T0295429

FROM CATCHMENT MANAGEMENT TO MANAGING RIVER BASINS

Current Directions in Water Scarcity Research

VOLUME 1

Series Editors

Robert Mcleman

Henry David Venema

FROM CATCHMENT MANAGEMENT TO MANAGING RIVER BASINS

Science, Technology Choices, Institutions and Policy

Edited By

M. DINESH KUMAR
Institute for Resource Analysis and Policy,
Hyderabad, Telangana, India

V. RATNA REDDY
Livelihoods and Natural Resource Management Institute,
Hyderabad, Telangana, India

A.J. JAMES
Environmental & Natural Resource Economist,
New Delhi, India

ELSEVIER

Elsevier
Radarweg 29, PO Box 211, 1000 AE Amsterdam, Netherlands
The Boulevard, Langford Lane, Kidlington, Oxford OX5 1GB, United Kingdom
50 Hampshire Street, 5th Floor, Cambridge, MA 02139, United States

© 2019 Elsevier Inc. All rights reserved.

No part of this publication may be reproduced or transmitted in any form or by any means, electronic or mechanical, including photocopying, recording, or any information storage and retrieval system, without permission in writing from the publisher. Details on how to seek permission, further information about the Publisher's permissions policies and our arrangements with organizations such as the Copyright Clearance Center and the Copyright Licensing Agency, can be found at our website: www.elsevier.com/permissions.

This book and the individual contributions contained in it are protected under copyright by the Publisher (other than as may be noted herein).

Notices
Knowledge and best practice in this field are constantly changing. As new research and experience broaden our understanding, changes in research methods, professional practices, or medical treatment may become necessary.

Practitioners and researchers must always rely on their own experience and knowledge in evaluating and using any information, methods, compounds, or experiments described herein. In using such information or methods they should be mindful of their own safety and the safety of others, including parties for whom they have a professional responsibility.

To the fullest extent of the law, neither the Publisher nor the authors, contributors, or editors, assume any liability for any injury and/or damage to persons or property as a matter of products liability, negligence or otherwise, or from any use or operation of any methods, products, instructions, or ideas contained in the material herein.

ISBN: 978-0-12-814851-8
ISSN: 2542-7946

For information on all Elsevier publications
visit our website at https://www.elsevier.com/books-and-journals

Publisher: Candice Janco
Acquisition Editor: Louisa Munro
Editorial Project Manager: Charlotte Rowley
Production Project Manager: Prem Kumar Kaliamoorthi
Cover Designer: Miles Hitchen

Typeset by SPi Global, India

Working together to grow libraries in developing countries
www.elsevier.com • www.bookaid.org

Contents

Contributors *ix*
About the Editors *xi*

1. Introduction **1**
M. Dinesh Kumar, V. Ratna Reddy and A.J. James
1.1. Introduction 1
1.2. Land use-hydrology interactions for catchment assessment and planning 2
1.3. Scale effects in water management 3
1.4. Traditional approach to water management decision making in river valley projects 5
1.5. From large river valley projects to development and management of micro watersheds 7
1.6. Current approach to integrated watershed management 7
1.7. Water accounting for exploring water management options 10
1.8. Inadequate use of scientific knowledge in catchment management decisions 12
1.9. Scope of the book 13
1.10. Contents of the book 13
References 17

2. Catchment management to basin management: International perspectives and overview of global experience **21**
M. Dinesh Kumar, Charles Batchelor and A.J. James
2.1. Introduction 21
2.2. Catchment assessment and planning for WSM: Theoretical and conceptual aspects 23
2.3. International literature on catchment assessment and planning 23
2.4. Catchment planning for watershed management programs in India 40
2.5. Major findings 46
References 50
Further reading 54

3. Watershed management in Afghanistan: Lessons from South Asia **55**
V. Ratna Reddy
3.1. Introduction 55
3.2. Evolution of WSM in South Asia 57

3.3. Watershed structures 64
3.4. Impacts and equity 67
3.5. Watershed institutions and implementation 74
3.6. Lessons for Afghanistan 79
3.7. Policy imperatives 81
References 82
Further reading 85

4. From watershed to sub-basins: Analyzing the scale effects of watershed development in Narmada river basin, central India 87
Jayesh Talati and M. Dinesh Kumar
4.1. Introduction 87
4.2. Objective, methodology, and scope 89
4.3. Results and discussion 90
4.4. Conclusion and policy implications 100
References 100

5. Catchment assessment and planning for improving effectiveness of watershed development interventions: A case study from South Gujarat 103
A.J. James, James Batchelor, Nitin Bassi and Grant Milne
5.1. Introduction 103
5.2. The issue 105
5.3. Study planning 107
5.4. Setting up and using the model 113
5.5. Understanding the catchment and its problems 117
5.6. Modeling the catchment characteristics 122
5.7. Scenarios of impact 128
5.8. Lessons for hydrology-based watershed management 139
5.9. Postscript 145
References 146

6. Water accounting for understanding water tenures: A case study of water rich Warna sub-basin of Krishna river basin, Maharashtra 149
M. Dinesh Kumar, K. Siva Ram Kishan, A.J. James and Nitin Bassi
6.1. Introduction 149
6.2. Hydrology and geohydrology of Warna river basin 150
6.3. Water resource system of Warna sub-basin and water supplies 155
6.4. Socioeconomic system of Warna sub-basin 157

6.5. Methodology and analytical framework for water accounting study 159
6.6. Data used for the analysis and sources 161
6.7. Analysis of water accounts of Warna sub-basin 162
6.8. Findings 174
6.9. Implications of water accounts for water tenure in the sub-basin 176
Acknowledgments 180
References 180
Further reading 181

7. Water accounting for Luni river basin, Western Rajasthan 183
M. Dinesh Kumar, Nitin Bassi and A.J. James

7.1. Introduction 183
7.2. Luni river basin: A bird's eye view 184
7.3. The basin hydrology and groundwater resources 186
7.4. Socioeconomic drivers of water use in the basin 190
7.5. Methodology and analytical procedure 193
7.6. Presentation of results 195
7.7. The basin water accounts 209
7.8. Conclusions 210
References 211
Further reading 212

8. Downscaling IWRM plans from basin to districts: IWRM plan for Pali district in Western Rajasthan 213
A.J. James and M. Dinesh Kumar

8.1. Introduction 213
8.2. District IWRM planning 213
8.3. Analyzing water demand 216
8.4. Analyzing water supply 223
8.5. Balancing demand and supply 229
8.6. District water management options 234
Annexure 1: Water use hydrology of irrigation 237
Annexure 2: Livestock population in Luni river basin (2007). 238
Annexure 3: Conveyance losses in irrigation and piped water systems in Pali district 239
Annexure 4: Microirrigation and mulching 240
References 243
Further reading 243

9. Water accounting for Kanpur metropolitan area in Ganges river basin **245**

Nitin Bassi, S.V. Suresh Babu and M. Dinesh Kumar

9.1. Introduction 245
9.2. Methodology 247
9.3. Kanpur metropolitan area 248
9.4. Status of water sources, water supply, wastewater and its treatment in KMA 252
9.5. Institutional framework for urban water management in KMA 260
9.6. Water accounts of KMA 261
9.7. Water supply management challenges 264
9.8. Major findings 270
9.9. Conclusion 273
References 273
Further reading 274

10. Addressing the scale issues in watershed development **275**

V. Ratna Reddy and Geoff Syme

10.1. Background 275
10.2. WSD scale and its evaluation 279
10.3. Socioeconomic implications 286
10.4. The way forward 294
References 295
Further reading 296

11. Operationalizing IWRM concepts at the basin level: From theory to practice **299**

M. Dinesh Kumar, Charles Batchelor and A.J. James

11.1. Introduction 299
11.2. General plan of action for implementing IWRM 301
11.3. Specific action plan for different regions 309
11.4. Conclusions 327
References 328
Further reading 329

12. Conclusions and areas for future research **331**

M. Dinesh Kumar, V. Ratna Reddy and A.J. James

12.1. Summary 331
12.2. Areas for future research 337
References 341

Index *343*

Contributors

Nitin Bassi
Institute for Resource Analysis and Policy, Delhi Liaison Office, New Delhi, India

Charles Batchelor
Water Resource Management Ltd., London, United Kingdom

James Batchelor
Digital Specialist, Atkins, London, United Kingdom

M. Dinesh Kumar
Institute for Resource Analysis and Policy, Hyderabad, Telangana, India

A.J. James
Environmental & Natural Resource Economist, New Delhi, India

K. Siva Ram Kishan
Institute for Resource Analysis and Policy, Hyderabad, Telangana, India

Grant Milne
Senior Natural Resource Management Specialist, World Bank, Washington, DC, United States

V. Ratna Reddy
Livelihoods and Natural Resource Management Institute, Hyderabad, Telangana, India

S.V. Suresh Babu
WWF-India, New Delhi, India

Geoff Syme
School of Business and Law, Edith Cowan, Hobart, TAS, Australia

Jayesh Talati
Formerly with IWMI-TATA, International Water Management Institute, Anand, Gujarat, India

About the Editors

M. Dinesh Kumar received his Ph.D. in Water Management in 2006. He has 28 years of professional experience in the field of water resources, undertaking research, consulting, action research, and training. He is currently the Executive Director of Institute for Resource Analysis and Policy in Hyderabad. He has nearly 170 publications to his credit, including books from internationally renowned publishers, several book chapters, many articles in peer-reviewed international journals, research monographs, and conference papers. He has authored six books and is the lead editor of four edited volumes. He has published extensively in many international peer-reviewed journals, such as *Water Policy*, *Energy Policy*, *Food Security*, *Water International*, *Journal of Hydrology*, *Water Resources Management*, and *Journal of Resources, Energy, and Development*. He is currently also Associate Editor of *Water Policy*, and Member of the Editorial Board of *International Journal of Water Resources Development*.

V. Ratna Reddy is an economist by training with experience in the fields of environmental economics and natural resource management, climate change, livelihoods, and institutions. He is an Alexander von Humboldt Fellow at South Asian Institute, Germany. He is Visiting Fellow at the Institute of Development Studies, Sussex, and the University of Leeds, UK, and served as consultant to organizations like The World Bank, ADB, UNDP, DFID, EC, WWF, IWMI, ICRISAT, UNU, etc. He had served as a working group member in the preparation of 11th and 12th Plans for the Government of India. He has published 12 books and more than 150 research papers in international and national peer-reviewed journals. He has a Ph.D. in Economics. Since 2009, he is Director of Livelihoods and Natural Resource Management Institute, Hyderabad.

A.J. James holds a Ph.D. in Economics (University College, London); has taught at St. Stephen's College and the Delhi School of Economics; and has been a Visiting Fellow at the Institute of Economic Growth, Delhi. He has over 20 years of research and work experience in developmental issues, including water and sanitation, water pollution, watershed development, irrigation and drainage, health, education, adaptation to climate change, natural resource management, agriculture, forestry, and poverty alleviation. Another area of specialization is monitoring and evaluation, wherein he had helped to develop innovative methodologies for community-level assessment of qualitative information. Besides India, he has worked in Nepal, Sri Lanka, Vietnam, Ethiopia, Tanzania, and Afghanistan. His clients are funding agencies (World Bank, DFID, UNDP, UNICEF, FAO, and ADB), research and development institutions (ENTRO, Addis Ababa), IRC International Water and Sanitation Centre (the Netherlands), Overseas Development Institute (London), Institute for Arable Crop Research, UK, and Natural Resources Institute, UK, consulting firms in India and abroad, nongovernmental organizations, and Indian government agencies. He has given several invited talks and seminars, besides editing books, contributing book chapters, and publishing articles in international peer-reviewed journals. He is an Honorary Visiting Professor at the Institute of Development Studies, Jaipur, since 2011.

CHAPTER 1

Introduction

M. Dinesh Kumar[a], V. Ratna Reddy[b], A.J. James[c]
[a]Institute for Resource Analysis and Policy, Hyderabad, Telangana, India
[b]Livelihoods and Natural Resource Management Institute, Hyderabad, Telangana, India
[c]Environmental & Natural Resource Economist, New Delhi, India

1.1 Introduction

As population grows, as food production increases, economic activities pick up and as societies become more affluent, demand for water burgeons. In very many places, demand has far outstripped supply–particularly so in seasons when supply is severely limited or in years of drought, or at times when demand is particularly high for large water-consuming sectors such as irrigation. Climate extremes compound the problem. As decisions on water allocation are going to be difficult under such situations (Molle and Vallée, 2009), water managers have to look for new management options, in a way that doesn't require compromises on social and economic development, ecological sustainability, and hydrological integrity of water systems.

Water management options for a locality or a region depend on three important factors, viz., hydrology, socioeconomic systems, and ecosystem (Mitchell, 1990; Voropaev and Zaitseva, 1996). Hydrological and topographical characteristics of the region determine the amount of water that is naturally available from the catchments or drainage basins that can be harnessed for meeting various water needs. Socioeconomic characteristics determine the demand for water for various human needs—domestic and productive. The ecological characteristics determine the amount of water required for maintaining the health of the ecosystem in the form of soil moisture storage, river flows and flow regimes, chemical and biological quality of river water, water storage and quality of water in wetlands, depth to groundwater table, groundwater flows, and wetland-groundwater interactions.

If the first two key parameters, i.e., availability of water resources within a hydrological unit and the demand for water various sectors of use within

Current Directions in Water Scarcity Research, Volume 1
https://doi.org/10.1016/B978-0-12-814851-8.00001-X
© 2019 Elsevier Inc.
All rights reserved.

that unit are known, the next important question, which confronts a water manager, is to know how much of the available water resources are already tapped through various interventions; how much of it is depleted, through various processes, and how much of the developed water resources in the basin remain un-utilized (or available for use); what are the unmet water demands, and how much of the water in the basin is untapped and joining the natural sink.

The interventions to tap the water to increase its utilization will comprise in situ water harvesting through field bunds and farm bunds, watershed treatment activities, small water harvesting systems, tanks, and large reservoirs and diversion systems. On the other hand, the processes, which cause depletion of water, are beneficial consumptive use (evapotranspiration, consumptive use by humans and animals), nonbeneficial consumptive use (evaporation from water bodies, barren soil, cropped land not covered by canopy, etc.), and nonbeneficial nonconsumptive use (return flows into saline formations, nonrecoverable deep percolation of water) and beneficial nonconsumptive use (recharge to groundwater and return flows from irrigated fields to streams) (Molden et al., 2003; Perry, 2007).

Addressing these questions is crucial from the point of view of identifying the nature of future water management interventions (i.e., whether supply augmentation or end use conservation or both) possible in the basin, and analyzing the extent to which these interventions could help achieve a balance between water availability and water demand. Failure to do so would lead to suboptimal water management for the basin. If large amount of water in the basin is already developed through reservoirs, but capacity of the infrastructure created to transfer the water to the points of demand is insufficient, some parts of the basin might experience water shortages. A solution to this problem lies in expanding the water conveyance systems, rather than building new water storage facilities such as reservoirs.

1.2 Land use-hydrology interactions for catchment assessment and planning

As regards the first decision variable, i.e., hydrology of the catchment and basins, the land use and land cover have a significant bearing. Since both keep changing with time, in order to determine the amount of water that is naturally available from them, information on the following is crucial: (1) how the type of catchment land use influences the impact of increased vegetation cover on stream flows (including water quality), in different agro ecologies (Idol, 2003); (2) how the nature of vegetation (whether shallow rooted grasses and shrubs or deep rooted trees) determines the impact of increased vegetation cover on the consumptive use of water from the soil profile and groundwater system of the catchment, and how these impacts can change across agroecologies (Oliveira et al., 2005; Idol, 2003); (3) the hydraulic interdependence between groundwater and surface water in a catchment and therefore the impact of change in groundwater withdrawal on stream flows downstream (Winter et al., 1998). Often, the sources of sedimentation can often be quite localized such as newly constructed roads, stream-bank cultivation, and movement of livestock around, to or from a water point.

Hence, there is a need for developing clear understanding of the catchment characteristics, current land use in the catchment, and the hydrological regime before interventions are planned for changing their hydrological regime. Of particular interest are the current stream flow regimes; the extent of base flow contribution to the stream flows; geological and geohydrological environment of the catchment, particularly the depth to water table and the thickness of the vadoze zone; area under natural forests, their condition, type of forests; area under cultivation, type of crops, their seasonality, and their geographical spread within the catchment; groundwater use in the catchment, its seasonality and location of wells; and amount of committed flows from the catchment for downstream uses.

A large number of catchment experiments conducted all over the world clearly demonstrated that deforestation of catchment implies an increase of water yield from it and, conversely, the establishment of a forest cover implies a decrease of water yield (Sahin and Hall, 1996). The application of this knowledge to designing sustainable water management practices, although necessary in water-stressed regions, has been largely delayed because of difficulties inherent to the change of any scientific paradigm, the limited experience on the hydrological consequences of land cover changes in large territories, and the disconnection between policy and science (Falkenmark, 2004; Calder, 2002).

1.3 Scale effects in water management

Scale effects in hydrology are well documented (Merz et al., 2009; Wood et al., 1988). The nature of hydrological interactions (surface water-groundwater interactions, infiltration-groundwater recharge) and hydrological relationships (rainfall-runoff relationships, rainfall-base flow, etc.)

within a catchment depends on the spatial unit of analysis. The rainfall-runoff relationship for a small (mini micro) catchment of say, 10 ha area, can be remarkably different from that of a hilly/mountainous catchment of say, 10 km^2 area, to which it becomes a part. The runoff coefficient for the latter will be much higher than that of the former, as in the latter case, it will also include base flows along with surface runoff unlike the former.

Similarly, the rainfall-runoff relationship for a small hilly/mountainous watershed, say of area, 10 km^2 can be different from a large watershed or river basin to which it becomes a part, even if there is no spatial difference in the rainfall. This can be because of many factors. First, the land use conditions could change drastically as one moves from the hilly/mountainous upper catchments to the plains. With the intensity of crop cultivation increasing toward the lower plains, a greater proportion of the runoff generated would be captured in situ, reducing the runoff coefficient. Another factor, which plays a crucial role in changing hydrological interactions spatially, is the topography. As lower catchments of basins will be relatively flat, the chances of natural outflow of groundwater to surface streams would reduce, with the result that the runoff coefficient will be less. The third factor is the climate. Generally, upper catchments of river basins are cold and humid, as compared to the lower catchments, which are hot and arid. This difference in temperature and relative humidity can significantly influence the rate of moisture depletion from the top soil through barren soil evaporation and ET losses from grasses and trees, and therefore infiltration and runoff generation.

As regards return flows, the effect of runoff from irrigated field and wastewater outflows from industrial areas and urban centers on the surface flows appears only in the lower parts of the basin and not in the watersheds and catchments from which the water is diverted for irrigation and other uses, essentially because of the terrain and drainage characteristics. But, return flows from irrigated fields to shallow groundwater from irrigated fields might appear in the lower part of the same agricultural watershed, though unlikely in the well of the farmer who irrigates.

As regards water demands, as one moves from upper catchment to lower catchments, not only does the water demand for competitive uses increase in aggregate terms, the nature of demand also varies. Generally, the upper catchments of river basins have low proportion of its land under crop cultivation due to the presence of forests, grass land, etc. This is compounded by cold and humid climate, which reduces the water demand for agriculture significantly. Generally, upper catchments of river basins are less populated, and less urbanized as compared to lower basin areas. This means, the concentrated water demands for domestic and municipal uses are generally low. Further, barring the exception of mining, very few industries are located in upper catchments of basins that are generally hilly and mountainous, not ideal for manufacturing, which require heavy and quick transportation.

The situation changes dramatically as one moves to the lower basin areas, which are hot and arid and closer to the coastal plains. These areas are more heavily populated, with less area under forest cover and other natural vegetation, characterized by commercial farming, and display high degree of urbanization and industrialization. All these factors put together not only mean new areas of water demand (such as industrial, urban, and even thermal power and navigation), but also increased water demand in individual use sectors such as farming. Also, there can be significant differences in the pattern of water demand. The farms that are located close to urban areas would require water round the year to produce fruits, vegetables, and flowers, as compared to those which are in the remote villages located in the upper watersheds.

Water requirements for maintaining ecosystem health, over and above the water which is directly consumed by the ecosystem in the form of evaporation, will be higher in lower basin areas, as compared to upper basin areas that are generally characterized by thick natural vegetation and pristine stream flows. The relatively higher temperature and aridity and the presence of large build-up areas, and the significant alterations in flow regime of the rivers during their course, owing to the upstream diversions, increase the ecological water demand, comprising water for growing trees in urban areas, watering lawns and gardens and parks, maintaining the environmental flows in rivers and assimilating pollution. Added to this is the pollution of river water, which takes place during its course from the upper catchment to the lower plains, due to disposal of trade effluents, urban wastewater, and return flows from irrigation.

1.4 Traditional approach to water management decision making in river valley projects

The foregoing analysis suggests that our ability to arrive at effective water management decisions for any region would depend heavily on: (1) quantitatively analyzing the land use-hydrological interactions and surface water-groundwater interactions within a catchment or river basin, which is critical for assessing the amount of water supplies available for meeting various needs, and its quality; and (2) choosing the right spatial scale for analysis of water availability and water demands, so as to capture the different types of interactions that are at play, affecting the effective water availability and water demands in different sectors and their spatial patterns. While the first is crucial for planning water resources management interventions, the second becomes important for all three activities, viz., water resources management, water resources development, and water allocation.

Historically, large water resource development projects, especially river valley projects in most developing countries, particularly in South Asia and Africa, were based on assessment of catchment water flows (yields), and normative demand for water in some other part of the basin, which experienced scarcity of water for meeting various needs, especially irrigation and domestic water supplies. This approach of considering a selected catchment as a source of water supply to meet the demand in another site lacked integrated basin-wide assessment of effective water availability and demands at any given point of time (Molle et al., 2010). The major technical challenge was in estimating the dependable (runoff) yield of the designated catchment, and sediment load, in lieu of the fact that for many rivers, stream-gauging data were not available for long time periods. This was by and large a piecemeal approach. The earlier criticism of the approach was on the inability of hydrologists to make realistic estimates of soil erosion from catchments, as in many situations the observed siltation rate was much higher than the values considered for reservoir design. Application of this catchment-based approach often led to creating conditions of water abundance in certain parts of the basin, and water scarcity, especially environmental water stress in certain other parts.

The catchment-based approach of basin development to meet water demands is segmented and often led to overappropriation of runoff in river basins. It also left no scope for exploring the range of alternatives in water management, particularly conjunctive management and water demand management. For instance, it is possible that within a basin, a large amount of water from a catchment is available as return flows to the shallow aquifers underlying the irrigation command area. Because of this, there could be build-up of water table over a period of time. Hence, it is important that well irrigation be encouraged in the region to prevent water logging, which would enable holding back a lot of water from the reservoirs and allocating it for meeting water requirements in other parts of the basin. Such an approach can, to a great extent, escape the need for investing in new water development projects. But, for embarking on it, we need to know what the annual groundwater storage change in the basin is. Similarly, if we know how much of the developed water is lost in nonbeneficial consumptive uses in the basin in sectors such as irrigation (in the form of evaporation from the harvested fields), measures can be adopted to prevent them. But to gain insights into the changing dynamics of water availability and use across the basin, we need to do basin-wide resource evaluation and planning.

1.5 From large river valley projects to development and management of micro watersheds

In India and in many other developing countries of South Asia and Africa, watershed management is not merely a soil and water conservation work meant for protecting the catchments of large reservoirs, but a major developmental activity for the rural areas, aimed at improving rural livelihoods through agricultural productivity enhancement. India, which is a pioneer in the field, has a long history of watershed development and management programs, with schemes from three Ministries, viz., Ministry of Agriculture, Ministry of Rural Development, and Ministry of Environment and Forests of the Govt. of India operational since mid-1990s.

By as far back as 2005, India had invested around 3.1 billion dollars for watershed development programs, covering a total area of around 45.0 m. ha of rain-fed areas in the country (Parthasarathy Committee Report, 2006). But, these interventions had very limited success in terms of productivity impacts on agriculture. The problems identified vis-à-vis impacts include limited and temporary productivity gains; poor revegetation of common land; and fast dissipation of groundwater recharge benefits (Joshi et al., 2005; Kerr et al., 2004). There have been problems with watershed planning approach due to flawed criteria used for selection of catchments for watershed treatment (Kumar et al., 2017); and limited scientific inputs in planning catchment management interventions (James et al., 2015). This is compounded by poor implementation of work (Reddy, 2006). These are policy issues, which need urgent attention for investments in watershed development programs to yield sufficient returns.

1.6 Current approach to integrated watershed management

The current approach of integrated watershed management has a clear emphasis on improving the conditions of large majority of the people living in rain-fed areas who are dependent on land and water for their livelihoods.

The thrust is on implementing activities, which can be done by the local communities with minimum outside support on technical matters (Farrington et al., 1999; Kerr, 2002; Hope, 2007). The thrust is also on taking up physical activities, which will have immediate as well as medium and long-term impacts. In that respect, local employment generation is also given sufficient emphasis. These are the strengths of the program.

Improvement in water use efficiency and equity in catchment-wide distribution of water are two key long-term objectives of watershed management programs. In that respect, lack of sufficient information for catchment assessment and watershed planning is a major issue with the current approach. While the management is based on microwatersheds as the unit for planning and operation of watershed interventions, the hydrological data of the catchment are not available at that scale, particularly the annual stream flows, groundwater recharge and flow gradients, base flows and withdrawal, and the uncommitted flow from the catchment (James et al., 2015). Many small basins in India are ungauged, and in many large basins, gauging is done for large tributaries. In none of the small river basins, and smaller tributaries of large basins, flow measurements are done by the agencies concerned (Kumar et al., 2006).

Further, information is not readily available to the PIAs on the characteristics of the catchment. Generally, a decision to implement watershed management programs in a village is taken on the basis of limited data available from Survey of India toposheets. The information on cropping pattern and crop types, available at the village level, is never available on spatial reference and hence cannot be transposed on watershed boundaries. In the absence of such data, watershed planning as attempted by some NGOs purely on PRA approaches has not yielded required results. While good hydrologists with orientation for information technology might be able to generate data on land use/land cover and runoff for microcatchments, the issue is such trained professionals are very few in India. Even when they are available, the financial capability of PIAs to hire such skilled professionals is open to question (James et al., 2015).

The current approach to planning WSM interventions is based on a strong notion that if comprehensive treatment is done, it would improve the water availability and or soil moisture regime within the microwatershed along with reducing soil erosion, and that it can even increase the effective water availability for the downstream communities during the lean season. The underlying assumption is that a large amount of runoff during the monsoon goes uncaptured and eventually gets wasted as it joins the natural sink of sea, ocean, or swamps. Since the geographical unit of planning is too small, the PIAs are not concerned with what happens to the runoff, which used to flow out of the microcatchment, and the social, economic, and environmental values it generated (Calder et al., 2008; Kumar et al., 2006, 2008). The approach tends to ignore the fact that there could be different values and interests, which communities within a large catchment attach to water but are all legitimate and need to be recognized (Jakeman and Letcher, 2003; Mitchell and Hollick, 1993).

The net result is that often there is overdoing of interventions within a watershed in terms of creating tree cover, building vegetative barriers to reduce the speed of runoff water, reduce soil erosion, and improve soil moisture conservation. Small water bodies are built to harvest monsoon runoff, without due consideration to storage efficiency. New crops are planned to utilize the augmented groundwater in the catchment, without due consideration to the water use efficiency ($\$/m^3$) of the newly introduced crops. In "closed" catchments, this reduces the overall economic viability of the interventions, and efficiency of water use at the catchment scale as downstream uses is adversely affected (Kumar and van Dam, 2013). While such practices are subject to criticism (Batchelor et al., 2002; Kumar et al., 2008), the current approach does not encourage proper assessment of changes in water use efficiency at the catchment scale.

Under the current approach, certain treatment activities within the catchment are promoted on the premise that they do not take much water from the hydrological system, and therefore, economic losses due to their adverse impacts are not taken cognizance of in the planning decisions (Batchelor et al., 2003). Such decisions tend to lose sight of the fact that within the same catchment, often there are downstream water systems like tanks and ponds, which depend on this runoff for uses such as domestic needs, supplementary irrigation of crops, and fisheries.

When the basins become closed, hydraulic interconnectedness of surface water and groundwater becomes marked (Molle et al., 2010). But, lack of recognition of the hydraulic interconnectedness of aquifers and streams in the catchment is another issue in watershed development programs. Groundwater recharge is promoted within the catchment under watershed management as a "positive value" with the assumption that it would increase the base flows, thereby making streams flowing in the lower catchment perennial through base flows during lean season. But, hardly any attention is paid to the fact that this activity is followed by indiscriminate drilling of wells by farmers in the area, which ultimately leads to increased draft, threatening even the existing natural discharge of groundwater into streams and wetlands. But, there is virtually no control on groundwater abstraction planned under watershed management (James et al., 2015).

Lack of integration between micro and macro and local and regional watersheds is another major issue. The current approach to watershed management is also highly decentralized (Calder et al., 2008; Syme et al., 2012). Hence, the decision of an individual PIA about the degree and extent of treatment in a particular watershed is driven by "what is optimal for that watershed," with the result that often the aggregate of the activities planned for all watersheds together is suboptimal for the large encompassing catchment.

Finally, the current approach is devoid of any measures for regulating the use of land and water resources within the catchment, particularly agricultural intensification and groundwater abstraction. Participation of the catchment communities is sought only for planning and implementing various physical interventions. The community organizations formed in the watersheds have no role, whatsoever, either in regulating land and water use or in allocating water among various uses.

1.7 Water accounting for exploring water management options

In developing countries, decisions to execute large-scale water resource development projects and projects to improve water use efficiency in agriculture are not based on realistic assessment of the opportunities that really exist at the basin-scale to augment the supplies and reduce water demands. In the case of water resource development, the individual catchment is the unit of analysis, rather than the basin. In many situations, the new developments such as construction of water harvesting structures and local groundwater mining, are undertaken in "closed basins," inducing hydrological externalities for the existing ones located downstream (source: based on Kumar et al., 2008; Molle et al., 2010). To prevent such negative outcomes, in the case of water supply augmentation projects, the basin manager should be concerned with the amount of uncommitted flows in the basin and what fraction of it can be harnessed.

As regards projects to improve water use efficiency in agriculture (such as introduction of micro irrigation technologies), they are normally based on the concept of field level water saving, which considers the possible reduction in the amount of water applied against the crop water requirement, through the use of those technologies, which help minimize the deep percolation of water applied in the fields (Perry, 2007). But in many situations, the deep percolation "losses" resulting from inefficient irrigation practice is available for reuse by well irrigators in the same catchment, and microirrigation technologies only reduce this beneficial, nonconsumptive use. Hence, using such concepts can lead to overestimation of water-saving benefits (Allen et al., 1998; Perry, 2007). Hence, to determine the "real water saving," the basin manager should be concerned with reduction in the amount of "water depleted" (Kumar and van Dam, 2013).

Water accounting studies, based on an analysis of data related to various components of the hydrological system (e.g., rainfall, stream flows, groundwater recharge, evaporation from water bodies and swamps, storage changes in reservoirs and land use and water use) at appropriate hydrological units (e.g., basin, catchment or watershed-level) can give a range of useful information about potential water-related extreme events, how much of water in the basin is used in different sectors (drinking and domestic uses, municipal use, industrial use, livestock use, power generation, crop consumptive uses, and various in-stream uses), how much of the basin's water goes uncaptured into natural sinks (e.g., saline formations, seas) and how much is lost through nonbeneficial uses (Molden et al., 2003).It could also help the water administration active in the basin to plan actions to mitigate the adverse impacts of hydrological stresses on the socioeconomic system, such as drinking water shortages for human and livestock populations in affected areas, crop losses, and reduced income and livelihoods of affected populations, and damage to water infrastructure.

From the point of view of water governance and management, water accounting studies is useful in the sense that it can provide insights into: (1) optimum level of investments in water augmenting measures (e.g., watershed development programs, water harvesting systems, large reservoirs and diversion systems); (2) possible measures to (a) increase effective water availability (e.g., reducing evaporation from surface reservoirs and preventing soil moisture depletion from rain-fed crop land), (b) reduce water demand (e.g., reducing nonbeneficial (both consumptive and nonconsumptive) use of water in irrigated fields) and (c) convert other nonbeneficial uses of water in the basin (e.g., evaporation from barren land) into productive uses; and, (3) better intersectoral water allocation, including from lower-value to higher-value uses.

1.8 Inadequate use of scientific knowledge in catchment management decisions

Most of the microwatershed management programs in India are grounded on inadequate knowledge of the relationship between changes in land use/land cover and catchment yield and soil erosion. They mainly concern differential impacts of various specific interventions on soil conservation and sediment control; differential impacts of grass buffer strips and retention ponds on sediment control; impacts of replacing crop land by tree cover, clearance of native trees from forested catchments, and replacing forest cover by crops on seasonal and annual yield of catchments; differential impact of grass and tree cover on erosion control and runoff reduction; and impact of use of efficient irrigation technologies on overall water use efficiency in the agricultural catchments and effective water availability for other uses in the catchment. Often, there are wide misconceptions, concerning their functions in different agro ecologies (James et al., 2015).

The knowledge on the dynamics of interaction between a particular land use and land cover and water use in the upper parts of the catchment, and the hydrology and ecosystems of a given catchment can provide pointers on the way in which the former needs to be modified to produce social, economic and environmental outcomes that are widely acceptable among the catchment communities. But, which land use or land cover-based intervention needs to be taken up and to what extent they need to be changed (scale analysis) to achieve the optimum outcomes in terms of water yield, sediment load reduction, meeting water quality standards and reduction in soil loss, etc. can only be assessed using complex mathematical models, which simulate the hydrological and biophysical processes. Such models basically integrate those used for prediction of soil erosion from the catchment; crop growth; rainfall-runoff; sediment transport; and groundwater flow. At the same time, the models used for predicting hydrological changes in catchments due to land use changes should be such that they can simulate the effects of irrigation efficiency improvement measures on degree of return flows from irrigated fields to groundwater and streams, and changes in groundwater withdrawals on magnitude of base flows from aquifers to streams.

In nutshell, scientific understanding of the relationship between catchment land use and land cover changes and hydrological and ecological processes is essential for identifying watershed management interventions that can achieve desired water management outcomes in social, economic, and environmental terms—improved livelihoods, drinking water security, enhanced agricultural production, and protection of hydrological integrity of watersheds. The key research questions are: (i) what is the level of treatment on land—tree cover, vegetative barriers, crop rotations, soil and moisture conservation measures—, water harvesting and irrigation efficiency improvement measures that can ensure optimum runoff yield and sediment load reduction, while ensuring water for various competitive and in-stream uses within the catchment; and, (ii) what is the economic viability of these interventions?

1.9 Scope of the book

The book discusses the theoretical and operational aspects of integrated catchment management and integrated water management in river basins, through reviews of international best practices and case studies. While there are a set of concepts and principles in IWRM that are widely known, there is lack of clarity on what practices (both physical and institutional) work for achieving its goals in different situations.

The book synthesizes key scientific facts that are crucial for catchment assessment and planning, and river basin water accounting, from extensive reviews of international literature on catchment hydrology, forest hydrology, and other hydrological processes such as groundwater-surface water interactions. It discusses not only the science of catchment assessment and planning, but also the catchment planning process. It documents several of the positive international experiences with integrated catchment management and integrated basin management and distils the key learning from them, which are important from the point of view of shaping water management paradigms for the developing world. It presents detailed case studies from India and other parts of South Asia with integrated watershed/catchment management and illustrate how lack of scientific planning lead to a myriad of unintended consequences. It then presents pilot studies on integrated catchment management planning, and basin water accounting to show what kind of interventions would be required in the basins of those specific regions for enhancing that the social, economic, and environmental outcomes of ICM and integrated water resources management.

1.10 Contents of the book

Following this introduction, the second chapter of the book by Kumar, Batchelor and James presents a strategic review of international perspectives and global experience with catchment management and basin management. It deals with various conceptual and theoretical aspects of catchment management, defining the principles and practice underlying it, as available from scientific research in the different parts of the world. It then discusses the technical aspects of catchment assessment and planning, including the use of hydrological simulation models. It then goes on to discuss the experience of some of the developed countries in operationalizing catchment management and river basin management concepts, on the institutional front.

The third chapter of the book by VR Reddy discusses the experience of Afghanistan in watershed management. It provides a basic policy and implementation framework for watershed management in that country based on a critical review of experiences in South Asian countries. The chapter reviews the evolution of watershed management in south Asian countries; examines the implementation and institutional modalities adopted in these countries; assesses the impacts (environmental and socioeconomic) of watershed interventions in varying agroclimatic conditions of these countries; identifies the implementation bottlenecks for sustainable watershed management; and draws lessons for Afghanistan.

The fourth chapter by Talati and Kumar illustrates the scale effects in watershed development using a case study from one of the sub-basins of Narmada river basin in central India, which experienced intensive and decentralized watershed development and water harvesting activities during the late 1990s to the early 2000s. The case study used analysis of field data on the geohydrological parameters at the level of microwatersheds, and the impact was analyzed by comparing sample (with watershed treatment) and control watersheds (with no watershed treatment). For the sub-basin level impacts of intensive watershed development activities in the upper catchment, secondary (hydrological) data were analyzed on a time scale in relation to rainfall. The impact analysis involved estimation of recharge coefficients for the treated and untreated watersheds, and analysis of rainfall-runoff relationship for the tributary draining the sub-basin, which experienced intensive watershed development.

The fifth chapter by James et al. discusses the findings of a catchment assessment and planning study carried out in South Gujarat, western India, using a simulation model based on SWAT (Soil and Water Assessment Tool). The model was calibrated using historical data of stream flow discharge from the catchment and validated to test its robustness. The validated model simulated the impact of various physical processes such as changes in rainfall magnitude and pattern, and land use/land cover on the catchment runoff; and changes in rainfall and groundwater withdrawal on surface water-groundwater interactions, including the occurrence of lean season flows in a catchment. The chapter presents the findings of the study vis-à-vis how scope for and catchment wide impacts of various water management interventions would change from a wet year to a dry year.

The sixth chapter by Kumar & others discusses an innovative methodology developed by the authors for basin water accounting for a river basin with different competing uses of water, but with limited measurement of inflows and outflows. It also presents the results of a water accounting exercise carried out for a small, water-surplus river basin. The analytical tool for water accounting, which considered the factors such as inflows (renewable water resources), outflows (consumptive use in various sectors of use), and annual change in storage, was applied to Warna sub-basin of Krishna river basin in peninsular India. The water accounting study used data on reservoir mass balance, annual groundwater recharge and storage change, consumptive water use in agriculture and other sectors (such as domestic, livestock, and industry), and stream discharge at the last drainage point of the sub-basin to estimate the water accounts. The methodology helped quantify the total amount of renewable water resources generated in the basin, and the amount of water actually depleted for beneficial and nonbeneficial uses.

The methodology for water accounting will have to change as basin characteristics, defined by rainfall, climate, morphology, and geohydrology change. The seventh chapter presents the findings of a basin-level water accounting study done for one of the most arid and water-scarce river basins of India, i.e., Luni river basin in western Rajasthan, which suffers severely from lack of reliable data onstream flows. There is only one stream gauging station in the entire drainage area of 69,000 km^2, which is located near the last drainage point of the basin. Overappropriation of surface water, groundwater mining, and heavy transmission losses in the surface runoff are unique features of this basin. The aim of basin water accounting was to evolve integrated water resource Management Plan for the basin. Estimating the consumptive use of water in agriculture was one of the biggest challenges in the hyperarid river basin. The seventh chapter also discusses the viable technical options for improving water balance of the basin.

River basins are the most appropriate unit for planning most IWRM interventions, as the physical systems affecting water availability and socio-economic system affecting the demand for water can be integrated at that scale (Jønch Clausen, 2004). However, the present situation in most countries is that these interventions have to be executed by agencies, which are not organized around river basins or their smaller hydrological units such as sub-basins and catchments. They operate at the state and district level or sometime even at the level of smaller administrative units. Therefore, interventions that are found to be effective at the basin scale for achieving IWRM goals in the Luni has to be downscaled, to arrive at the "scale" at which each intervention has to be carried out in the district. The eighth chapter discusses the analytical method and result of the process of downscaling IWRM plan for Pali, one of the districts of Luni River Basin.

The ninth chapter by Nitin Bassi and M Dinesh Kumar also uses a water accounting approach to assess the water security challenges of Kanpur Metropolitan Area (KMA), a city located in Ganga River Basin, which is under tremendous stress due to overappropriation and large-scale pollution. The chapter particularly focuses on the following: development of methodology for doing water accounting for cities; preparing water accounts for KMA; and identifying the present and future water-related risks resulting from the inadequacy of the quantity and quality of water supply for domestic and industrial consumers, and ineffectiveness of the regulatory framework related to water quantity and quality management in the KMA.

The scale effects in watershed management, particularly the downstream effects of upstream interventions due to land-water interactions, are well documented (Darghouth et al., 2008). Chapter 10 by Reddy and Syme discusses about addressing the scale issues in watershed development. One of the aims of using the IWRM approach is to internalize such effects.

That said, many of the concepts and principles underlying integrated water resources management are largely recognized and appreciated by basin managers. The challenge is in operationalizing the concept of IWRM through appropriate coordinating institutions at the river basin level. As discussed by Biswas (2008, p. 12), "... operationally it has not been possible to identify a water management process at a macro- or meso-scale which can be planned and implemented in such a way that it becomes inherently integrated, however this may be defined, right from its initial planning stage and then to implementation and operational phases." There is also a lot of confusion about what an IWRM plan should contain in terms of interventions in different situations. As noted by Jønch Clausen (2004), there is no one-size-fits-all solution in IWRM. The solutions would change with the typology of areas, defined by the hydrological and socioeconomic reality, and the political and institutional setting.

In the eleventh chapter, the authors make an earnest beginning in this direction. The chapter suggests the following:

(i) A framework for identifying the most vital interventions, which need to be carried out in a river basin for achieving IWRM goals, for different typologies;

(ii) The tools for planning the interventions at the basin scale;

(iii) The key technical and institutional interventions necessary for executing the physical interventions; and,

(iv) The institutional mechanisms and processes that have to be put in place to monitor the effectiveness of IWRM interventions.

The chapter would also illustrate how the potential outcomes of some widely known water management interventions on supply and demand side could change at the basin level, as the basin characteristics change.

The concluding chapter summarizes the key findings and inferences from the previous chapters and discusses the areas for future research. It will also discuss some of the policy interventions that are required at the national and provincial level for promoting integrated catchment management and integrated basin management.

References

Allen, R.G., Pereira, L.S., Raes, D., Smith, M., 1998. Crop evapotranspiration. Guidelines for computing crop water requirements. FAO Irrigation and Drainage Paper 56, Food and Agriculture Organization of the United Stations, Rome, Italy. 1999.

Batchelor, C., Singh, A., Rao, M.S.R.M., Butterworth, J., 2002. Mitigating the potential unintended impacts of water harvesting. In: Paper Presented at the IWRA International Regional Symposium 'Water for Human Survival', 26–29 November, 2002, Hotel Taj Palace, New Delhi.

Batchelor, C., Rama Mohan Rao, M.S., Manohar Rao, S., 2003. Watershed development: a solution to water shortages in semi-arid India or part of the problem? Land Use Water Resour. Res. 3 (2003), 1–10.

Biswas, A.K., 2008. Integrated water resources management: is it working? Int. J. Water Resour. Dev. 24, 5–22. https://doi.org/10.1080/07900620701871718.

Calder, I.R., 2002. Forest valuation and water; the need to reconcile public and science perceptions. In: Verweij, P.A. (Ed.), Understanding and Capturing the Multiple Values of Tropical Forests, Tropenbos Proceedings. Tropenbos International, Wageningen, pp. 49–62.

Calder, I., Gosain, A.K., Rama Mohan Rao, M.S., Batchelor, C., Garratt, J., Bishop, E., 2008. Watershed development in India. 2. New approaches for managing externalities and meeting sustainability requirements. Environ. Dev. Sustain. 227–240. https://doi.org/10.1007/s10668-006-9073-0.

Darghouth, S., Ward, C., Gambarelli, G., Styger, E., Roux, J., 2008. Watershed Management Approaches, Policies and Operations: Lessons for Scaling Up. Water Sector Board Discussion Paper Series, Paper No. 11, The World Bank, Washington, DC.

Falkenmark, M., 2004. Towards integrated catchment management: opening the paradigm locks between hydrology, ecology and policy-making. Int. J. Water Resour. Dev. 20 (3), 275–281.

Farrington, J., Turton, C., James, A.J., 1999. Participatory Watershed Development: Challenges for the Twenty-First Century. Oxford University Press, New Delhi, India.

Hope, R.A., 2007. Evaluating social impacts of watershed development in India. World Dev. 35 (8), 1436–1499.

Idol, T., 2003. Hydrologic Effects of Changes in Forest Structure and Species Composition. Deptartment of Natural Resources and Environment Management, University of Hawaii-Manoa.

Jakeman, A.J., Letcher, R.A., 2003. Integrated assessment and modelling: features, principles and examples for catchment management. Environ. Model. Software 18 (2003), 491–501.

James, A.J., Kumar, M.D., Batchelor, J., Batchelor, C., Bassi, N., Choudhary, J., Gandhi, D., Syme, G., Mile, G., Kumar, P., 2015. Catchment Assessment and Planning for Watershed Management. vol. 1. World Bank Group. Main Report.

Jønch Clausen, T., 2004. Integrated Water Resources Management (IWRM) and Water Efficiency Plans by 2005. Global Water Partnership, Stockholm, Sweden.

Joshi, P.K., Jha, A.K., Wani, S.P., Joshi, L., Shiyani, R.L., 2005. Meta-analysis to assess impact of watershed program and people's participation. Comprehensive Assessment Research Report 8. Comprehensive Assessment Secretariat, Sri Lanka.

Kerr, J., 2002. Watershed development, environmental services, and poverty alleviation in India. World Dev. 30 (8), 1387–1400.

Kerr, J., Pangare, G., Pangare, V., 2004. Watershed development projects in India: an evaluation. Research Report 127. International Food Policy Research Institute, Washington, DC.

Kumar, M.D., van Dam, J., 2013. Drivers of change in agricultural water productivity and its improvement at basin scale in developing economies. Water Int. 38, 312–325. https://doi.org/10.1080/02508060.2013.793572.

Kumar, M.D., Ghosh, S., Patel, A., Singh, O.P., Ravindranath, R., 2006. Rainwater harvesting in India: some critical issues for basin planning and research. Land Use Water Resour. Res. 6 (1), 1–17.

Kumar, M.D., Patel, A.R., Ravindranath, R., Singh, O.P., 2008. Chasing a mirage: water harvesting and artificial recharge in naturally water-scarce regions. Econ. Pol. Wkly 43 (35), 61–71.

Kumar, M.D., Reddy, R., Narayanamoorthy, A., Bassi, N., James, A.J., 2017. Rainfed areas: poor definition and flawed solutions. Int. J. Water Resour. Dev. 34, 278–291. https://doi.org/10.1080/07900627.2017.1278680.

Merz, R., Parajka, J., Bloschl, G., 2009. Scale effects in conceptual hydrological modeling. Water Resour. Res. 45 (9), 1–15.

Mitchell, B., 1990. Integrated water management. In: Mitchell, B. (Ed.), Integrated Water Management: International Experiences and Perspectives. Belhaven Press, London and New York.

Mitchell, B., Hollick, M., 1993. Integrated catchment management in western Australia: transition from concept to implementation. Environ. Manage. 17 (6), 735–743.

Molden, D., Murray-Rust, H., Sakthivadivel, R., Makin, I., 2003. A water productivity framework for understanding and action. In: Kijne, J. et al., (Ed.), Water Productivity in Agriculture: Limits and Opportunities for Improvement, Comprehensive Assessment of Water Management in Agriculture. CABI Publishing in Association with International Water Management Institute, UK.

Molle, F., Vallée, D., 2009. Managing competition for water and the pressure on ecosystems. In: Water in a Changing World, United Nations World Water Development Report 3. UNESCO Publishing, Earthscan, London, UK.

Molle, F., Wester, P., Hirsch, P., 2010. River basin closure: processes, implications and responses. Agric. Water Manag. 97 (2010), 69–77.
Oliveira, R.S., Bezerra, L., Davidson, E.A., Pinto, F., Klink, C.A., Nepstad, D.C., Moreira, A., 2005. Deep root function in soil water dynamics in cerrado savannas of central Brazil. Funct. Ecol. Br. Ecol. Soc. 19, 574–581.
Parthasarathy Committee Report, 2006. From Hariyali to Neeranchal: Report of the Technical Committee on Watershed Programmes in India. Department of Land Resources, Ministry of Rural Development, Government of India, New Delhi.
Perry, C.J., 2007. Efficient irrigation, inefficient communication, flawed recommendations. Irrig. Drain. 59, 367–378. https://doi.org/10.1002/ird.323.
Reddy, V.R., 2006. Getting the implementation right: can the proposed watershed guidelines help? Econ. Polit. Wkly XLI (40), 4292–4295.
Sahin, V., Hall, M.J., 1996. The effects of afforestation and deforestation on water yields. J. Hydrol. 178, 293–309.
Syme, G., Reddy, V.R., Pavelik, P., Croke, B., Ranjan, R., 2012. Confronting scale in watershed development in India. Hydrgeol. J. 20 (5), 985–993.
Voropaev, G.V., Zaitseva, N. (Eds.), 1996. Environmental and Socio-Economic Consequences of Water Resources Development and Management, International Hydrology Programme. UNESCO, Paris.
Winter, T.C., Harvey, J.W., Franke, O.L., Alley, W.M., 1998. Groundwater and surface water a single resource. US Geological Survey Circular 1139, Denver, Colorado.
Wood, E.F., Sivapalan, M., Beven, K., Band, L., 1988. Effects of spatial variability and scale with implications to hydrologic modeling. J. Hydrol. 102 (1–4), 29–47.

CHAPTER 2

Catchment management to basin management: International perspectives and overview of global experience

M. Dinesh Kumar[a], Charles Batchelor[b], A.J. James[c]
[a]Institute for Resource Analysis and Policy, Hyderabad, Telangana, India
[b]Water Resource Management Ltd., London, United Kingdom
[c]Environmental & Natural Resource Economist, New Delhi, India

2.1 Introduction

Watershed development is an approach to raise agricultural productivity, conserve natural resources, and improve rural livelihoods in the regions suffering land degradation, which are often characterized by high levels of food insecurity and income poverty (Hope, 2007; Farrington et al., 1999). Watershed development has emerged as an important policy instrument for rural development in many developing countries, including India (Hope, 2007). Since the 1970s, India has invested significantly in watershed development as a driver of rural development (Joshi et al., 2005), partly in an attempt to scale up successes from a handful of well-known village-level watershed projects, such as Pani Panchayat, Ralegaon Siddhi, and Sukhomajri (Turton et al., 1997). But, the focus of watershed development in India has evolved over the last 25 years from soil conservation to water conservation to now include a more participatory planning approach (Hope, 2007). In fact, the importance of institutions of collective action for implementing and managing watershed programs was recognized in the GOI guidelines formulated in 1994–95 (Reddy et al., 2004).

A principal assumption underlying watershed development and management project is that there is sufficient amount of un-utilized water flowing out of the agricultural watersheds (Kumar et al., 2006, 2008), which, if can be captured during the wet season, may be made available in dry periods

Current Directions in Water Scarcity Research, Volume 1
https://doi.org/10.1016/B978-0-12-814851-8.00002-1
© 2019 Elsevier Inc.
All rights reserved.

offering several potential benefits (Farrington et al., 1999). Socioeconomic benefits commonly associated with watershed development include improved agricultural yields and farmer returns, increased access to domestic water, and new employment opportunities (Kerr et al., 2002). However, these benefits will vary for different resource user groups located across watersheds. However, emerging global evidence suggests that there are limits and tradeoffs to modifying watersheds due to complex hydrological and social systems' interactions (Calder, 2005). This is partly because watershed development interventions modify land use impacts on water resources, which in turn may alter downstream water access, while augmenting upstream water supplies (Batchelor et al., 2003; Gosain et al., 2006).

Evident are dramatic changes in the hydrology of many dry areas of India in recent years as a result of increased groundwater-based irrigation, watershed development, water harvesting, and land use change. Although current watershed development programs bring a range of benefits, they may also change the temporal and spatial pattern of water availability and use at the scale of large catchment in which these watersheds fall (Batchelor et al., 2003; Calder, 2005; Kumar et al., 2008). This can result in significant negative tradeoffs such as more unreliable lean season flows, particularly during low rainfall or drought years, drying up of tanks and large reservoirs, and damage to riverine ecology.

But, the political preference for watershed development remains high in India, generally for supply-side interventions. Given the fact that nearly Rs. 2200 crore (US$340 million) per year is allocated to watershed projects in India at present (Reddy, 2012), there is a need to ensure enhanced performance of the watersheds from social, economic, and environmental perspectives and to identify and mitigate perverse outcomes and negative externalities (Batchelor et al., 2003; Calder et al., 2008). There is increasing recognition among scholars of the need to take a look at the hydrology and land use in the larger catchment, in order to decide on the degree /scale of agricultural intensification and rainwater harvesting in order to avoid undesirable effects of watershed management (Calder et al., 2008). But, while doing this, water managers should recognize that hydrological characteristics of catchment are not necessarily stationary, and instead are constantly altered by changes in regional land use and hydrological structures (Moench and Dixit, 2004).

This chapter provides a review and puts together catchment assessment and planning aspects of watershed management, that are practiced under international programs and projects; the catchment/watershed management programs undertaken by different countries of the world; and, the integrated watershed management program of government of India, and other donor-supported projects.

2.2 Catchment assessment and planning for WSM: Theoretical and conceptual aspects

Watershed management professionals generally view that increasing the thickness of vegetative cover in the upper catchments would reduce erosion and sediment load in the runoff, and increase the base flows, irrespective of the agroecology. Often international agencies and scholars, aggressively promoting small-scale rainwater harvesting in catchments, view rainwater as separate from groundwater and surface water (COMMANS, 2005; Stockholm Environment Institute, 2009). Such views ignore the following: (1) how the type of catchment land use influences the impact of increased vegetation cover on stream flows (including water quality), in different agroecologies; (2) how the nature of vegetation (whether shallow-rooted grasses and shrubs or deep-rooted trees) determines the impact of increased vegetation cover on the consumptive use of water from the soil profile and groundwater system of the catchment (Oliveira et al., 2005) and the stream flows (Bosch and Hewett, 1982), and how these impacts can change across agroecologies; (3) the hydraulic interdependence between groundwater and surface water in a catchment and therefore the impact of change in groundwater withdrawal on stream flows downstream (Leblanc et al., 2012); and, (4) rainwater (precipitation) is the source of surface water, soil moisture, and groundwater in the catchment. But, this knowledge is extremely important for catchment management measures.

2.3 International literature on catchment assessment and planning

2.3.1 Catchment assessment

2.3.1.1 State-of-the-art knowledge of relationship between changes in land use/land cover and catchment yield and soil erosion

The traditional perception that forests increase water resources has long been questioned by the results of scientific forest hydrology since the early 20th century (e.g., Calder, 2002). A large number of catchment experiments conducted all over the world clearly demonstrated that the deforestation of a catchment implies an increase of water yield from it and, conversely, the establishment of a forest cover implies a decrease of water yield (Sahin and Hall, 1996). The application of this knowledge to designing sustainable water management practices, although necessary in water-stressed regions, has been largely delayed because of difficulties inherent to the change of any scientific paradigm, the limited experience on the hydrological consequences of land cover changes in large territories, and the disconnection between policy and science (Falkenmark et al., 2004; Calder, 2002).

Most of the microcatchment management programs in India are grounded on inadequate knowledge of the relationship between changes in land use/land cover and catchment yield and soil erosion and wide misconceptions concerning their functions in different agroecologies. Therefore, scientific understanding of the relationship between catchment land use and land changes and hydrological and ecological processes is essential for effective catchment planning.

Such scientific understanding is developed elsewhere through experimental field monitoring of data on soil erosion, sediment yield and catchment runoff, and the use of simulation models. Some of the most fascinating ones, including those available from modeling studies, are relative effectiveness of field-scale soil conservation measures taken in reducing both soil loss and sediment yield over grass buffer strips and retention ponds (Verstraeten et al., 2002); negative impact of increase in forest cover in dry areas of Spain on river flows in Spain (Gallart and Llorens, 2003); rising groundwater levels and increase in soil salinity as a result of clearance of native vegetation for agricultural use in Murray Darling basin (Leblanc et al., 2012); combined effects of a combination of socioeconomic developments and land use changes occurring historically on river water quality in Scotland (Pollard et al., 2001); the importance of analyzing the spatial and temporal patterns in rainfall and land use in the catchments over and above the aggregate scenarios, in explaining the spatial and temporal variations in runoff occurring in the catchment in North West England (Orr and Carling, 2006); the negative impact of replacing paddy fields by forests on catchment yield, with just an opposite impact from replacement of forests by crop land, and positive impact of irrigated paddy on watershed hydrology in terms of quantum of flow in Chi river basin of Thailand (Homdee et al., 2011); and distinction between notional water saving and real water saving through efficient irrigation technologies, and importance of scale in deciding the water-saving impact of water use efficiency improvements in agriculture (Wallace and Batchelor, 1997).

2.3.1.2 Knowledge use in catchment management decisions

As mentioned in Chapter 1, simple knowledge about the nature of impacts that land use and land cover can cause on hydrology of the catchment and water quality parameters will not be sufficient for taking catchment management decisions. Rather, a quantitative understanding of the impacts of these impacts, which finally decide whether the social, economic, and environmental outcomes at the catchment level are optimal or not, is needed. This can only be obtained using complex mathematical models, which simulate the hydrological and biophysical processes.

What is important to note is that while changing land use and land cover and construction of vegetative bunds in the upper parts of the catchment as part of the ICM approaches could change the catchment yield along with soil loss, withdrawal of either surface water or groundwater in the catchment to affect such changes could also cause variations in water yield received in the lower parts of the catchment. This in turn can bring about drastic ecological changes in those areas in terms of changes in nature and extent of vegetation that the river plains support. Hence, these models have to be used in an integrated way to understand the cumulative effects.

As indicated by Verstraeten et al. (2002), when soil conservation measures such as changing the crop rotation sequence or terracing or contour bunds are taken up at catchment scale, the effect of sediment control measures such as grass buffer strips or retention ponds on sediment yield from the catchment would be less visible. Hence, there is an optimum level of treatment, which needs to be carried out in the catchment beyond which intensifying the treatment would not produce additional benefits. In certain situations, such activities can also lead to negative social or environmental outcomes. Integrated catchment models could help identify the optimum level of various interventions to maximize the overall benefits[a] in physical terms at catchment scale.

In the case of water use efficiency (WUE) improvement measures, the knowledge about the difference between notional water saving and real water saving from efficient irrigation systems is crucial, while applying knowledge derived from farms to the catchments for planning. Sufficient insights into the likely impact of using efficient irrigation systems in terms

[a] The overall benefits in physical terms could be sum of the individual benefits of reduction in soil loss reduction and sediment yield, flood control, increased water availability of drinking, irrigation and other competitive uses, and water availability for ecological flows during lean season.

of real water saving would help avoid unintended consequences of depleting more water rather than freeing water from agriculture (Wallace and Batchelor, 1997; Kumar and van Dam, 2013; Howell, 2001; Molle and Turral, 2004; Perry et al., 2009). Since the hydrological process involved is quite complex and dynamic, in order to get an empirical understanding of the catchment scale impacts of farm-level WUE improvement measures, modeling is required. The model used for catchment assessment and management planning should be such that it is able to simulate the dynamic hydraulic interaction between surface irrigation and groundwater.

The fundamental question is how the knowledge, which has been in existence for several decades in the public domain, gets ignored in policy formulation. The key issue is with regard to the ability to convert this knowledge into "proper evidence" to convince the policy makers and counter the reductionist and misinformed view of "what constitutes evidence" (Whitty and Dercon, 2013).

2.3.2 Main components of integrated catchment management in major international programs/projects

2.3.2.1 Physical/technological aspects

There are a wide range of technologies used in catchment management worldwide, depending on the agroecology. Ideally, the type of intervention should vary from humid, high-rainfall regions to arid low-rainfall regions; it should also vary from temperate climates to more tropical climates; the interventions for an area with mild slopes would be vastly different from those used in a hilly watershed.

The treatment technologies adopted in catchment management programs are terracing; contour bunding for preventing erosion and soil & water conservation; contour trenching for moisture conservation; grass plantation for soil conservation; gully plugs for preventing gully erosion due to runoff; gabien structures; check walls for reducing erosive force of the runoff and to retain sediments; check dams for reducing erosive force of water and to impound some water; vegetative barriers on stream banks for preventing bank erosion; earthen dams for sediment control; sediment retention ponds; mulching for moisture conservation; and afforestation (Buhl and Sen, 2006).

In addition, there are several regulatory measures, which can be used to bring about changes in the performance of the catchment, in terms of yield, runoff quality, condition of the wetlands, etc. For instance, reducing area under rain-fed crops might help increase the runoff. Similarly, reduction in area under irrigated cropland and size of livestock might help reduce the nitrate load in the runoff, occurring as a result of leaching of fertilizers and organic waste from dairy farms in areas (McDonald et al., 1995).It can also prevent nitrate contamination of groundwater in humid high-rainfall areas with shallow water table conditions (Kraft and Stites, 2003). Augmenting river flows during low flows and high temperature through increased release of water from reservoirs in the upper catchment is another widely practiced regulatory measure for catchment management, particularly for reducing algal growth, protecting birdlife, and increasing fish breeding (Collins et al., 2007).

But over the last two decades, several new technologies and practices were added to the watershed treatment technology package activities in India to address the local needs. Most notable among them is structures for groundwater recharge, adopted in response to depletion and degradation of groundwater in the semiarid and arid areas. The recharge restructures try to create the saturated conditions in the unsaturated zone whereby the hydraulic conductivity of the formation increases sharply. Most of them are aimed at augmenting groundwater resources in the watersheds (COMMANS, 2005).

2.3.2.2 Institutional, legal/regulatory measures

For successful implementation of integrated catchment management programs, institutional structures that facilitate the involvement of stakeholders in the development and implementation of appropriate natural resource management strategies and policies need to be developed (Batchelor, 1999). There are some instances where developed countries have implemented integrated catchment management programs through legal, institutional, and policy approaches. The fundamental change brought about in water management in these countries through the adoption of ICM approach has been organizing water resource management around hydrological boundaries. Examples are Australia, Britain, South Africa, and France (based on Johnson et al., 1996; Bellamy et al., 2002; Buller, 1996; Batchelor, 1999; Connell, 2012; European Environmental Bureau (EEB), 2010; Herrfahrdt-Pähle, 2010; Mitchell and Hollick, 1993; Schreiner and Van Koppen, 2002).

However, many ICM strategies have not been able to bring about improvements in resource management at the catchment scale mainly due to lack of delivery mechanisms and enabling policies that generate the interest and trigger the participation of local institutions and communities (Batchelor, 1999). In certain cases, catchment management programs were also implemented effectively without legislative support. But, in such cases, the success of the initiatives is very much dependent on involvement of community leaders and resource agencies (Johnson et al., 1996). Several institutional models are tried in ICM around the World with varying degrees of success, beginning in Europe (Britain and France in the early 90s, though the most common among them being tried in developing countries is decentralized, community-based institutions for implementing watershed management program at the level of microcatchment (Darghouth et al., 2008).

The nature and functions of these institutions vary from country to country and in a few instances from province to province. They range from: (1) the Catchment Management Plans in Britain, which form the basis for the actions of its National River Authority (NRA) in the respective basins; (2) the two-tier tradition of water management with SAGE (*Schemas d aménagement et de gestion des eaux*) and SADGE (*Schémasdirecteurs d'aménagementet de gestion des eaux*), implemented through local management commissions and a higher-order River Committees, respectively, in France; (3) the River Basin Management Plans (BMPs) developed by the member states of European Union under its flagship legislation of European Water Framework Directive; (4) the Basin Plan, which provides a new management framework for a transboundary, river-catchment-level management of water resources in the Murray Darling Basin, encompassing four basin states, being implemented by the newly constituted MDB authority; (5) Integrated Catchment Management program of western Australia, which enjoys, legislative, policy, administrative, and financial support since the early 90s; and (6) Catchment Management Agencies (CMAs) in South Africa that are created for ensuring poor people's access to water for domestic and productive purposes, being facilitated the DWAF under the National Water Act.

While the CMPs of Britain remain wholly consultative with respect to the planning process, with no binding force on land users, developers, and local planning authorities, the NRA, which is the higher-level institution, possesses a wide range of regulatory powers over the private and public bodies that directly impact on the water environment. While SDAGE has statutory obligations, SAGE is discretionary. With the passing of the European Water Framework Directive in 2000, more than 10 years of planning and consultation went into development of River Basin Management Plans (RBMPs) by member countries. These plans are meant to be the main instruments for realizing the environmental objectives of protection of water bodies from nutrient pollution. 14 out of the 27 member countries have come out with RBMPs for the basin districts in their respective countries. In Australia, the Basin Plan, which is legally enforceable, defines sustainable limits for groundwater and surface water; basin-wide environmental objectives; roles for a basin-wide water trading system; requirement of subplans for each one of the four states to implement the Basin Plan objectives; measures to improve the security of water entitlement holders. In South Africa, the process of establishment of Catchment Management Agencies began with the passing of the National Water Act in 1998, but only two out of the nine proposed CMAs are fully established and operational so far.

Table 2.1 summarizes the institutional arrangements for catchment management and the management instrument administered by them in selected countries worldwide.

2.3.3 Catchment planning tools and processes

2.3.3.1 Integrated crop, rainfall runoff, soil erosion, and economic models

There are multiple social, economic, and environmental objectives underlying catchment management approaches, and the interventions needed for achieving one of them could work at cross purposes with the other two.

As a result, value of the marginal returns from ICM interventions owing to incremental benefits to some sectors may not be higher than the marginal costs, which will be the sum of the incremental cost of the interventions and the reduction in benefits to some other sectors, at every scale of intervention. The net marginal returns can become negative at some scale of a given management activity. Therefore, it is important to analyze the tradeoffs. This would involve quantification of each one of the impacts from the selected ICM interventions, and assessment of the economic value of the net benefits accrued, for various scales of interventions, and choosing the one which gives highest overall benefits. Only such integrated assessments of resource use and resource management can provide good guidance to resource managers on decision making with respect to catchments.

But, planning for catchment or watershed management in India is not driven by considerations of the costs and benefits associated with the economic, social, and environmental outcomes. This is partly because of the lack of scientific data defining the physical relationship between catchment interventions on land use and land cover, and their hydrological outcomes, and the economic imperatives (Kumar et al., 2006). Another reason is the drive to address the concerns of equity in distribution of benefits of agricultural productivity enhancement programs to rain-fed areas where large number of poor people live (Kerr, 2002). Ultimately, what gets implemented as a catchment management plan has more to do with the current political economy rather than what is optimal from social, economic, and environmental angles.

Table 2.1 Institutions for catchment management and management instruments.

Name of institution	Year of establishment-continued till	Nature	Characteristics	Objective functions	Outcomes
River Basin Organizations, France	1970–1992	Water financial investment institutions with no statutory powers	Intermediary institution between central and local government; and, between water consumers and water industry	Water resource management in major river basin, by promoting efficient water use and pollution reduction through financial incentives to local agencies and private players	**(i)** Introduced charges for water abstraction and discharge, by various users in the basin, under the 1964 Water Act, and became financially powerful **(ii)** Effective in case of larger & identifiable point polluters, but not for widespread non-point source pollution **(iii)** Unsuccessful in implementing "polluter pays principle"
Regional Water Authorities (RWA), England and Wales	1973–1989	Regulatory and management functions	Centralized body	**(i)** Flood control **(ii)** Organized water supplies **(iii)** Water resources management	**(i)** No active community participation in catchment management **(ii)** Distortion in catchment development pattern due to inconsistency between local land-use planning authorities and RWAs in their goals **(iii)** Had a dual role both in policing water pollution (regulatory function) and being polluters themselves

Integrated Catchment Management Coordinating Group (ICMCG) and Community Catchment Groups (CCG), Western Australia	1987	ICMCG is to plan for ICM, its implementation, and advice CCG CCG is expected to participate in planning process	ICMCG is a centralized body with representation from various State government agencies responsible for land and water management CCG is a community-based institution formed under guidance of ICMCG	**(i)** ICMCG provides forum for interagency coordination of policies and activities **(ii)** CCG was expected to involve local people in defining problems and developing strategies for catchment management	**(i)** ICMCG played a key role in developing a draft state wetlands policy and provided a platform to develop coordinated action on the Swan River coastal plain **(ii)** Reporting structure of ICMCG created resentment in some senior officers who thought it to be questioning their judgment and activities **(iii)** Formation of ICMCG led to uncertainty about the role of other agencies such as the WA Water Resources Council and the Soil and Land Conservation Council, land care groups, etc.
National Rivers Authority (NRA) and Catchment Management Plans (CMP), England and Wales	1989	NRA is a regulatory institution CMP forms the basis for NRA action within each catchment	NRA is a centralized agency with range of regulatory powers on private and public bodies CMPs are consultative with respect to the planning process, with no ability to enforce on land-users, developers or local planning authorities	**(i)** Protection of water environment **(ii)** Establish management protocols and monitoring programs **(iii)** Facilitate better coordination between land use and water policy makers	**(i)** Highlighted the divergence between water management and land-use planning functions **(ii)** However, NRA did little to incorporate water management decisions in land-use planning as it had limited formal authority to restrict land use development **(iii)** CMP had no immediate impact on statutory authorities

Continued

Table 2.1 Institutions for catchment management and management instruments.—cont'd

Name of institution	Year of establishment-continued till	Nature	Characteristics	Objective functions	Outcomes
Catchment Co-ordinating Groups (CCG) and Catchment Care Committees (CCC), Queensland, Northern Australia	1991	CCG and CCC were formed to provide forum for public discussion on catchment management	No legislative basis, voluntary involvement of stakeholders	**(i)** Promote stake-holders partici-pation in catchment management **(ii)** CCC identifies and prioritizes natural resource issues; and develops, pro-motes, and facil-itates implemen-tation of CM strategies **(iii)** CCG take part in local action and planning	**(i)** ICM in Queensland has fos-tered co-ordination between landholders, community action groups, and govern-ment agencies **(ii)** Under this institutional arrangement, ICM is dependent on the support of prominent community leaders and resource agencies **(iii)** ICM is not successful in solving the land use conflicts **(iv)** Financing remains a major issue for effective function-ing of CCC
SAGE and SDAGE in France	1992	SAGE is a precise planning document for small local catchments SDAGE is a forward planning document for	SDAGE has statutory obligation SAGE is discretionary Local Management Commission to	**(i)** SAGE is to influ-ence land-based activities that have direct impact on water resources and aquatic environment	**(i)** A step toward involving all possible stakeholders in catchment management **(ii)** Spatial and institutional fragmentation of local water responsibility continues in France

		major river basins Both draw on elaborate participation of water users, consumers, regulators, and policy makers	implement SAGE and a higher-level institution of River Committees to implement SDAGE		**(iii)** This institutional set up was driven more by political rather than environmental considerations, and is not suitable to achieve sustainable water management objectives
Catchment Management Agencies (CMA), South Africa	1998	Implementing National Water Act and management of water at local level	Participatory; but are statutory bodies Governing Board of CMA to have representation from present and potential future water users, local and provincial government, & environmental interest groups	**(i)** Planning of WRM at catchment scale **(ii)** Registration of water users **(iii)** Water charges collection **(iv)** Water authorization and licensing **(v)** Ensure poor communities' access to water for domestic and productive needs	**(i)** Shift from a centralized management approach to a decentralized participatory approach based on equity **(ii)** Water users from poor communities got opportunity to discuss directly with the concerned agency on WRM **(iii)** Only two CMAs became fully operation in 12 years **(iv)** Delegation of water management functions to the CMAs has only been partially implemented **(v)** Poor cooperation between the CMAs and the water supply agencies, due to the misfit between catchment boundaries and local administrative boundaries

Continued

Table 2.1 Institutions for catchment management and management instruments.—cont'd

Name of institution	Year of establishment-continued till	Nature	Characteristics	Objective functions	Outcomes
River Basin Management Plans (RBMPs) under the European Water Framework Directive	EWFD was adopted in 2000 by EU members States	EWFD is a legislative framework RBMPs are instruments for realizing the new environmental objectives	Preparation of RBMPs is a consultative process which each EU member State has to follow for realizing the ecological objectives set under EWFD	**(i)** Water quality protection (both surface and groundwater), especially to control nutrient pollution **(ii)** To assess environmental quality **(iii)** RBMPs is for water management based on river basins	**(i)** Till 2010, only 14 member States had adopted RBMPs **(ii)** Lack of transparency and robustness in assessment of environmental quality by the member States **(iii)** Measures suggested in the RBMPs to restore specific water quality elements such as nutrient conditions were useless **(iv)** Only 1/3rd of water bodies were targeted for restoration in next few years **(v)** Limited progress in dealing with negative impacts of nutrient pollution on aquatic ecosystem

Murray Darling Basin Authority (MDBA), Australia	2007	Authority for developing and implementing Basin Plans Basin plan has legislative backing and is an enforceable document	Centralized body with community participation through basin community committee	**(i)** To limit surface and groundwater diversions from the basin for environmental sustainability **(ii)** Ensure that sub-plans for each state are in line with the basin plan **(iii)** To promote water trading to maximize the social and economic benefits, while ensuring hydrological integrity of the basin	**(i)** MDBA replaced the existing institutional arrangements at the state level, which were unable to achieve environmental sustainability and resource security targets **(ii)** Shift in responsibility for high-level policy making for basin management from the states to the national government **(iii)** Creation of environmental water holder at the basin level **(iv)** Provision of water entitlements for environment created resentment among irrigation stakeholders as it was bound to reduce the overall size of allocation for that sector

Continued

Table 2.1 Institutions for catchment management and management instruments.—cont'd

Name of institution	Year of establishment-continued till	Nature	Characteristics	Objective functions	Outcomes
Catchment Management Authorities in New South Wales (NSW), Australia	2003	They replaced the existing Regional Catchment Coordination Committees & Catchment Management Boards, Trust, and local land-care groups	Legal entities set up after the enactment of Native Vegetation Act-2003; Catchment Management Authorities Act-2003 and Natural Resource Commission Act-2003 13 CMAs exist in NSW	**(i)** Responsible for implementing integrated approach to catchment management, with community participation **(ii)** Develops Catchment Action Plans	**(i)** Effective mechanism for supporting land owners to voluntarily manage their land for public and private benefit **(ii)** Gives regional communities a direct say in the complex task of reconciling their needs with ecosystem health **(iii)** Very good, well-designed projects were delivered throughout NSW; the activities had strong links with the expected outcomes; over 90% achieved expected short-term outputs (as per the Audit carried out in 2010)

Source: Based on Johnson et al. (1996), Bellamy et al. (2002), Buller (1996), Connell (2012), European Environmental Bureau (EEB) (2010), Herrfahrdt-Pähle (2010), Mitchell and Hollick (1993), Schreiner and Van Koppen (2002).

Internationally, in many cases, water resource management projects have concentrated only on "physical control of water." In other cases, economic aspects were attended to, but, environmental and social effects were at best given token consideration (Jakeman and Letcher, 2003).

The reason for this is that the decision on how much water is needed for meeting environmental flows demands answering questions like when, how often, and for how long river flows are needed to protect various river ecosystem goods and services. But, the concepts, theories, and methods of assessment of environmental flow and their practical applications are largely new in most developing countries. Moreover, environmental flow is a relatively new concept for the water sector and there is lack of awareness among the general public on the concept and its application (Kashaigili et al., 2005). Integrated assessment tools try to address these concerns by allowing efficient scenario generation for different catchment management choices, in particular assessment of the resultant tradeoffs among indicators of environmental, economic, and social outcomes (Jakeman and Letcher, 2003).

2.3.3.2 Planning processes in catchment management

Catchment management is about managing the multiple perspectives and stakeholder interests, and the conflicts emerging out of it (Prato and Herath, 2007). But to begin with, planning for integrated catchment management requires scientific understanding of the interaction between the catchment flows (both surface water and groundwater) and the human systems, which depend on it; and the catchment flows and water dependent ecosystems such as wetlands, swamps, and biodiversity reserves. It is really important to understand two things: (1) The hydrological processes that determine flows and fluxes with respect to space and time; and, (2) how the hydrological gain in one location in the catchment would mean a loss for a downstream location. It is also important to base decisions on empirical evidence, rather than expert opinions and anecdotes (Whitty and Dercon, 2013). When such evidence does not exist in amble measure, it will have to be generated, sometimes through complex modeling tools, which are also referred to as Decision Support Tools (DST), which are used to decide on ICM interventions to enhance the catchment functions. As seen earlier, catchment-wide management of land and water resources has to be based on multiple objectives and criteria (Prato and Herath, 2007), and therefore involve tradeoffs (Prato and Herath, 2007). For the tradeoffs to be socially acceptable, multistakeholder dialogs is an essential component (Falkenmark, 2004; Warner, 2006).

There are a variety of tools and techniques for facilitating informed decision making for improved catchment management, in situations of multiple values and objectives that exist among stakeholders. Prato and Herath (2007) discussed Multicriteria decision making in catchment management. Collins et al. (2007), using case studies of three catchment management projects in the United Kingdom, illustrated on the extent to which a systemic approach to understanding multiple perspectives and stakeholder interests could improve policy and practice. Ewing et al. (2000) illustrated the potential of a decision support system called Adaptive Environmental Assessment and Management (AEAM) in facilitating Integrated Catchment Management (ICM) processes in the context of Western Australia, through a case study of Blackwood River Catchment.

Besides the scientific and technical aspects of catchment planning process, there are other important aspects in the planning process, and which precede the former. They concern identifying community desires and objectives with respect to development of catchment; developing strategies to meet the objectives; and carrying out audits. This requires institutional platforms like the CMPs in England and Wales, SAGE in France, CMAs in Western Australia.

Institutional development process, which is often long drawn, is integral to the planning process, as illustrated by the experience of setting up Catchment Management Agencies in South African river catchments (Schreiner and Van Koppen, 2002).

Community participation in the management of river basin areas is explicitly stipulated in Article 14 of the European Water Framework Directive (EWFD) (European Environmental Bureau (EEB), 2010), which states that the general public should be consulted in the formulation of basin management plans. However, there are indications that the "Directive" accepts public involvement, which is reduced to mere "consultation" and "information provision" rather than meaningful participation. Again, if such consultation processes are dominated by the powerful groups like the State water agencies, the communities will not have much influence on the process (Mostertman, 2005), communities can easily get frustrated into thinking that "nothing ever changes" and therefore will find the entire exercise meaningless. Since participation involves considerable opportunity cost for the stakeholder groups, the benefits have to outweigh these costs (Warner, 2006).

2.3.4 Outcomes of implementing CM plans: Experience from select countries

A few countries or territories have implemented the concept of ICM programs and projects in terms of management instruments called "catchment management plans," and also created institutional arrangements for their implementation. Australia has been a pioneer in implementing ICM, often basin wide, using formal institutional structures or otherwise, covering large geographical areas. Many countries in the European Union had implemented ICM, sometimes within the country's territories or at the scale of international river basins. But, studies that scientifically documented the outcomes and impacts of such initiatives on water resources are extremely limited. Limited evidence seems to suggest a mixed experience.

The United Kingdom had implemented environmental management policies and institutional structures for achieving the goals of ICM in early 1990s. But, as per the recent monitoring data across England and Wales, percentage of rivers that have nitrate levels exceeding 30 mg/L changed from 29.9% in 1995 to 28.3% in 2005 and the percentage of rivers with phosphate levels exceeding 0.1 mg/L increased from 50.3% to 51.5%, during the same period. Such unexpected outcomes are the result of failure to apply "systems perspective" while using the well-established scientific knowledge about physical processes in policy formulation. What this suggests is that catchment management requires natural and social scientists to work more closely, to provide robust analysis of water management outcomes of environmental policies that fully considers the biophysical, social, political, and economic settings (Macleod et al., 2007).

But, there are notable positive experiences with catchment-wide water trading and reallocation of water for environment in the case of Murray Darling Basin and integrated catchment management in New South Wales in Australia, even prior to the implementation of the National Water Act of 2007. Water trading has been progressively introduced in the southern MDB since the late 1980s, along with further reforms in the mid-1990s and again in the mid-2000s. It was based on the premise that trading provides economic benefits to buyers and sellers, and to society as a whole. Trading of water rights enabled increased water use efficiency in irrigated production, through the movement of water from low-valued uses to high-valued uses (National Water Commission, 2010).

In South Africa, through the implementation of the NWA-1998, attempt was made to improve government and management of water resources by organizing them around hydrological boundaries under the institution called CMAs. But, several issues still need to be sorted out. Water-related services such as drinking water and industrial water supply are being managed by Water Service Agencies, which are within the administrative control of local governments such as municipalities. The Water

Services Development Plans being developed by the WSAs within the local governments are based on information provided by the CMAs on water availability within the basin. But, these plans do not take into water demands in agriculture, and concern only water supply and sanitation and urban and industrial water supply (Herrfahrdt-Pähle, 2010).

There are problems of poor cooperation between the newly established CMAs and the WSAs, due to the misfit between catchment boundaries and administrative boundaries of local governments under which the line agencies of water supply work. There are also problems of lack of coordination within the DWA between the divisions dealing with water resources management and water-related services, and between DWA and the water service agencies (Herrfahrdt-Pähle, 2010). There are currently no specified procedures and rules that guide cooperation between CMAs and the local governance institutions. Cooperation is based on capacity and levels of understanding of legislation and strategies by individuals within these institutions (Mazibuko and Pegram, 2006).

Whereas in the NSW, the 13 catchment management authorities created under the ICMP have become effective mechanism for supporting land owners to voluntarily manage their land better for both public and private benefits. It has given the regional communities a more direct say in the complex task of reconciling community needs with ecosystem health. The audits of the CMAs done, during the recent droughts, showed improvement in resource condition in half of the cases (source: Report of the Commissioner, New South Wales Natural Resources Commission, Level 10/15 Castlereagh Street, Sydney, NSW 2000, Australia. John.Williams@nrc.nsw.gov.au).

2.4 Catchment planning for watershed management programs in India

2.4.1 State of knowledge about catchment hydrology for WSM programs in India

In India, the current state of knowledge about the catchment processes is extremely inadequate. One major issue is that the other key factors, which determine the impact of land use and land cover on catchment hydrology, viz., rainfall regime, geohydrology, climate, soils and topography, vary remarkably from region to region, making it difficult to apply the limited knowledge available for select catchments from other parts of the world, for the hundreds of thousands of country's catchments, which have varying environmental conditions. This means, in the Indian context, such scientific data need to be generated for typical catchments.

A major lacuna arises from the lack of adequate scientific knowledge of catchment characteristics, hydrology, hydrogeology, stream flows, and groundwater-surface water interactions. To start with, hydrological data are not available for many catchments. Many small basins in India are ungauged for stream flows, and some do not even have adequate number of rain gauges to capture the spatial variation in rainfall across the basin. Also, the data from non-IMD stations such as those managed by agricultural universities, state irrigation department, and forest department are of poor quality. While rainfall-runoff relationship can be established with known values of daily rainfall, data on soil type, land use, and land cover, such datasets are also not available for small catchments on spatial scale. This makes it practically impossible to analyze how changes in land use and land cover in the past had affected the catchment hydrology and the likely future changes in the same.

Improved access to online terrestrial and data on rainfall, soil moisture, etc. acquired using wireless sensors and properly calibrated, is making it a lot easier to understand catchment hydrology, especially rainfall-runoff relationships, when supported by field evidence of catchment outflows and inflows into tanks and other wetlands (Campbell and Newfoundland, 2010).

Groundwater-surface water interactions are important, but more significant for hilly and mountainous catchments, wherein groundwater outflows contribute to stream flows not only during monsoon but also during lean season (Kumar et al., 2005). By the very nature, these hilly and mountainous become priority areas for catchment management initiatives. But, water resource assessment, especially groundwater resource assessments, does not provide any separate data on groundwater outflows into surface streams (base flow) in such regions.

2.4.2 Planning for catchment management in the IWMP

The planning for watershed management activities in India under the IWMP is highly decentralized, while the funds for undertaking the program is from the central plan allocation, administered by the Ministry of Rural Development, Government of India. The plan allocation is driven by the consideration of the number of districts and the total land area falling under rain-fed category in each state. The underlying rationale is that areas that are rain-fed—areas where the total net irrigated area is less than 25% of the cultivable area—attract much less government investment in agriculture as compared to irrigated areas. Also, since most rain-fed areas in the semiarid and arid parts of the country face problem of groundwater overexploitation (Kumar, 2007), the program is viewed as a tool for arresting it through resource augmentation.

In developed countries, the primary focus of integrated catchment management was of reallocating water from agriculture to other sectors and to make agricultural land use more ecologically sustainable. But, in India, the primary objective has been to improve the productivity of agricultural land by making more water or soil moisture available to the farming communities, and arrest degradation of common land (Farrington et al., 1999; Hope, 2007). The key concern is improving soil moisture regime in the agricultural and nonagriculture land and augmenting groundwater recharge. The unit for planning catchment management interventions is a small microwatershed of 500–1000 ha size, which makes it difficult for communities within the watershed to foresee other potential uses of water, which flows out of their catchments.

The central funds, which are allocated to the states, are routed through the district rural development agencies, which in turn allocate the funds to a wide range of agencies, to finally reach the watershed communities. The project implementing agencies (PIA), which receive the funds from against the proposals that specify the watersheds and the area to be treated, carry out watershed planning in their respective areas with the involvement of the stakeholder communities, through an institution of the communities, which are promoted by them, called Village Watershed Committees.

Watershed planning in India under IWMP generally does not involve any scientific considerations of the catchment hydrology, including available runoff and uncommitted runoff from the catchment (Batchelor et al., 2003). The program focus is on soil and water conservation measures and water resources development. Since the unit for planning is very small, for which hydrological data are not available (except for experimental watersheds), hydrological planning is hardly attempted. The KAWAD project is a clear illustration of this poor planning.

It also does not consider the demand for water in the watershed (from irrigation, livestock drinking, domestic uses, and water for environment) against the utilizable surface and groundwater. While water demand far exceeds the supplies in most semiarid and arid regions of India (Biggs et al., 2007; Batchelor et al., 2003; Kumar et al., 2008), with increasing number of basins either closed or on the verge of "closure" (Biggs et al., 2007; Kumar et al., 2008), the program by design does not involve components, which address demand management issues (Batchelor et al., 2003).

Instead, the accent is on building more water-harvesting structures, on the premise that it would increase the local water supplies, an idea supported by most researchers and academicians. While environmental flow considerations are gaining greater acceptance in water resource decision making, this is hardly visible in watershed management programs and policies in India (Smakhtin and Anputhas, 2006).

While the approach is participatory planning at the microwatershed level (James, 2000), the Village Watershed Committees are not supported by any scientific data on the catchment slopes, soil types, the drainage lines, land use, and land cover, which are very crucial for scientific planning for effectiveness from both physical and economic perspectives. Generally, the implementing agency does not have resources available for generating such information in the project funds. Nearly 85%–90% of the funds allocated for are to be used for physical interventions. The remaining funds are used for covering the overheads of the PIA to support social engineering. The knowledge about how various land uses affect catchment hydrology is very tentative and hence the impact of the proposed interventions.

Very often, conflicting claims are made about the potential impact of the interventions. No doubt, a well-designed and executed watershed development intervention works in terms of improved recharge in the treated watershed and reduced soil erosion. But these benefits accrued often come at the cost of undesirable hydrological changes like reduced stream flows downstream. While one of the socioeconomic indicators of success of IWMP has been increase in area under cultivation and irrigation in the treated watershed, which meant increased consumptive use of water, the claim of "increased base flow during lean season" is also used simultaneously to defend extensive WH work carried out under the program by the PIA. But, the point is that increased consumptive use of water in the watershed has to come with a reduction in runoff from the watershed or base flow or both.

Several misconceptions drive the choice of technologies and practices under watershed management. It is widely believed by the planners as well as PIAs that tree planting is very good for improving the flow regime of the catchment if not augmenting the flow itself, irrespective of the climate. The application of scientific knowledge of forest hydrology is hardly visible. Similarly, it is believed that adoption of microirrigation technologies would eventually lead to reduction in water use in agriculture.

In India, the agroecology of the rain-fed area ranges from hot and arid tropical areas with very low rainfall and flat topography to cold and humid subtropical areas with excessively high rainfall and mountainous topography.

Interestingly, rain-fed areas exist under both the conditions. While in the first case, the crops remain largely rain-fed due to lack of sufficient water for irrigation, whereas in the second case, the crops do not require irrigation water for most parts of the year due to excessive rainfall over long time periods and soil moisture. The kind of issues relating to catchment management in the former would be drastically different from that in the latter. While in the former, the focus should be on introducing limits on water diversion for biomass production and improving the efficiency of use of the water allocated to that sector, in the latter plenty of scope for regulating the monsoon flows, reducing soil erosion, controlling sediment load in run-off and augmenting local surface water storage might exist.

Hence, the ICM interventions will have to change as we move from one agroecology to another. But, the uniform guidelines for watershed management and the standard norms on financing based on area treated do not encourage development of innovative approaches that would make hydrological and economic sense.

2.4.3 Physical and socioeconomic impacts of watershed management programs in India

The focus of watershed development in India has modified over the last 25 years from soil conservation to water conservation to now include a more participatory planning approach, evaluation studies estimating the distribution or magnitude of social impacts from watershed development are often unclear or disputed (Hope, 2007; Kerr et al., 2002; World Bank, 2004).

Very few studies, based on catchment monitoring and research, are available to help evaluate the impact of catchment management interventions on their environmental conditions. They pertain to the following: the hydrological and biophysical impacts of the interventions at the level of microcatchments in terms of changes in runoff collection efficiency; increase in vegetation cover; soil loss from the catchment (Panwar et al., 2012; Garg et al., 2012); and groundwater recharge (Garg et al., 2012). They all show positive impacts at the local level (Panwar et al., 2012; Pathak et al., 2013). But, these microlevel studies are not representative of the hydrological conditions such as rainfall, slope, geohydrology, soil type, land use, and land cover, of the large catchments they are part of. In nutshell, the impact assessments do not factor in the "scale effects." In other words, the hydrological processes that matter for microwatersheds or the mathematical relationships representing the hydrological processes will not be same as that for the large catchment in which these microwatersheds fall. This makes it difficult to draw useful inferences of catchment-wide interventions for the large catchment in question.

Several researchers have provided empirical evidence to illustrate the "scale effects" in small water harvesting (Batchelor et al., 2002; Gupta, 2011; Kumar et al., 2008; Ray and Bijarnia, 2006; Talati et al., 2005) and watershed development (Batchelor et al., 2002; Syme et al., 2012).

Economic valuation of watershed management programs also suffers on many counts: they fail to link the measurable indicators of watershed impacts to planned interventions (Hope, 2007); and they are not able to capture the negative externalities (World Bank, 2004), which are very significant as shown by several empirical studies across the country Batchelor et al., 2003; Calder, 2005; Hope, 2007; Kumar et al., 2006, 2008). Over and above, there is failure to capture the lasting effects of project interventions on the social and natural systems (Barron and Noel, 2011). Failure to quantify the hydrological gains from the structures to compute the economic value created from the use of the additional water stored in the catchment is another pressing problem. Findings of studies on socioeconomic impact of watershed development programs from different authors, particularly those relating to poverty impacts, often contradict (source: based on Hope, 2007; Buhl and Sen, 2006; Kerr, 2003), while the skewed distribution of benefits from projects, which achieved resource conservation productivity gains, toward large land holders, is visible (Kerr, 2003).

2.4.4 Sustainability of watershed management institutions at the microlevel

In the countries where catchment-wide management of land and water resources has been attempted, the need for the same has been felt by the community groups living within the catchment. Such needs have arisen due to conflicts arising out of excessive diversion of water for agriculture or excessive use of nitrogenous fertilizers in the farm land or disposal of effluents in the rivers by users who constituted a small share of the watershed population. Such uses had impacted several of the downstream users of water, including environmental uses, fisheries, and recreation and quality of drinking water for a large section of the population (Prato and Herath, 2007; Collins et al., 2007).

The relationship between the nature of land use and the resultant impact on the ecological and economic services provided by the catchment was quite visible. More importantly, the communities could value enhanced ecological services such as increased hot weather flows and reducing coastal salinity, high-quality water for drinking water supplies and management of aquatic ecosystem including protection of rare species of fish, which could be derived from such initiatives (Collins et al., 2007; Connell, 2012; Johnson et al., 1996; Prato and Herath, 2007). In certain other cases, it came out of the realization that there is high degree of fragmentation of water management institutions as found in Britain and France (Buller, 1996) or the policies and activities carried out by various agencies in the catchments need to be coordinated and communities need to be involved in the program as in Western Australia (Mitchell and Hollick, 1993), or as the aftermath of a legislation to ensure poor communities' access to water for domestic and productive needs as in South Africa (Schreiner and Van Koppen, 2002).

But, the situation in most of the treated watersheds in India differs drastically from what is illustrated here. In general, the opportunity cost of participation outweighs the benefits, creating a tendency to self-exclude. First: there are clear winners of an effectively implemented watershed program, while there are losers (Johnson et al., 2001). Second: the opportunity cost of nonparticipation in the program does not exist for the individual members of the communities, including those who are likely to be adversely affected. The reason is that the program mostly produces nontangible public goods and the newly created institutions do not perform any role for distribution of benefits accrued from enhanced watershed performance, either for improving the poor's access to groundwater or ensuring sufficient water in the local aquifer for domestic uses during the lean season (Shiferaw et al., 2006). Third: the local institutions are not internally driven, but instead are driven by external support, and receive too little support for their own capacity building (Singh et al., 2012), fall in the capability trap, and ultimately collapse once the donor withdraws from an area (Sinha and Sinha, 1996).

2.5 Major findings

1. The science of catchment hydrology is well developed internationally, with sufficient knowledge and understanding of the manner in which the hydrology changes in response to changes in land use and land cover in the catchment. But, there is also sufficient evidence of delayed use of this knowledge in catchment management programs, particularly in developing countries like India.

2. Modeling tools exist for simulating the complex hydrological processes in catchments, which, if used correctly, have the potential to predict hydrological outcomes for projected changes in land use and land cover in terms of runoff, soil loss, and sediment transport. Models, which incorporate economic outcomes of catchment management interventions into the hydrological and biophysical models, also exist.
3. But, given the variety of uses and users of land and water in the catchments, catchment management decisions will be based on multiple objectives and criteria, which are social, economic, environmental, and political. Often, there could be strong tradeoff between maximizing economic outcomes and meeting environmental and social goals, and vice versa. Making the right management choice is about minimizing the tradeoffs. But to what extent this tradeoff could be minimized depends on which stakeholder has the political influence or is powerful. The challenge is to ascertain the weightage to be given to each one of the criteria.
4. Catchment management plans should offer a vision for the catchment and its communities for the future. Therefore, it is important that that they accept the rules and framework of actions broadly defined by such plans. Often, the communities are not really convinced about the influence of individual actions on catchment functions.
5. Hence, catchments cannot be managed merely on the basis of scientific knowledge of hydrological and ecological processes. There is a need to recognize the fact that individual actions of the community members are not governed by scientific practices that promote good catchment functioning, but other considerations. This recognition only can foster the awareness of interdependencies between individual actions and catchment functioning and optimize the individual actions. Facilitating the dialog among the stakeholder groups in the catchment, the scientists and policy makers would help the latter understand the rationale behind the individual actions. Awareness of the interdependencies can be created through appropriate practical initiatives that provide a systemic awareness of the context in which the individual actions and catchment functions are positioned.
6. In countries where ICM practices are attempted, the need for catchment-wide management of land and water resources had mostly arisen in response to the prevailing or perceived future conflicts over land and water use so as to ensure water for environmental flows to prevent coastal salinity, augmenting lean season flows for ecology, and preservation of aquatic life and protection of water quality for drinking. The objectives for catchment management varied across countries and situations though.

7. Catchment management planning is a process, outcomes of which would be determined by who initiates and facilitates the process. Different countries follow different processes. While in the MDB of Australia, which has a long history of catchment management, the activities of various catchment management agencies are to be now regulated through a Basin Plan, which is a legally enforceable document. In South Africa, under the National Water Act, the DWAF is to facilitate the process of setting up CMAs within each basin, statutory bodies for basin-wide management of water resources. The CMAs are expected to come out with their own technical proposals for managing their catchments with the larger goal of participatory basin management.
8. In Britain, the CMPs form the basis for the NRA's actions within each catchment. The French SAGE and the SDAGE follow a two-tier tradition in catchment management. SDAGE is a forward planning document at the regional level, concerning major drainage basins, whereas SAGE is a more precise planning document at the local level for small local catchments (around 100–200 km^2). In Western Australia, the Community Catchment Groups are formed to define the principles and rules for management of their own catchments, under political, administrative, and financial support from the Office of Catchment Management.
9. In India, watershed management has evolved as a participatory approach for managing both private and common degraded lands for enhancing agricultural productivity in rain-fed areas using watershed as the unit for treatment of land, with the "ridge to valley" concept. The entire focus is on soil and water conservation and water resources development at the microcatchment level. The "scale effects" are not considered in the planning.
10. There is little evidence of hydrological considerations in catchment planning. Given the fact that scientific data on stream flows, groundwater resource availability, groundwater-surface water interactions in the catchment, and the catchment characteristics are largely absent for microwatersheds chosen for implementation, the tendency is to overestimate the resource availability so as to enhance the scope for water resource development interventions.
11. The program is implemented using a decentralized approach, through village watershed committees. They are financially supported by the Ministry of Rural Development through the dept. of rural development of the respective states and their district-level arms. The physical interventions are planned by the VWCs with the help of the PIAs, which

are in turn supported by state-level resource agencies. The monitoring and evaluation of the watershed activities are carried out by the State line departments. The DOLR undertakes periodic monitoring of the projects, midterm evaluations, and special impact assessment studies.

12. The program by design promotes resource development and exploitation and does not envisage any mechanism either for sharing the augmented resources equitably or for promoting sustainable and efficient use of water and land. Intensive water-harvesting activities carried out under the program supported by groundwater overexploitation help improve economic returns for the farmers. But, they also simultaneously cause negative effects downstream in terms of reduced flows into tanks and other wetlands and drinking water shortage.

13. Though there are multiple uses and users of water, biomass and land in the treated watersheds, the evidence to the effect of involving them in designing management strategies is absent. The heavy emphasis on water harvesting means that the program is primarily targeted at benefiting the agricultural communities in terms of improved land productivity and water access, while the cost is being borne by the landless, who are denied access to the forest land and pastures due to ban on grazing and tree cutting.

14. Watershed management programs across the country follow the same norms and guidelines vis-à-vis the technical strategy and the expenditure for treatment activities. However, the agroecology of rain-fed areas varies drastically from hot and arid semiarid tropical climate with plains to cold and humid, subtropical climate in the mountainous areas. Accordingly, the land and water-related problems also vary. While in the former, the problem is of excessive use of water for agriculture with resource depletion, in the latter the problem is of poor soils, and excessive soil erosion and sediment load in water along with poor utilization of the available renewable water resources. But the focus everywhere is on one type of solutions, i.e., soil and water conservation and water resources development, indicating a lack of clarity of purpose.

15. There is some evidence available from scientific research and monitoring on the positive hydrological impacts of watershed management interventions at the local level in terms of improved runoff collection efficiency, soil loss reduction, sediment control, and augmentation of groundwater recharge when interventions are designed on the basis of scientific data. But, the schemes do not appear to be economically viable at catchment scale. Evidence available from studies in watersheds of Karnataka, Andhra Pradesh, Gujarat, and Madhya Pradesh shows negative impacts on areas outside the watersheds in terms of reduced inflows into reservoirs.

16. Social impact assessment is fraught with many methodological challenges, owing to the difficulty in establishing ideal "controls," and the practical issues in nullifying the effect of larger socioeconomic changes. The available independent studies involving robust research designs and methodologies do not show very encouraging outcomes in terms of agricultural productivity growth, improvement in drinking water supply situation, and reduction in rural poverty, though evaluation studies of some of the widely acclaimed watershed programs do show some positive changes in terms of poverty reduction and environmental sustainability.

17. Even in situations where projects were successful in achieving resource conservation and productivity enhancement objectives, skewed distribution of benefits toward large land owners is apparent. In any case, there is no evidence of watershed management programs contributing toward addressing the concerns of inequity in access to groundwater.

18. The projects that followed proper microplanning taking into consideration of the existing resource condition both for private land and common land, using scientific as well as participatory assessment tools seem to be doing good in terms of realizing the intended outcomes and impacts. Also, such projects are also found to have made special provisions in their design for landless communities to derive benefits out of the project.

References

Barron, J., Noel, S., 2011. Valuing soft components in agricultural water management interventions in meso-scale watersheds: a review and synthesis. Water Alternatives 4 (2), 145–154.

Batchelor, C., 1999. Improving water use efficiency as part of integrated catchment management. Agric. Water Manage. 40 (1999), 249–263.

Batchelor, C., Singh, A., Rao, M.S.R.M., Butterworth, J., 2002. Mitigating the potential unintended impacts of water harvesting. In: Paper Presented at the IWRA International Regional Symposium 'Water for Human Survival', 26–29 November, 2002, Hotel Taj Palace, New Delhi.

Batchelor, C., Rama Mohan Rao, M.S., Manohar Rao, S., 2003. Watershed development: a solution to water shortages in semi-arid India or part of the problem? Land Use Water Resour. Res. 3 (2003), 1–10.

Bellamy, J., Ross, H., Ewing, S., Meppem, T., 2002. Integrated catchment management: learnings from the Australian Experience for the Murray Darling Basin. Final Report, January 2002.

Biggs, T.W., Gaur, A., Scott, C.A., Thenkabail, P., Rao, P.G., Gumma, M.K., Acharya, S., Turral, H., 2007. Closing of the Krishna Basin: Irrigation, Streamflow Depletion and Macroscale Hydrology. Research Report 111, International Water Management Institute, Colombo, Sri Lanka.

Bosch, J.M., Hewett, J.D., 1982. A review of catchment experiments to determine the effect of vegetation changes on water yield and evapo-transpiration. Hydrology 55, 3–23.

Buhl, S., Sen, R., 2006. A Preliminary Poverty Impact Assessment. Indo-German Development Cooperation, Natural Resource Management, GTZ, Delhi.

Buller, H., 1996. Towards sustainable water management catchment planning in France and Britain. Land Use Policy 13 (4), 289–302.

Calder, I.R., 2002. Forest valuation and water; the need to reconcile public and science perceptions. In: Verweij, P.A. (Ed.), Understanding and Capturing the Multiple Values of Tropical Forests, Tropenbos Proceedings. Tropenbos International, Wageningen, pp. 49–62.

Calder, I.R., 2005. The Blue Revolution, Integrated Land and Water Resources Management. Earthscan Publications, London.

Calder, I., Gosain, A.K., Rama Mohan Rao, M.S., Batchelor, C., Garratt, J., Bishop, E., 2008. Watershed development in India. 2. New approaches for managing externalities and meeting sustainability requirements. Environ. Devel. Sustain, 227–240. https://doi.org/10.1007/s10668-006-9073-0.

Campbell, P.D., Newfoundland, L., 2010. Application of Earth Observation Technology to Improve Water Resource Management in Newfoundland and Labrador. Hydrologic Modelling Section Water Resources Management Division, Department of Environment and Conservation.

Collins, K., Blackmore, C., Morris, D., Watson, D., 2007. A systemic approach to managing multiple perspectives and stake holding in water catchments: some findings from three UK case studies. Environ. Sci. Policy 10 (2007), 564–574.

COMMANS, 2005. Managing groundwater resources in rural India: communities and beyond. British Geological Survey Commissioned Report CR/05/36N.

Connell, D., 2012. Catchment management across borders in the Murray–Darling basin. Int. J. River Basin Manag. 11, 167–173. https://doi.org/10.1080/15715124.2012. 727827.

Darghouth, S., Ward, C., Gambarelli, G., Styger, E., Roux, J., 2008. Watershed Management Approaches, Policies and Operations: Lessons for Scaling Up. Water Sector Board Discussion Paper Series, Paper No. 11, The World Bank, Washington, DC.

European Environmental Bureau (EEB), 2010. 10 Years of Water Framework Directive: A Toothless Tiger? Snapshot Assessment of EU Environmental Ambitions. European Environmental Bureau.

Ewing, S.A., Grayson, R.B., Argent, R.M., 2000. Science, citizens, and catchments: decision support for catchment planning in Australia. Soc. Nat. Res. Int. J. 13 (5), 443–459.

Falkenmark, M., 2004. Towards integrated catchment management: opening the paradigm locks between hydrology, ecology and policy-making. Int. J. Water Resour. Dev. 20 (3), 275–281.

Falkenmark, M., Gottschalk, L., Lundqvist, J., Wouters, P., 2004. Towards integrated catchment management: increasing the dialogue between scientists, policy-makers and stakeholders. Int. J. Water Resour. Dev. 20 (3), 297–309.

Farrington, J., Turton, C., James, A.J., 1999. Participatory Watershed Development: Challenges for the Twenty-First Century. Oxford University Press, New Delhi, India.

Gallart, F., Llorens, P., 2003. Catchment management under environmental change: impact of land cover change on water resources. Water Int. 28 (3), 334–340.

Garg, K.K., Karlberg, L., Barron, J., Wani, S.P., Rockström, J., 2012. Assessing impacts of agricultural water interventions in the Kothapally watershed, Southern India. Hydrol. Process. 26 (3), 387–404.

Gosain, A.K., Sandhya, R., Debajit, B., 2006. Climate change impact assessment on hydrology of Indian river basins. Curr. Sci. 90 (3), 346–353.

Gupta, S., 2011. Demystifying 'tradition': the politics of rainwater harvesting in rural Rajasthan. Water Alternatives 4 (3), 347–364.

Herrfahrdt-Pähle, E., 2010. Introducing Catchment Management: The Case of South Africa. Discussion Paper 3/2010, German Development Institute, Bonn.

Homdee, T., Pongput, K., Kanae, S., 2011. Impacts of land cover changes on hydrological responses: a case study of Chi River basin, Thailand. Annu. J. Hydraul. Eng. (JSCE, Vol. 55).

Hope, 2007. Evaluating social impacts of watershed development in India. World Dev. 35 (8), 1436–1499.

Howell, T., 2001. Enhancing water use efficiency in irrigated agriculture. Agron. J. 93, 281–289.

Jakeman, A.J., Letcher, R.A., 2003. Integrated assessment and modelling: features, principles and examples for catchment management. Environ. Model. Software 18 (2003), 491–501.

James, A.J., 2000. MPA: A Methodology for Participatory Assessments, Waterlines. October Issue.

Johnson, A.K.L., Shrubsole, D., Merrin, M., 1996. Integrated catchment management in northern Australia: from concept to implementation. Land Use Policy 13 (4), 303–316.

Johnson, N., White, A., Perrot-Maître, D., 2001. Financial Incentives for Watershed Management: Issues and Lessons for Innovators. World Resources Institute, Washington, DC.

Joshi, P.K., Jha, A.K., Wani, S.P., Joshi, L., Shiyani, R.L., 2005. Meta-Analysis to Assess Impact of Watershed Program and People's Participation. Comprehensive Assessment Research Report 8, Comprehensive Assessment Secretariat, Sri Lanka.

Kashaigili, J.J., Kadigi, R.M.J., Lankford, B.A., Mahoo, H.F., Mashauri, D.A., 2005. Environmental flows allocation in river basins: exploring allocation challenges and options in the Great Ruaha river catchment in Tanzania. Phys. Chem. Earth 30 (2005), 689–697.

Kerr, J., 2002. Watershed development, environmental services, and poverty alleviation in India. World Dev. 30 (8), 1387–1400.

Kerr, J., 2003. Price policy, irreversible investment, and the scale of agricultural mechanization in Egypt. In: Lofgren, H. (Ed.), Research in Middle East Economics, Volume 5: Food, Agriculture and Economic Policy in the Middle East and North Africa. JAI Press, Oxford, UK, pp. 161–185.

Kerr, J., Pangare, G., Pangare, V.L., 2002. Watershed development projects in India—an evaluation. Research Report 127, International Food Policy Research Institute (IFPRI), Washington, DC.

Kraft, G.J., Stites, W., 2003. Nitrate impact on groundwater from irrigated-vegetable systems in a humid north-central US sand plain. Agric. Ecosyst. Environ. 100 (1), 63–74.

Kumar, M.D., 2007. Groundwater Management in India: Physical, Institutional and Policy Alternatives. Sage Publications, New Delhi.

Kumar, M.D., van Dam, J., 2013. Drivers of change in agricultural water productivity and its improvement at basin scale in developing economies. Water Int. 38, 312–325. https://doi.org/10.1080/02508060.2013.793572.

Kumar, M.D., Ghosh, S., Singh, O.P., Ranade, R., Ravindranath, R., 2005. Changes in Groundwater Ecology and its Implication for Surface Flows: Studies from Narmada River Basin, Madhya Pradesh, India. Water Policy Research Highlight # 25, IWMI-Tata Water Policy Research Program, Anand, Gujarat, India.

Kumar, M.D., Ghosh, S., Patel, A., Singh, O.P., Ravindranath, R., 2006. Rainwater harvesting in India: some critical issues for basin planning and research. Land Use Water Resour. Res. 6 (1), 1–17.

Kumar, M.D., Ankit Patel, A., Ravindranath, R., Singh, O.P., 2008. Chasing a mirage: water harvesting and artificial recharge in naturally water-scarce regions. Econ. Pol. Wkly 43 (35), 61–71.

Leblanc, M., Tweed, S., Van Dijk, A., Timbal, B., 2012. A review of historic and future hydrological changes in the Murray–Darling basin. Global Planet. Change 80–81, 226–246. https://doi.org/10.1016/j.gloplacha.

Macleod, C.J.A., Scholefield, D., Haygarth, P.M., 2007. Integration for sustainable catchment management. Sci. Total Environ. 373 (2007), 591–602.

Mazibuko, G., Pegram, G., 2006. Evaluation of the Opportunities for Cooperative Governance Between Catchment Management Agencies and Local Government. Water Research Commission, Gezina.

McDonald, P., Edwards, R.A., Greenhalgh, J.F., Morgan, C.A., 1995. Animal Nutrition, fifth ed. Addison Wesley Longman, UK. 353 pp.

Mitchell, B., Hollick, M., 1993. Integrated catchment management in western Australia: transition from concept to implementation. Environ. Manag. 17 (6), 735–743.

Moench, M.H., Dixit, A. (Eds.), 2004. Adaptive Capacity and Livelihood Resilience: Adaptive Strategies for Responding to Floods and Droughts in South Asia. Institute for Social and Environmental Transition, International and Nepal: Institute for Social and Environmental Transition, Boulder.

Molle, F., Turral, H., 2004. Demand management in a basin perspective: is the potential for water saving over-estimated? In: Paper Prepared for the International Water Demand Management Conference, June2004, Dead Sea, Jordan.

Mostertman, I., 2005. Paying for the commons: an investigation into a participatory decision making process in implementing a bulk water pricing system in the Paraiba do Sul river basin. MSc Thesis, Irirgation and Water Engineering Group, Wageningen University.

National Water Commission, 2010. The Impacts of Water Trading in the Southern MURRAY–Darling Basin: An Economic, Social and Environmental Assessment. NWC, Canberra.

Oliveira, R.S., Bezerra, L., Davidson, E.A., Pinto, F., Klink, C.A., Nepstad, D.C., Moreira, A., 2005. Deep root function in soil water dynamics in cerrado savannas of central Brazil. Funct. Ecol., Br. Ecol. Soc. 19, 574–581.

Orr, H.G., Carling, P.A., 2006. Hydro-climatic and land use changes in the river lune catchment, north west England, implications for catchment management. River Res. Appl. 22 (2006), 239–255.

Panwar, P., Sharmistha, P., Bhatt, V.K., Tiwari, A.K., 2012. Impact of conservation measures on hydrological behaviour of small watersheds in the lower Shivaliks. Indian J. Soil Conserv. 40 (3), 257–262.

Pathak, P., Kumar, A.C., Wani, S.P., Sudi, R., 2013. Multiple impact of integrated watershed management in low rainfall semi-arid region: a case study from eastern Rajasthan, India. J. Water Resour. Prot. 5, 27–36.

Perry, C., Steduto, P., Allen, R.G., Burt, C.M., 2009. Increasing productivity in irrigated agriculture: agronomic constraints and hydrological realities. Agric Water Manag 96, 1517–1524.

Pollard, P., Devlin, M., Holloway, D., 2001. Managing a complex river catchment: a case study on the River Almond. Sci. Total Environ. 265 (2001), 343–357.

Prato, T., Herath, G., 2007. Multiple-criteria decision analysis for integrated catchment management. Ecol. Econ. 63 (2007), 627–632.

Ray, S., Bijarnia, M., 2006. Upstream vs downstream: groundwater management and rainwater harvesting. Econ. Polit. Wkly. July 10.

Reddy, V.R., 2012. Environment and employment in rural India: moving towards 'win-win' strategies. Indian J. Labour Econ. 55 (1), 201–215.

Reddy, V. Ratna, Malla Reddy, YV and Soussan, John (2004).Water and poverty: a case of watershed development in Andhra Pradesh (jointly). Water Nepal. 11, 1 (August 2003–January 2004).

Sahin, V., Hall, M.J., 1996. The effects of afforestation and deforestation on water yields. J. Hydrol. 178, 293–309.

Schreiner, B., Van Koppen, B., 2002. Catchment management agencies for poverty eradication in South Africa. Phys. Chem. Earth 27 (2002), 969–976.
Shiferaw, B., Bantilan, C., Wani, S., Sreedevi, T.K., Rao, G.D.N., 2006. Collective action for integrated community watershed management in semi-arid India: analysis of multiple livelihood impacts and the drivers of change. In: Paper Presented at the International Association of Agricultural Economists Conference, Gold Coast, Australia, August 16–18, 2006.
Singh, S., James, A.J., Reddy, V.R., Marothia, D.K., Gardener, J., 2012. IWRM and Local Level Planning in Rajasthan, Special Studies Series, Rajasthan Study 2, European Union state Partnership Programme. The European Union.
Sinha, F., Sinha, S. (Eds.), 1996. From Indifference to Active Participation: Six Case Studies of Natural Resource Development Through Social Organisation. EDA Rural Systems, Gurgaon, India.
Smakhtin, V., Anputhas, M., 2006. An Assessment of Environmental Flow Requirements of Indian River Basins. International Water Management Institute, Colombo, Sri Lanka. 42 p (IWMI Research Report 107).
Stockholm Environment Institute, 2009. Rainwater Harvesting: A Lifeline for Human Well-Being. Report Prepared for the UNDP, Stockholm Environment Institute, Stockholm.
Syme, G., Reddy, V.R., Pavelik, P., Croke, B., Ranjan, R., 2012. Confronting scale in watershed development in India. Hydrgeol. J. 20 (5), 985–993.
Talati, J., Kumar, M.D., Ravindranath, R., 2005. Local and Sub-Basin Level Impacts of Local Watershed Development Projects: Hydrological and Socio-Economic Analysis of Two Sub-Basins of Narmada. Water Policy Research Highlight 15, IWMI-Tata Water Policy Research Program, Anand, Gujarat, India.
Turton, C., Birardar, N., Ramamurthy, V., Singh, J.P., Misra, A.K., 1997. Opportunities and Constraints for Fodder Development in Rainfed Areas of India. Developing Linkages With Smallholder Farmers. Indian Grassland and Fodder Research Institute and Overseas Development Institute, London.
Verstraeten, G., Van Oost, K., Van Rompaey, A., Poesen, J., Govers, G., 2002. Evaluating an integrated approach to catchment management to reduce soil loss and sediment pollution through modelling. Soil Use Manage. 19 (2002), 386–394.
Wallace, J.S., Batchelor, C.H., 1997. Managing water resources for crop production. Philos. Trans. R. Soc. Lond., B, Biol. Sci. 352, 937–947.
Warner, J.F., 2006. More sustainable participation? Multi-stakeholder platforms for integrated catchment management. Int. J. Water Resour. Dev. 22 (1), 15–35.
Whitty, C., Dercon, S., 2013. The Evidence Debate Continues, Chris Whitty and Stefan Dercon Respond from DFID. People, Spaces, Deliberation. Available at http://blogs.worldbank.org/publicsphere.
World Bank, 2004. Managing Watershed Externalities in India. The World Bank, Washington, DC.

Further reading

Alansi, A., Amin, M., Halim, G., Shafri, H., Aimrun, W., 2009. Validation of SWAT model for stream flow simulation and forecasting in Upper Bernam humid tropical river basin, Malaysia. Hydrol. Earth Syst. Sci. Discuss. 6, 7581–7609.
NWDPRA, 1991. Watershed Areas Rainfed Agricultural Systems Approach (WARASA) Guidelines. Ministry of Agriculture, Department of Agriculture and Cooperation, Government of India, New Delhi.

CHAPTER 3

Watershed management in Afghanistan: Lessons from South Asia

V. Ratna Reddy
Livelihoods and Natural Resource Management Institute, Hyderabad, Telangana, India

3.1 Introduction

Land degradation is increasing over years, extending even to the less densely populated mountain regions like Afghanistan, Bhutan, Nepal, etc. The initial perception that anthropogenic factors are responsible for degradation has changed with the availability of more scientific information on geology, steepness, soils, rainfall intensity, etc. These factors coupled with the local land and water management factors are identified as the main reasons for land degradation. Such an awareness has prompted actions promoting soil conservation technologies as well as government policies and programs, viz., forest management, watershed management, and protected area management (for a review see Tiwari et al., 2008). Of these watershed management is the most widely adopted technology as it can contain water and wind erosion effectively. Watershed technology is also suitable to protect and enhance soil fertility. More importantly, it caught the policy attention in some countries due to its potential to conserve soil and water in the water-constrained (low rainfall and low irrigation potential) regions. For, these regions account for more than 75% of the cropped area in the world and a third of the developing world's population (World Bank, 2008).

3.1.1 Watershed management

A *Watershed* is a topographically delineated area that is drained by a stream. Watershed development (WSD) is a land-based technology that helps conserve and improve in situ soil moisture, check soil erosion, and improve water resources, especially groundwater. Watershed Management (WSM) deals with adoption of watershed technologies in specific watersheds

Current Directions in Water Scarcity Research, Volume 1
https://doi.org/10.1016/B978-0-12-814851-8.00003-3
© 2019 Elsevier Inc.
All rights reserved.

(Reddy et al., 2010a). The approach of watershed management ought to be holistic and integrated involving hydrological, biophysical, and socioeconomic systems. Often adoption of watershed technologies requires collective or cooperative strategies among individual households and communities as watersheds cover large areas of private and public lands, transcending villages. People's participation, thus, is recognized as a key factor in effective implementation of the program. Thus, WSM is a combination of "science and art"/"technology and philosophy" (Tiwari et al., 2008; Reddy et al., 2010a). Throughout this chapter, the terms WSM and IWSM are used interchangeably.

Approach and purpose of WSM varies across countries of South Asia. It ranges from soil and water conservation (most countries) to rangeland protection (Afghanistan and Pakistan) protecting rivers/reservoirs (irrigation) (Afghanistan, Pakistan, etc.) to protecting reservoirs (hydel plants) (Nepal and Bhutan) to enhancing ecosystem services (Sri Lanka) and to improving land productivity in the water-constrained regions (India). In terms of scale, some of the South Asian countries have shifted to micro (500 ha) watersheds, others continue to treat areas above 5000 ha. India (2009 onwards) also adopted the mega watershed (5000–10,000 ha) approach under the integrated watershed management program (IWMP). Afghanistan, after a long-drawn unrest, is attempting to implement the watershed development program in order to enhance the livelihoods of the local communities.

Afghanistan is an arid to semiarid country receiving erratic rainfall over the years. Rainfall varies from 75 mm in Farah to 1170 mm in south Salang. Southern part of Afghanistan receives the least rainfall (100–300 mm/year) followed by western (200–400 mm) and northern parts. Rainfall occurs mostly in the winter months, particularly in the February–April period. Afghanistan is drought prone with severe droughts occurring once every 10–15 years. Cropped area accounts for only 12% of land area, while 45% of the land area is under range lands, which supports livestock. About 41% of the crop land is irrigated. Livestock contributes substantially to the livelihoods in Afghanistan though the relative importance differs between provinces. Afghanistan has a typical problem of crop season not coinciding with rainy season. As a result, it faces spatial as well as seasonal water scarcity, making water storage necessary for protecting the crops. Both crop lands and range lands are degraded heavily due to destruction of forests, excessive grazing, fuel wood collection, and encroachment of pasture lands for cultivation. Apart from soil erosion and low biomass production on rangelands, incidence of flash flood has increased (Favre and Kamal, 2004). Watershed development in Afghanistan could potentially transform agriculture.

This chapter aims at providing a basic watershed policy and implementation framework for Afghanistan. This framework would be based on a critical review of experiences of WSM initiatives in different countries of South Asia. The idea is to identify appropriate methods and approaches in terms of technology and institutions that are being adopted in different countries. These methods and approaches need to be tuned to suite the specific conditions in Afghanistan.

3.2 Evolution of WSM in South Asia

Watershed interventions may be considered as 5000 years old (FAO, 2006), though the potential of watershed technology started to be fully exploited in Europe between the 16th and 17th centuries. This coincided with the advent of new cultivation techniques (slow drainage and abundant fertilization), emergence of new crops (maize, potato, etc.) and implements, private farming, and higher yields. Sustaining these yield rates to meet the needs of growing population called for major public investment in irrigation, land reclamation, and watershed management works (FAO, 2006).

In the context of South Asia, WSM evolved in two phases. The first phase was triggered by the construction of big irrigation schemes and hydropower dams during 1950s and 1960s. Fast sedimentation of reservoirs, increased incidence of flash floods in downstream areas has given rise to WSM planning. This has received further fillip with the environmental concerns coming into fore during the 1970s. The decade of 1970s, however, was occupied with the promotion of green revolution technologies and food security shadowing every other aspect related to agriculture. WSM was back into policy agenda during the early 1980s with the saturation of yield gains from green revolution, which was limited to irrigated regions. As productivity growth in the green revolution areas was showing signs of slowing down (Pingali and Rosegrant, 2001), future growth in agricultural production and food security is likely to depend on improving the productivity in the rain-fed areas. From an economic angle too, there was evidence indicating that marginal returns to investments would be substantially higher in these regions when compared to the irrigated regions where potentials for productivity growth have already been exploited

(Fan and Hazell, 2000). Improving the natural resource base and livelihoods of the communities in the poorly endowed regions has also become a policy priority. In fact, the "agenda 21" stresses that successful watershed management must be based on local stakeholders' informed participation in natural resource management, economic growth, and social change (FAO, 2006).

In most of the south Asian countries, evolution of WSM had the beginnings during the 1980s. The purpose of WSM varies across south Asia depending on their national priorities (Table 3.1). Some countries (India, Nepal, and Bhutan) have accorded high/medium policy priority for the WSM programs, while in others (Pakistan, Afghanistan), watershed interventions are limited to mountainous terrains, range lands, and forest areas. The difference between India and other countries is the emphasis on crop lands. This differential coverage or focus has a natural division of covering private lands (crop lands) and public or common lands. Nepal also has to deal with treating private and common lands together due to the terrain. This has given rise to the need for people's participation for better implementation, as number of private parcels of land are to be treated under the watershed. Thus, India and Nepal are among the first to introduce participatory WSM in the region and others adopted it later.

The level of participation and involvement is dependent on the benefits directly accruing to the local communities. Direct benefits to the communities are more in the case of interventions on crop lands, range lands, and forest areas. On the other hand, interventions such as catchment protection, controlling soil erosion and protecting reservoirs, protection of hilly areas, etc. provide indirect benefits to the communities. In order to achieve active involvement of people, Nepal has integrated livelihoods components with watershed interventions. In some cases, the interventions take place far from the communities. For instance, while communities in the upstream areas (where the interventions take place) may get limited benefits, downstream communities benefit more. In such cases, upstream communities have little incentive to participate.

3.2.1 India

The earliest example of soil conservation measures in undivided India is the Bombay Plan during the early decades of 20th century, which introduced systematic adoption of dry farming techniques. These were known as soil conservation methods in drought-prone regions. These methods were adopted on a piecemeal basis, more as relief measures (Shah, 1998). This piecemeal

Table 3.1 Evolution of modern watershed management in South Asia.

Country	Period	Biophysical conditions	Purpose	National priority	Approach
Afghanistan	1980s	Low rainfall and arid; high proportion of range lands; heavy soil and range land degradation	Range land protection and catchments	Low (mostly external)	In the initial stages
Bangladesh	1970s	High rainfall; flood prone	Check soil erosion and forest protection in Hilly regions	Low only hilly regions (mostly internal)	Forest and catchment focus (integrated watershed management)
Bhutan	1980s	High rainfall; degradation in Mountain regions and catchments	Protection of reservoirs (Hydel)	Medium only mountainous (mostly external) areas	Integrated watershed management
India	1970s	Arid and semiarid with varying rainfall; large proportion of rain-fed; heavy soil and water resource degradation	Enhancing the productivity and reducing yield vulnerability in the rain-fed regions	High. Covering crop and forest lands in all terrains of rain-fed regions (mostly internal)	Participatory and integrated watershed management with a focus on enhancing livelihoods in rain-fed regions
Nepal	1970s	High rainfall; degradation of mountain regions	Protection of reservoirs (Hydel) forest protection and livelihoods	Medium to high. Covering mountainous regions (mostly external)	Participatory and integrated watershed management
Pakistan	1980s	Low rainfall; large proportion of range lands; degradation of range lands; mountain regions and catchments	Protection of reservoirs (irrigation); protection of rangelands	Low covering mountainous regions (mostly external)	Participatory and integrated watershed management
Sri Lanka	1980s	High rainfall; degradation of soil and water resources	Protection of reservoirs (irrigation). upstream soil protection and water quality	Low covering upstream areas of the catchment (mostly external)	Participatory watershed management at the basin scale reservoir protection (irrigation)

Source: Based on the studies quoted in the text.

approach had continued till 1980s when the integrated National Watershed Development Programme for Rainfed Agriculture (NWDPRA) was launched. Successive Indian Governments have recognized the value for WSM and have often sought to promote it, though approaches and guidelines kept changing. With one of the highest land use intensities in the World, with very little amount of land for grazing, the main focus of WSM has been on croplands.

Since 1991, three guidelines were issued in order to improve the performance of the program. Of these, the 1994 guidelines are landmark in WSM in many respects. These guidelines formalized participation through making the involvement of all sections of the community as a prerequisite. Involvement of NGOs in the implementation was encouraged. These guidelines also set forth systematic implementation modalities at the district level and below (Reddy et al., 2010a,b). While 1994 guidelines remained the basis, in 2006 *haryali* guidelines were brought in giving the village panchayats central role in implementing the program. Cost norms were increased Rs. 4000 (US$ 100) to Rs. 6000 (US$ 150) per hectare. The *haryali* guidelines were replaced by the common guidelines in 2008, which introduced six major changes.[a]

3.2.2 Nepal

Soil and water conservation interventions were initiated for the first time in Nepal during the third plan period (1965–70). A systematic approach to WSM was first adopted in 1966, when the United Nations Development Programme (UNDP), FAO, and the Government of Nepal jointly initiated a pilot project called Survey and Demonstration for the Development and Management of the Trishuli Watershed, which was implemented through the Department of Forests. This project lasted for 4 years and carried out diversified activities with an aim to serve as a demonstration scheme for identifying the most suitable approach to mountain area development (Singh et al., 2004). With the establishment of the Shivapuri Watershed Area Development Board in 1976, soil and water conservation extension and

[a] The common guidelines have introduced six major changes, viz., (i) creation of national rain-fed authority (NASDORA); (ii) enlarging the size of watershed and making the micro-watersheds part of a mili watershed (4–10 thousand ha); (iii) adding a livelihoods component with financial allocations and making it a comprehensive rural development program; (iv) increasing the duration of the program from 5 to 8 years; (v) special provision for women and weaker sections by providing 50% and 33% quota, respectively, in Village Watershed Committee (VWC) and also the creation of women watershed councils; and (vi) doubling the financial allocation, i.e., from Rs. 6000 to 12,000 per hectare.

education programs were promulgated across the country and 14 watershed projects were implemented.

People's participation was introduced during early 2000s with the introduction of Forest Sector Policy in the year 2000. Nepal has attempted to adopted a holistic approach to WSM integrating forestry, agriculture, pasture and water management, with an objective of sustainable resource management (Pandit et al., 2007). In the present context, watershed management is not only for managing or conserving natural resources in a holistic manner, but also for involving local people for betterment of their lives. In Nepal, watershed management is more focused on local people's livelihood needs rather than water quality and supply, which is the case in the developed countries (Tiwari et al., 2008).

3.2.3 Pakistan

Two-third of land in Pakistan is used only for grazing by domestic animals. Range lands support a substantial portion of the yearly forage demand ranging between 50% in Hazara region to over 70% in Baluchistan. These range lands are highly degraded and produce hardly 10%–50% of their original productivity potential. Deterioration in range land productivity has naturally resulted in reduced quantitative and qualitative production of range livestock and their products (Sheikh, n.d.). Though the soil and water conservation department was established in 1951, during the third 5-year plan (1965–70), an area of 2000 sq. mile was covered under watershed implementation (Khattak, n.d.). Pakistan Agricultural Research Council (PARC) and Pakistan Poverty Alleviation Foundation (PPAF) fund localized water storage, water conveyance, and water distribution systems, yet there is a dire need for holistic adoption of watershed management approaches, based on resource integration concept and practices.

The initial WSM programs were initiated in the catchment areas of irrigation reservoirs, viz., Mangala/Tarbela/Rawal/Shahpur Dam WSM programs (Ahmad, 2001) to check rapid siltation of reservoirs. The main focus has been protection of mountain regions from soil erosion to prevent silting of reservoirs. In the process, the interventions provide socioeconomic and livelihood benefits as well.

3.2.4 Bhutan

WSM in its modern or comprehensive form is relatively of recent origin in Bhutan, though it is part of a number of legislations since 1974 (National Forest Policy). Even the Constitution of Bhutan mentions about watershed management (article 5) followed by most of the recent policies and acts (Bhutan 2020; Land Act, 2007; Bhutan Water Vision 2025; Bhutan Water Policy, 2008 and Bhutan Water Act 2010). But none of them have provided any clear guidelines and objectives or institutional framework for implementing the program. Watershed management was part of many activities and programs implemented by different departments. And there is no specific allocation of resources for watershed management and lack of emphasis on ground-level coordination (Tsering, 2011).

The WSM Division (WMD) was established under the Department of Forests and Park Services. WMD is the nodal agency at the national level for operationalizing the WSM programs as part of a larger initiative of river basin management in the country. The main objectives of the program include sustaining the flows of natural resources and better stewardship of watersheds providing the country and its people with goods and services in ways that maintain the long-term productive capacity of natural resources without damaging the environment (Tsering, 2011). Prior to 2009, WSM projects were implemented with donor support. These include the Wang WSM project (WWMP) supported by European commission and Lingmutey Chhu watershed project (LCWP) supported by the International Development Research Centre (IDRC) and the Swiss Agency for Development and Cooperation (SDC) (Singh and Karki, 2004). These programs did not have any common objective or framework.

3.2.5 Sri Lanka

WSM approach was not inbuilt in Sri Lanka's natural resource management policies till 1970s, though soil and water conservation interventions date back to precolonial period. The soil conservation act was introduced in 1951 (Gunawardena, n.d.). Most of the investments were made toward rehabilitating the tank systems. This has resulted in the neglect of their catchments. This led to increase in runoff and siltation of tanks, canal systems, and the major reservoirs. An immediate and significant impact was on the reduction in the capacity of the country's hydropower generation capacity (80% of installed capacity depends on hydropower). In fact, a severe drought in 1995 had reduced the GDP growth by 1.5% largely due to reduced electricity generation in the hydropower plants. The degradation of watersheds and the siltation it causes also reduce the capacity of irrigation canals and the availability of water.

The regulatory measures adopted to control soil erosion coupled with limited interventions failed to control soil erosion due to the increasing demands on land. This has led to a policy shift toward the broader issue of environmental protection and management. This led to the development of a more comprehensive watershed-based action program (Murray and Little, n.d.). Thus, only during the late 1970s, the first holistic, broad-based watershed management projects were initiated in Sri Lanka. Initially these projects were implemented by the State institutions and later shifted to more participatory approaches (Gunawardena, n.d.). Besides, Indian government currently spends US$300 million per year on integrated WSM incorporating small-scale farmer managed soil and water harvesting devices implemented in collaboration with community-based institutions. The WSM policy was introduced in 2004 and number of projects were taken up with the support of donors.

3.2.6 Bangladesh

Given the geographical and biophysical features of Bangladesh, WSM can be a developmental priority in the upland regions of Chittagong region. However, the development plans of Chittagong did not have a watershed component. Though some of the departments associated with forest department (Bangladesh Forest Research Institute and the Bangladesh Forest Academy) carry out training courses on WSM, they are not involved in implementing the watershed projects. The Chittagong Hill Tracts Development Board (CHTDB) too has no specific WSM program though it worked in collaboration with the International Centre for Integrated Mountain Development (ICIMOD) on establishing run-off and erosion monitoring plots (Misbahuzzaman, n.d.). Of late, WSM has evolved to respond to the complex challenges of natural resource management in the Chittagong hills. These projects focus on enhancing agricultural productivity and livelihoods through soil and water conservation (SWC) interventions. Watershed communities along with local government institutions/nongovernment organizations (NGOs), etc. are recognized as stakeholders (Misbahuzzaman, n.d.).

3.2.7 Afghanistan

Afghanistan is at early stages of planning and designing a developmental path after decades of war and civil disturbances. Land and water resources are degraded, affecting agriculture as well as livestock production. It is observed that during the last two decades 70% of the forest area was lost, livestock population was reduced by 60%, and agriculture production was halved (Renner, 2010). Possibly with the American and other international communities' collaboration/assistance, the importance of WSM was recognized soon after returning to normalcy from war. Afghanistan has prepared a watershed Atlas for the entire country in 2004, which reflects the importance accorded to WSM. While the importance of WSM is well recognized at policy level, it has a long way to go in terms of developing capacities, institutional arrangement, resource allocation, etc. At present, number of donor-funded WSM programs are being implemented.

3.3 Watershed structures

The technical interventions of WSM include soil conservation practices, vegetative cover, afforestation, contour bunds, water harvesting structures (farm ponds, check dams), field *bunds* (raised edges), ridge bunds, etc. All these interventions are expected to facilitate higher land productivity through improved overall ecological conditions such as moisture and water availability for agriculture. Nature and intensity of interventions vary according to the hydrogeology and biophysical conditions of the region or watershed. The interventions can be grouped under three broad groups (i) soil conservation; (ii) in situ moisture conservation and (iii) water-harvesting structures.

3.3.1 Soil conservation

Majority of soil conservation programs instituted in the 1970s and 1980s were dominated by engineered systems. These systems were originally developed for large land holdings in temperate regions (Doolette and Smyle, 1990). Most popular ones in south Asia include: (a) rock and spill drains—graded drain, which is usually built along contours and acts by capturing runoff in small stilling ponds, which allows the rain water to infiltrate slowly; (b) stone walls—constructed along contours in fields and provide irregular form of terracing; (c) bench terraces—construction of large benches on steep slopes and cultivation may be carried out on slightly downward or back sloping surfaces; (d) bunds-artificial embankments constructed to intercept rainfall and sediments and lead runoff away from the cultivated land.

These interventions are widely used across South Asia though they are mostly implemented on the common lands and steep slopes. Besides, these approaches are commonly used in the centrally managed tea estates in Sri Lanka, where land is not in short supply and valuable topsoil can be sacrificed for the construction of bunds and terraces (Carson, 1989; Doolette and Magrath, 1990). Pebble bunding is popular in the regions where soil quality is poor and crop lands are covered with pebbles. Clearing the crop lands of pebbles and using them to make bunds across gullies and small streams observed to serve double purpose of improving the quality of land and checking soil erosion (Reddy et al., 2010a).

Maintaining the vegetative cover is an effective way of reducing rain water runoff and soil erosion. In fact, it is observed that vegetative cover is more effective in soil erosion control as well as in reducing catchment yields. Calder (1991) has estimated that afforestation will reduce river flows in the order of 200 mm/year (or 2000 cubic meters/ha/year) in Sri Lanka. Afforestation is also widely adopted in the watershed management, especially at the ridge locations. Afforestation and forest conservation measures are extensively used in the hill terrains of Nepal, Bangladesh, and Pakistan. Their popularity in South Asia is mainly due to the small size of holdings. Either these interventions are capital intensive or land intensive or both. Often governments provide incentives or subsidies to promote on-farm soil conservation activities. But they do not sustain once the incentives dry up.

3.3.2 In situ moisture conservation

Number of interventions are adopted to increase in situ soil moisture content. These activities are most widely promoted and adopted measures in India, especially during the initial years of 1970s and 1980s. These include farm and contour bunding to increase soil moisture within plots of land by reducing the run off. These interventions help retain the rain water within the farmers' field in the arid and semiarid zones, where annual rainfall is low (200–600 mm). These interventions, however, are land intensive and hence are not favored by small and marginal farmers. For, in the case of very small plots, the loss of area offsets the yield gains. With the declining farm size, these measures are least preferred by the farming communities. But for the land constraint, these interventions are observed to be very effective in the arid and semiarid regions. In India, these interventions are implemented with 90%–95% subsidy.

Agronomically, there are a number of farming practices that can help increase soil moisture content and enhance crop yields. They include vegetative soil cover, mulching, use of simple tillage practices, contour cultivation, ridging across the slope, vegetative barriers, ripped furrows, land leveling, level pans, and terracing. These practices, however, are only recommendations and not implemented as part of watershed program. Number of research studies have shown that these practices enhance soil quality and yield rates (for detailed review see Doolette and Magrath, 1990).

3.3.3 Rain water harvesting

Of late, rain water harvesting (RWH) is gaining popularity among local communities as well as policy makers. Often, this specific component draws a substantial share of the fund allocation. For instance, in India, the density and size of RWH structures has been on the rise during the last two decades, despite the fact that they are effective only during good rainfall years. The structures include check dams; farm ponds; percolation tanks or pits; recharge wells; and injection wells. All these interventions help recharge of groundwater if geology is favorable. While check dams and farm ponds are part of water-spreading methods, which serve the purpose of recharge as well as direct use of water, percolation pits/tanks, recharge wells, and injection wells are used for recharging the groundwater.

Adoption of these methods depends on the location (upstream/downstream) within the watershed as well as the hydrogeology of the location. In a ridge to valley context of watersheds, it is often recommended that in situ moisture conservation measures and water spread methods (farm ponds and check dams) in the middle and downstream locations. From the hydrogeology perspective, moderate-to-deep weathering and fracturing zones are suitable for artificial recharge through water-spreading methods; areas with deep fractures suitable for artificial recharge by injection methods; and areas with very shallow basement are not suitable for any intervention (Reddy and Syme, 2015). Minipercolation and percolation tanks are effective on the first- to third-order streams, and when the topography becomes plain, check dams are more effective. The interventions that are useful in the weathered zone are check dams, percolation tanks, and farm ponds.

The RWH structures produce more perceptible impacts. The visible collection of water at the check dams and the impact on the surrounding wells attract local community's attention. As a result, farmers demand more of these structures, as these structures can't be built by individual farmers due to their high capital costs. Besides, check dams are constructed on natural streams, which are common property. Even the implementing agencies are leaning toward creating more of these structures owing to the demands from the communities and their quick and visible impacts. However, the impact of these structures is location specific and hence is not equitable in terms of spatial distribution of benefits. Moreover, their impacts could be very limited when they are constructed in the absence of hydrogeological information of the watershed.

3.4 Impacts and equity

Measuring the impacts is becoming more complex in recent years as watershed development programs have transformed from a soil and water conservation program to a comprehensive rural development and livelihoods program. Hitherto, watershed impact assessment studies have focused on the socioeconomic and natural resource impacts (Reddy et al., 2010a; Joshi et al., 2004). With the introduction of the livelihoods component, impact studies have started using the sustainable livelihoods (SL) framework to assess the impacts (Reddy et al., 2004; Reddy et al., 2008). SL framework incorporates the social, human, and physical (capitals) dimensions of poverty, which are long term in nature. But, not much attention has been paid to assess the resilience aspects of watershed development in the literature. Nature of impacts and benefits varies across the countries of south Asia, as their priorities and interventions differ (Table 3.2).

3.4.1 India

An overwhelming majority of the studies have endorsed these programs in terms of costs and benefits in India. Some of them also highlight the less quantifiable ecological benefits (see for review Reddy et al., 2010a). Positive impact of watershed development programs on crop yields, cropping intensity, and cropping pattern changes is mainly highlighted though there are variations in the magnitude of impact across regions and crops. Recent years have revealed stabilization of B-C ratios at around 1.75.

Although the economics of watershed technology is unambiguously in its favor, the magnitude of its impact differs across regions and locations (Joshi et al., 2004). The impact of watershed technology is observed to be more effective in scarcity regions when compared to assured rainfall regions (Deshpande and Reddy, 1991), though the absolute benefits are more in the assured rainfall regions (Joshi et al., 2004). On the other hand, adoption of the technology itself might be a difficult proposition in extreme scarcity conditions, as poor households living on the margin can hardly afford to follow conservation practices such as halting grazing their animals (Singh, 1991). The overall picture indicates that the success rate in some of the states is as low as 20%–25%.

Table 3.2 Impacts of watershed management interventions across South Asia.

Country	Natural resources	Economic	Social & institutional	Overall livelihoods impacts (magnitude)	Remarks
Afghanistan	Improved vegetative cover; number of tree species	Improved fodder availability	Community managed grazing (social fencing)	–	Very few and not comprehensive
Bangladesh	Reduction Jhum cultivation; improved forest protection	Improved livelihoods	Creation of village community forest communities	–	Very few studies
India	Improvement in the resource base, especially groundwater. Increased moisture availability and reduction in soil erosion	Positive benefit cost ratios; increase in net income and employment	Strengthening of participatory institutions. Improved gender participation	Good impact on natural, financial and social capitals	Wide variations across the regions. Overall success rate is very low (20%–25%)
Nepal	Checking forest degradation. Soil erosion and siltation declined Vegetative cover improved	Crop intensities increased; cropping changed to high value crops; yields increased	Education and health improved. Participation has increased; social cohesion increased	Improvement in all capitals Household income doubled in some cases	Not much variation across the studies
Pakistan	Reduction in soil erosion (21%) Reduced sediment (38%)	Improved yield rates of wheat, groundnut, fodder. Increase in income	Increased participation—including women	–	Not many studies. There could be selection bias
Sri Lanka	Reduced runoff and sedimentation in reservoirs	Improved power generation. High IRR	Low participation. Lack of institutions	Limited	The focus is not enhancing the livelihoods of the communities

Source: Based on the studies quoted in the text.

3.4.2 Nepal

WSM projects implemented in Nepal have focused on rural development as well as the protection of watersheds for hydropower development. In Kulekhani and Phewa Lake watersheds, watershed management activities have been linked with hydropower development. It is observed that halting deforestation and restoring the degraded natural environment were possible through community development activities for the upgrading of living standards of rural communities based upon their needs and initiative, promoting participatory decision-making processes, and paying due consideration to women and the rural poor (Acharya, 2000). WSM program has contributed to income growth, increased well-being, reduced vulnerability, improved food security, and more sustainable use of natural resources. It contributed to all livelihood assets of the local people (Thapa, n.d.). In the case of Phewa Tal, watershed fodder and fuelwood productivity has doubled when compared to formerly government-controlled protected forest. Agricultural productivity on terraced farm land has increased to provide sufficient food for a 36% increase in population between 1981 and 2001, with little or no increase in the area. More children, in particular girls, attend school than 30 years ago (Fleming and Fleming, 2009).

Participatory approach to WSM program appears to be contributing to the development of all the five capitals in rural areas (ICIMOD, 2007). Different components of WSM, viz., conservation farming, agroforestry, and conservation engineering have positive impacts. The individual farm approach in the Bagmati Integrated WSM Project (BIWMP) has raised the household income through integrated farming (multistory cropping, vegetable growing, and livestock rearing). In Udaipur, Siraha, and Saptari districts, poor and marginalized communities have benefited through agroforestry-based income-generating activities. Similarly, in Jhikhu Khola watershed in Kavre district, growing of grasses in private lands has reduced the workload of women and children for feeding their cattle (ICIMOD, 2007). Conservation engineering activities such as construction of water harvesting-catchment ponds, irrigation channel improvements, water source protection, river and stream embankments, and forestation on degraded land in the foothills and river flood plains have improved the groundwater water yield—both quality and quantity—in watershed program areas (Merz et al., 2003).

3.4.3 Pakistan

ICARDA (2010) using climate data from 2009 estimated that with protection structures have reduced the annual runoff by 28%. Retention of rainwater in the watershed has improved leading to increased crop yields. Average soil loss was reduced by 21% through soil conservation structures on agricultural land. Watershed interventions have helped increase the average yield of rain-fed wheat by 31%. Agronomic practices like raised bed sowing has resulted in 23% increase in wheat yields and net income by 30%. Similarly, net income from groundnut has gone up by four to seven times with improved practices under WSM. Yield rates of both summer and winter fodder have increased by more than 30%. Shift toward high-value crops was possible wherever water is available (ICARDA, 2010).

A comparative study of two watersheds in the subtropical scrub zone of northern Pakistan showed reduction in runoff and sediment yield by 55% and 78%, respectively (Hanif et al., 1990). In another comparative study, over a period of 9 years, there was 33% decrease in the runoff and 52% decrease in sediment release due to treatment of the watershed. Similar observations were made in the case of Rawal watershed (Bashir et al., 2013). A study to assess the interventions like construction of engineering structures, contouring, terracing, tree planting, and land management observed that the average net income rose from US$6.3 to 36.0 per hectare (Sheikh, n.d.). The Hilkot participatory WSM program led to increase in productivity and livelihoods of people. Economic improvements include doubling of wheat and paddy yields; increased employment opportunities, particularly for the landless and small farmers; and reduction in migration (Shah, 2008).

3.4.4 Sri Lanka

WSM programs in Sri Lanka have mainly focused on reducing soil erosion and siltation in the reservoirs. Changes in the adoption of practices, including engineering measures, tree planting, relevant market development, policy changes, etc., and their impacts could be taken as the intermediate benefits (Abernethy and Wijayaratna, 1998). Economic valuation of the Sri Lanka Upper Watershed Management Project (n.d.) highlighted the importance of protecting reservoirs for hydropower generation. The Project recommended the Government to regulate cultivation and agricultural activities on hills with a slope more than 60%. It was observed that the UMA-OYA upper WSM project has improved the educational levels of households resulted in the adoption of conservation practices. At the same time, conservation practices are not directly linked to farming (occupation) or land ownership (Jayaweera et al., 2004). It was observed that participatory

WSM is effective in achieving improved crop production and household incomes though number of them have struggled to get effective participation. This is mainly attributed to lack of proper institutional arrangements and coordination between line agencies and communities. On the contrary, in some of the projects, the participatory approach has increased people's expectations and dependence on line agencies (Wimalasinghe and Wijayaratne, 1998).

3.4.5 Afghanistan

WSM in Afghanistan is at an early stage of implementation. Some of the earlier initiatives with donor support are limited to specific interventions in the upper watershed areas, viz., reforestation and range land rehabilitation. While such initiatives are critical to long-term success in many areas, benefits may take several years to accrue. It is argued that smaller-scale projects with tangible short-term results would be more useful for the local communities (Groninger and Lasko, 2011). Another study on social fencing initiative with community collaboration facilitated by an NGO in the Pushtun Zarghun District has provided positive evidence in the context of integrated WSM. Vegetative cover increased from 5% to 48%, number of plant species increased from 10 to 31, and fodder availability went up from nil to 202 kg/ha. This has lowered surface wind speeds and significantly reduced wind erosion. A side effect has been an increase in wildlife and the Department of Agriculture has declared the zone as a nonhunting "Nature Reserve" and the blown sand problem had ceased in 4 years (Virgo et al., 2006).

3.4.5.1 Hydrological impacts

Although hydrological change is integral to WSM, hydrological impacts are not studied widely. Impact assessments have mostly focused on socioeconomic or biophysical aspects. Most of the countries in South Asia have focused on upstream WSM in terms of improving the vegetative cover in order to reduce runoff and sedimentation. Soil conservation and water-harvesting measures and their impacts on surface and subsurface hydrology are not given due importance in impact studies. As a result, watershed planning and designing have failed to integrate hydrogeological aspects. One reason could be the scale of watershed is not big enough to integrate hydrology and hydrogeology, as in some of the south Asian countries the focus has been on microwatersheds, especially after the 1990s. Moreover, the purpose of WSM in most of the countries is reduction in soil erosion and sedimentation. Though the aspect of groundwater improvement is integral to WSM in India, hydrogeology aspects are not taken into account while designing watershed interventions. This has given rise to issues pertaining to the nature and intensity of impacts. For, operating at the microwatershed scale does not necessarily aggregate up or capture upstream-downstream interactions.

WSM interventions may prove detrimental to downstream areas. Research has showed that the microwatershed approach may be producing hydrological problems that would be best addressed by operating at a macrowatershed scale. For example, in the Indian context, Batchelor et al. (2003) and Kumar et al. (2006) documented that successful water harvesting in upper watersheds came at the expense of lower areas within the river basin. Even the interventions like rain water harvesting structures could harm the interests of downstream communities, especially in the low rainfall regions (Ray and Bijarnia, 2006).

With the worsening of the groundwater conditions in downstream, more intensive drilling of wells is needed, which the poor could not afford, leading to inequitable distribution and use of water (Calder, 2005). Calder et al. (2006) cite this as "catchment closure," wherein upstream water harvesting accumulates groundwater locally and then intensive pumping depletes the shallow aquifer thereby depriving the downstream areas of both surface runoff and groundwater. As the size of watershed increases, the influence of land use on the upstream—downstream hydrology reduces while the influence of precipitation increases (FAO, 2006). A recent study at sub-basin scale has also observed that watershed interventions have caused water insecurity downstream, though they have benefited the onsite communities (Malano et al., 2015). Thus, while addressing socioeconomic considerations favors small microwatersheds as the unit of operation, approaching this hydrological problem calls for working in large macrowatersheds and the two may be inconsistent (Kumar et al., 2008; Ray and Bijarnia, 2006).

In the case of Sri Lanka, land cover changes (from forest to tea gardens and home gardens) have altered the flow regimes. The observed adverse environmental impacts were directly related to changes in flow regimes. Rapid runoff was responsible for high soil erosion, loss of land productivity, and frequent flash floods. The high rate of soil erosion has caused degradation of stream channels, increasing the likelihood of flash floods, reduced land productivity in rice fields with deposition of coarse material and silting of irrigation canals. During dry spells, relative droughts and irrigation water shortages have occurred. Reduced lean season flows in downstream have threatened the dependable supply of good quality water and increased the salinity intrusion at the river mouth, whereas increased high flows have aggravated the flooding (Elkaduwa and Sakthivadivel, 1998).

3.4.5.2 Who benefits

Participatory WSM implies a fundamental concern for ensuring equity in benefits. While it is desirable to involve all sections of the community, WSM by itself does not guarantee the distribution of benefit flows across the community. For, watershed is a land-based technology and hence most of the benefits accrue to landed—particularly those best placed to pump the augmented ground water. Though landless households are expected to benefit to some extent, through employment opportunities, the type and magnitude of benefits accruing to them are poorly understood. Even if the watershed covers the entire village lands, it might leave sections of dissatisfied households, as not all interventions produce private benefits. Under such situation, it is unlikely that they will participate or evince any interest in the program. Their apathy toward the program might jeopardize the sustainability of the program.

The distribution of economic benefits across socioeconomic groups holds the key to the success of WSM. Joy and Paranjape (2004) have observed that the distribution of direct and tangible benefits is often discriminatory and limited, even in some of the most "successful" projects. The evidence not only indicates little impact on equity; sometimes, they point to increased deprivation for women/poor in terms of their access to resources in addition to further workload for project activities (Shah, 2009).

Many scholars observed that WSM favors the landed and those having land in the lower reaches as well as those who have the ability to invest in wells and pumps (Arya and Samra, 1995; Adolph and Turton, 1998; Kerr et al., 1998; Reddy et al., 2001). A study by Reddy et al. (2010b) in the context of Rajasthan points toward a disturbing fact that benefits from WSM in poor and backward regions are not only low but are mostly cornered by large farmers resulting in the aggravation of inter- and intraregional inequalities. In a study of Orissa and Andhra Pradesh (Rout, 2013), participation was observed to have enhanced the benefit flows, but it could not ensure equitable distribution of the benefits. Equity issues arise even in the spatial context, i.e., downstream locations benefit more when compared to upstream interventions (Reddy and Syme, 2015).

The review of impacts and benefits of WSM clearly brings out that the economics of watershed technology is unambiguously in its favor across south Asia, though the magnitude of its impact varies across the countries and locations. The impact of watershed technology is observed to be more in scarcity (low rainfall and irrigation) regions when compared to assured rainfall regions and the absolute benefits are more in the medium rainfall zones. For, adoption of the technology itself might be a difficult proposition in extreme scarcity conditions, as poor households living on the margin can hardly afford to adopt conservation practices such as losing part of their land for contour bunding, etc. Adding to this is the long gestation in getting the benefits from the technology. This has led to the increasing demand for rain water harvesting structures to the neglect of soil conservation in countries like India. Thus, adopting appropriate and scientific designing and implementation is the key for ensuring its effectiveness.

3.5 Watershed institutions and implementation

Collective participation and action of the community is a critical ingredient for successful WSM. Given its relatively long gestation period of 5–7 years, it is necessary not only to involve the community but also show benefits (tangible economic) to all sections of the community even in the short run. Evolution of appropriate institutions plays a critical role in the implementation and management of watershed interventions. Institutions are required at various (policy to village) levels. Except in India and to some extent in Nepal, watershed institutions have not evolved much in South Asia, though most of the countries have established specific departments for implementing watershed programs at the national level. Institutional evolution is evident in some countries where watersheds were implemented through donor support.

3.5.1 India

Institutions are put in place at various levels to implement the program across the states. The present institutional arrangements are evolved over the years with successive guidelines recommending improvements over the earlier ones. While the central-level institutions were established in the form of ministries and departments during the 1980s, the state, substate, and village-level institutions have their systematic beginnings during the mid-1990s with the introduction of participatory watershed guidelines (GoI, 1994). At the national level, watershed development was brought under the ministry of rural development, though different states have put it under different departments, viz., agriculture, forestry, etc. Different institutional modalities were tried with limited success, mostly at the district and village levels between 1994 and 2006 (Reddy, 2006).

New guidelines (known as common guidelines) introduced in 2008 increased the size of the watershed (from 500 to 5000 ha) and tenure (4–7 years) of the program and renamed the program as integrated watershed management program (IWMP). IWMP is placed under a newly created central rain-fed authority, while keeping the program under the administrate control of the ministry of rural development (GoI, 2006). Besides, all the watershed programs that are being operated under different departments were brought under the department of land resources (DoLR). The enhanced size of the watershed required further changes in the institutions at the village level, as it extends beyond the village. The state governments are given the flexibility to adopt their own approaches. In 2014, IWMP was brought under an umbrella program named as the *Pradhan Mantri Sinchayi Yojana* (Prime Minister Irrigation Development Programme) along with other programs like irrigation and employment program (MGNREGA) at the central level. The state-level and district-level institutional arrangements remained as under IWMP.

Under the IWMP, a bottom-up approach of creating user groups in each watershed and also supporting the self-help groups (SHGs) is existing at the village level. While user groups (UGs) are part of the watershed committee, SHGs are part of the village organization that is being created for the overall development at the village level. Watershed committee takes on an active role in the implementation. Though the actual size of the watershed spreads over 8–10 villages, each village will have similar groups and the implementing agency would take their help in implementing the program in the respective village. This is known as cluster approach and most states adopt this approach. This approach makes the task of participation and collective action easy in each village, though coordinating between the villages is complex.

All the projects at the district level are coordinated by the district water management agency (DWMA). The state-level nodal agency (SLNA) is responsible for selecting the project implementing agencies and allocating the projects. SLNA is also responsible for capacity building of the implementing agencies and identifying new areas of intervention and research, if necessary. The state-level institutional arrangements are not very different from the earlier approach. The village-level cluster approach is new and could prove to be difficult in terms of coordinating the activities across number of villages. Compared to this, the institutional functioning is much smoother in the case of micro (one village) watersheds. There are also issues regarding the sustainability of these local institutions beyond the implementation cycle, which needs to be addressed at the policy level.

As far as implementation is concerned, project implementing agencies (PIAs) are selected by the SLNA. The PIAs could be the line departments of the state government or the NGOs.[b] PIA is responsible for implementing watersheds. PIAs will put up a team of experts including junior engineers, technical officer (institutions and capacity building), and a computer operator headed by a project officer. This team will support implementation of watersheds under the PIA in the district. At the watershed level, two technical officers (agriculture and engineering) would support implementation in each watershed. As per the guidelines, Rs. 12,000 (US$ 185) per ha. is allocated, which is to be spent over a period of 5–7 years. Guidelines also prescribe how this amount is to be spent on different components. While administration, institutional capacity building, monitoring and evaluations and preparation of detailed project report are provided with 22% of the budget, actual watershed works (NRM) get 56% of the budget. Livelihoods support and production enhancement components also get a substantial portion of 19%. While the watershed works and production enhancement components benefit the landed households, the livelihoods component benefits the landless.

3.5.2 Nepal

Nepal is spearheading in community-based natural resources management in the South Asia Region. Nepal's participatory forest management policies are well known. Participation is mandatory in government policy in all the natural resource management programs. It has helped in bringing large rural communities, including women, into the mainstream of development (Tiwari et al., 2008). Initiation and implementation of the forest policy and programs adopting the principle of community-based decentralization has changed the scenario of the natural resource management and community mobilization in the rural areas. The institutional development of watershed management in Nepal more or less adopted the Indian path as far as participatory WSM is concerned. The Department of Soil Conservation and Watershed Management (DSCWM) is the principal authority for undertaking WSM. Similarly, district-level DSCWM are established

[b] As per the guidelines, up to 25% of the projects can be given to NGOs for implementation.

(Sthapit, n.d.). During the Eighth Plan (1992–97), the DSCWM brought out program implementation strategies and people's participation guidelines, which made provision for partnerships with NGOs and CBOs. This provision emphasizes the formation of user groups (UGs) and the need to involve them in planning, implementing, and decision making. This plan also includes group savings and credit mobilization (Pandit et al., 2007).

Basically, three phases are involved in the implementation of the WSM programs. These include: (a) identification of primary users and formation of a committee as well as watershed planning; (b) implementation; and (c) monitoring of planned activities. District soil conservation officers (DSCOs) identify primary users, as well as their priority issues (areas) and programs, through field-level consultations with stakeholders and communities. Then, DSCO officials and user groups jointly prepare an implementation plan, which lays down the activities, level of participation from users, government contribution, work assignment, and time schedule. User group (CDG) implements the plan after receiving approval from local government bodies, including Village Development Committee (VDC) and District Development Committee (DDC). DSCOs provide necessary tools, equipment, and technical and financial support to the users to implement the plan and monitor its implementation. The CDG makes internal rules and regulations on maintenance and benefit sharing among members, whereas the community development committee (CDC) implements these rules and also arranges local people's participation in the program. Although implementation modalities vary between projects and agencies, generally the locally hired project staff facilitate program implementation.

The DSCWM has not been able to prepare participatory plans for areas larger than 2500 ha. While participatory planning for small areas is effective in addressing local issues, it fails to recognize the interests of stakeholders downstream and outside of these areas. Nonetheless, small watershed approach to WSM developed by the DSCWM has been found to be more successful. The DSCWM assesses land capability and encourages appropriate land use. It uses locally available, low-cost materials in the design and construction of structures, and it gains local cooperation for long-term protection of an area by increasing awareness of the need for resource conservation.[c] In most of the WSM projects, a range of small-scale income-generating activities have been launched to benefit the poor.

[c] Drawn from "Sector Assessment (summary): Watershed Management," Building Climate Resilience of Watersheds in Mountain Eco-regions (rrp nep 44214) (http://www.adb.org/sites/default/files/linked-documents/44214-024-nep-ssa.pdf).

In addition, savings and credit groups, mothers' groups and disadvantaged groups of the poor have been formed and economic program packages are introduced to strengthen their capacity and expand their economic activities (Pandit et al., 2007).

Though the importance of integrated WSM is well recognized at the policy level, institutional and operational capabilities are lacking. The human and financial constraints are hindering the coverage of watershed programs. Despite the anomalies in fund allocation and releases, implementation of activities by the communities provides an opportunity to spread the program over a larger part of the year by better utilizing unused labor available from the agricultural cropping calendar. Moreover, currently the program considers the political as well as social boundaries rather than hydrological. The activities are often carried out outside the drainage boundaries (Tiwari et al., 2008).

Except for India and Nepal, none of the other countries in the region have formulated systematic guidelines from national to village level. In fact, the absence of such institutional set up is the main reason for low adoption of WSM projects (Baloch and Tanik 2008; Gunawardena, n.d.). While Sri Lanka has a watershed policy and guidelines, Pakistan and other countries are yet to formulate their watershed policies. Sri Lanka though is moving toward people-oriented policies, and they are not backed by institutional arrangements. As per the national watershed management policy, the WSM committees under the chief provincial secretaries would coordinate the activities of all agencies at the rural, divisional, district, provincial, and national levels (Gunawardena, n.d.). Afghanistan has introduced participatory management of watersheds during 1990s but could not provide appropriate policy and institutional support (Ahmad and Wasiq, 2004).

Thus, countries in the region are at different levels of institutional arrangements for implementing WSM programs. Even in the countries with appropriate institutional arrangements, the implementation of watershed interventions is neither scientific nor comprehensive. All the countries are moving toward participatory approaches. Participatory approaches have to be backed by appropriate institutional framework in order to achieve the objectives. Implementation of the programs for enhancing the livelihoods of the communities needs to strengthen both technical and institutional aspects. While countries like India and Nepal are yet to strengthen the technical aspects in terms of integrating hydrogeology and biophysical aspects into watershed design, others need to look at these aspects as they move toward formulating watershed policies and watershed institutions.

3.6 Lessons for Afghanistan

Most of the South Asian countries adopted WSM to address problems of land degradation and reservoir protection earlier like the developed countries of Europe and Americas. However, India has adopted WSM with a different objective and approach. It has moved away from scientific management of catchments to protecting and enhancing the land productivity in the poorly endowed regions focusing on small hydrological units. Further, WSM is viewed as a technological strategy for these regions and accorded high policy priority and fund allocations. In the process, the scale issues of WSM are largely neglected to achieve smooth implementation, as participation of local communities in the program is recognized as critical for its success. Microwatersheds are taken up at the village level for better social organization (participation). The focus on social aspects of WSM has led to evolution of the program into a more comprehensive rural development approach of integrating livelihoods and production enhancement components. Though number of watersheds covering large areas are being identified and implemented, the technical aspects are not taken into account while selecting the watersheds for treatment. As a result, the effectiveness of these interventions has been limited despite better social or institutional outcomes. India's approach in WSM is considered effective in government policy circles as in addressing the rural livelihoods as well as resource degradation issues. The participatory and livelihoods approach to WSM is now being adopted in Nepal, Sri Lanka, and Pakistan. Bhutan continues to focus on basin/catchment management.

Thus, the experience of South Asia in WSM suggests that it can be adopted as an effective technology for addressing the broader aspects of soil and water conservation as well as productivity and food security in arid and semiarid regions. Given the high proportion of agriculture in these regions, WSM can provide a win-win situation when the broader aspects are integrated. Afghanistan is well placed, like India, to adopt such an approach to protect its soil and water resources and enhance it production and productivity.

Afghanistan is characterized by low rainfall in most parts. It faces regional as well as seasonal water shortages. Extent of land degradation is high (35% of its crop and range lands). Studies have identified watershed management as a cost-effective short- and medium-term solution to drought mitigation as well as sustainable resource management. It is observed that "watershed health and agricultural productivity issues are reducing opportunities for rural livelihood recovery in south-eastern Afghanistan" (Groninger and Lasko, 2011, p. 693). Hitherto emphasis has been on river basin scale with a focus on reforestation and range land rehabilitation. Therefore, there is immediate need for moving toward livelihood-oriented watershed management at smaller scale (less than 1000 ha) that can provide short-term benefits and increases the awareness among the rural communities (Groninger et al., 2010).

Though Afghanistan is unique in many respects, it can draw lessons from more than three decades of experiences of South Asian countries. The following aspects need to be considered in the context of Afghanistan:

- Given the low and medium rainfall, watershed interventions should be targeted toward enhanced livelihoods rather than improving the productivity alone. This would ensure broad-based benefit flows and strengthen the social acceptance and participation.
- Learning from the South Asian experience of aligning watersheds to hydrological boundaries would ensure sustainability and effective impacts. Afghanistan has already prepared the watershed atlas delineating the major watersheds and identifying smaller watersheds. The size of the watersheds needs to be smaller to make them socially feasible. Implementation of watersheds should be taken up in such a way that it would address both local (soil and water conservation, improved productivity and livelihoods) and basin-level issues (soil erosion due to wind).
- Participatory approaches need to be scientifically sound awareness building and improving their capacities to understand and adopt technically robust interventions rather than interventions that give quick and short-term benefits. Concerted efforts are required to increase awareness and gain acceptance for the interventions. Particular emphasis should be on hydrogeological and biophysical aspects.
- Resource sustainability and equity need to be built into WSM. While adopting scientific approach and choosing the right "scale" is necessary, it is not sufficient for addressing the sustainability. Any successful watershed intervention requires a combination of security/accessibility, clear land title and control, and appropriate geologic and climatologic conditions that will permit interventions to make a discernible difference (Groninger et al., 2010). Appropriate institutional arrangements for managing the resources need to be built around the existing institutions that are embedded in the culture of the land. For instance, Afghanistan has traditional irrigation systems like *Karez* and water institutions like *Water*

Waqils and *Shuras,* which are still functional. These institutions could form the basis for evolving watershed institutions.

- Afghanistan has limited capacities in understanding and implementing the WSM approaches. The implementing agencies in the government need capacity building at various levels.
- NGO sector needs to be encouraged to take part in implementing the WSM programs. Identifying right NGOs and encouraging them to participate in the process would be a good starting point. Selection of nodal NGOs at the provincial and national levels and capacitating them along with the key line departments should be the first priority. Networking of NGOs that are involved/interested in natural resource management and capacitating them is an effective way forward.

3.7 Policy imperatives

There is an immediate need for creating a policy environment and policies to initiate WSM program in Afghanistan. Providing a common framework for implementing the WSM programs should be the priority rather than leaving it to different funding agencies. Each agency comes with its own objectives and approaches. Often these agencies may strive to achieve diametrically opposite outcomes instead of working in tandem. A national policy with specific objectives could provide the direction so that goals can be achieved collectively. Given here are some policy imperatives for immediate attention:

- Formulating a national policy on WSM and preparation of guidelines for implementing the policy.
- Establishing appropriate institutions at the central level either by designating an existing department or by establishing an autonomous authority with roles and responsibilities to provide the policy directions, guidelines for institutional arrangements and implementation at the provincial level and below.
- Identifying the priority areas and allocating the resources is critical for achieving ecologically sound and economically efficient outcomes.
- Delineation of watershed boundaries is a necessary condition for implementing the program. This needs to be taken up at the national level.
- Establishing a national-level capacity building institution (in the lines of MANAGE, in India) or identifying some of existing institutions and strengthening them is a policy priority.

References

Abernethy, C., Wijayaratna, C.M., 1998. Evaluation of the impacts of projects to reduce human-induced soil losses in watersheds. In: Proceedings of the symposium on Modelling Soil Erosion, Sediment Transport and Closely Related Hydrological Processes, Vienna, July, pp. 287–295. IAHS Publ. no. 249.

Acharya, B., 2000. Watershed Management in Nepal: Recent Experiences and Lessons. Office of Agriculture and Rural Development, United States Agency for International Development Kathmandu, Nepal.

Adolph, B., Turton, C., 1998. Community Self-Help Groups and Watersheds. Partnerships and Policies Programme Series, Overseas Development Institute, London.

Ahmad, S., 2001. Watershed management in Pakistan: achievements and issues. In: Paper Presented in the National Seminar on "Water Resources". FFC, Lahore October.

Ahmad, M., Wasiq, M., 2004. Water Resource Development in Northern Afghanistan and Its Implications for Amu Darya Basin. World Bank Working Paper No. 36, The World Bank, Washington, DC.

Arya, S., Samra, J.S., 1995. Socio-Economic Implications and Participatory Approach of Integrated Watershed Management Project at Bunga. Technical Bulletin No. T-27/C-6, Central Soil and Water Conservation Research and Training Institute, Research Centre, Chandigarh, India, p. 112.

Baloch, M.A., Tanik, A., 2008. Development of an integrated watershed management strategy for resource conservation in Baluchistan province of Pakistan. Desalination 226, 38–46. https://doi.org/10.1016/j.desal.0000.00.000.

Bashir, S., Baig, M.A., Ashraf, M., Anwar, M.M., Bhalli, M.N., Munawar, S., 2013. Risk assessment of soil erosion in Rawal watershed using geoinformatics techniques. Sci. Int., Lahore 25 (3), 583–588.

Batchelor, C., Singh, A.K., Rao, C.H., Mohan, R., Butterworth, C., 2003. Watershed development: a solution to water shortages or part of the problem? Land Use Water Resour. Res. 3, 1–10.

Calder, I.R., 1991. Sri Lankan Link Programme: Hydrology Component. Draft Tour Report ODG 9014.

Calder, I.R., 2005. Blue Revolution. Integrated Land and Water Resource Management, second ed. Earthscan, London and Sterling, VA.

Calder, I., Gosain, A., Rao, M.S., Mohan, R., Batchelor, C., Snehalatha, M., Bishop, E., 2006. Planning rainwater harvesting in India: (1) biophysical and societal impacts. In: National Workshop on Priorities for Watershed Management in India. Bangalore, 22–23 May. Government of Karnataka and the World Bank.

Carson, B., 1989. Soil Conservation Strategies for Upland Areas of Indonesia. Occasional Paper No. 9, East-West Environment and Policy Institute, Honolulu.

Deshpande, R.S., Reddy, V.R., 1991. Differential impact of watershed based technology: some analytical issues. Indian J. Agric. Econ. XLVI (3), 261–269.

Doolette, J.B., Magrath, W.B. (Eds.), 1990. Watershed developments in Asia: strategies and technologies. In: World Book Technical Paper No. 127. Washington, DC.

Doolette, J.B., Smyle, J.W., 1990. Soil and moisture conservation technologies: review of literature. In: Doolette, J.B., Magrath, W.B. (Eds.), Watershed Developments in Asia: Strategies and Technologies. World Book Technical Paper No. 127. Washington, DC.

Elkaduwa, W.K.B., Sakthivadivel, R., 1998. Use of Historical Data as a Decision Support Tool in Watershed Management: A Case Study of the Upper Nilwala Basin in Sri Lanka. Research Report 26, International Water Management Institute, Colombo, Sri Lanka.

Fan, S., Hazell, P., 2000. Should developing countries invest more in less favoured areas? An empirical analysis of rural India. Econ. Polit. Wkly. 35 (17), 1455–1464.

FAO, 2006. The New Generation of Watershed Management Programmes and Projects. FAO Forestry Paper 150, Food and Agriculture Organization of the United Nations, Rome.

Favre, R., Kamal, G.M., 2004. Watershed Atlas of Afghanistan. In: Working Document for Planners (Parts I & II)), first ed. Ministry of Irrigation. Water Resources and Environment and the Food and Agriculture Organization of the United Nations (FAO), Kabul.

Fleming, B., Fleming, J.P., 2009. A watershed conservation success story in Nepal: land use changes over 30 years. Waterlines 28(1).

GoI, 1994. Guidelines for Watershed Development. Ministry of Rural Development, Government of India, New Delhi.

GoI, 2006. From Hariyali to Neeranchal: Report of the Technical Committee on Watershed Programmes in India. Department of Land Resources, Ministry of Rural Development, Government of India.

Groninger, J.W., Lasko, R.J., 2011. Water for agriculture: challenges and opportunities in a war zone. Water Int. 36 (6), 693–707.

Groninger, J., Ruffner, C.M., Walters, S.A., 2010. Sustaining rural Afghanistan under limited central government influence. Stability Int. J. Secur. Dev. 2 (2), 1–13. https://doi.org/10.5334/sta.bk, Part 24.

Gunawardena, E. R. N., n.d. Evolution of watershed management: the Sri Lankan experience. In: Proceedings of the Asian Regional Workshop on Watershed Management (Chapter 14) (ftp://ftp.fao.org/docrep/fao/009/a0270e/A0270E11.pdf).

Hanif, F., Subhan,M.,Megahan,W.F., 1990.Watershed restoration reduces runoff and sedimentation from comparative watersheds in Pakistan's subtropical scrub zone research needs and applications to reduce erosion and sedimentation in tropical steep lands. In: Research Needs and Applications to Reduce Erosion and Sedimentation in Tropical Steep lands Proceedings of the Fiji Symposium. June 1990. IAHS-AISH Publ. No. 192.

ICARDA, 2010. Integrated watershed development for food security and sustainable improvement of livelihoods in Barani, Pakistan. Final Report, International Center for Agricultural Research in the Dry Areas, Austrian Development Agency, Austria.

ICIMOD, 2007. Good Practices in Watershed Management Lessons Learned in the Mid Hills of Nepal. International Centre for Integrated Mountain Development (ICIMOD), Kathmandu, Nepal.

Jayaweera, J.M.U.I., De Zoysa, M., Kariyawasam, D., 2004. Uma-Oya upper watershed management project (Uwmp) in Sri Lanka: the impacts at household level. In: Baumgartner, D.M. (Ed.), Proceedings of Human Dimensions of Family, Farm, and Community Forestry International Symposium, March 29–April 1, 2004. Washington State University, Pullman, WA. Washington State University Extension MISC0526.

Joshi, P.K., Jha, A.K., Wani, S.P., Joshi, L., Shiyani, R.L., 2004. Meta-Analysis to Assess Impact of Watershed Programme and People's Participation. Comprehensive Assessment Research Report No. 8, International Crops Research Institute for the Semi-Arid Tropics (ICRISAT), Patancheru, Andhra Pradesh, India.

Joy, K.J., Paranjape, S., 2004. Watershed Development Review: Issues and Prospects. Technical Report, Centre for Interdisciplinary Studies in Environment and Development (CISED), Institute for Economic and Social Change, Bangalore.

Kerr, J.M., Pangare, G., Lokur-Pangare, V., George, P.J., Kolavalli, S., 1998. The Role of Watershed Projects in Developing India's Rainfed Agriculture. Report submitted to the World Bank, IFPRI, Washington.

Khattak, G.M., n.d. Watershed Management in Pakistan. http://lib.icimod.org/record/25325/files/c_attachment_419_4611.pdf.

Kumar, M.D., Ghosh, S., Patel, A.R., Singh, O.P., Ravindranath, R., 2006. Rainwater harvesting in India: critical issues for basin planning and research. Land Use Water Resour. Res. 6 (2006), 1–17.

Kumar, M.D., Patel, A., Ravindranath, R., Singh, O.P., 2008. Chasing a mirage: water harvesting and artificial recharge in naturally water-scarce regions. Econ. Polit. Wkly. 43, 61–71.

Malano, H., Davidson, B., George, B., Ryu, D., Kumar, K., Anshuman, Pavelic, P., Kulkarni, A., Deshpande, N., Patwardhan, S., Venkat, R., Teluguntla, P., Nune, R., Lagudu, S., Chatti, D., Grover, S., Nayal, A., Honnungar, V., 2015. Impacts of Climate Change and Watershed Development on Whole of-Basin Agricultural Water Security in the Krishna and Murray-Darling Basins. Australian Centre for International Agricultural Research, Australia. Final report.

Merz, J., Nakarmi, G., Shrestha, S.K., Dahal, B.M., Dangol, P.M., Dhakal, M.P., Dongol, B.S., Sharma, S., Shah, P.B., Weingartner, R., 2003. Water: a scarce resource in rural watersheds of Nepal's middle mountains. Mt. Res. Dev. 23 (1), 41–49. https://doi.org/10.1659/0276-4741(2003)023[0041:WASRIR]2.0.CO;2.

Misbahuzzaman K., n.d. Problems and Prospects of the Hilly Watersheds in Bangladesh: Priorities for their Conservation. Institute of Forestry and Environmental Sciences, Chittagong University, Chittagong, Bangladesh. http://lib.icimod.org/record/12176/files/3539.pdf.

Murray, F., Little, D., n.d. Towards Improved Management of Living Aquatic Resources in Watersheds of the Dry Zone, Sri Lanka. Institute of Aquaculture, Stirling University, Scotland.

Pandit, B.H., Wagley, M.P., Neupane, R.P., Adhikari, B.R., 2007. Watershed management and livelihoods: lessons from Nepal. J. For. Livelihood 6 (2), 67–75.

Pingali, P.L., Rosegrant, M., 2001. Intensive food systems in Asia: can the degradation problems be resolved? In: Lee, D.R., Barrett, C.B. (Eds.), Trade-Offs or Synergies?. Agricultural Intensification, Economic Development and the environment. CABI Publishing.

Ray, S., Bijarnia, M. 2006. Upstream vs Down Stream: Groundwater Management and Rain Water Harvesting. Econ. Polit. Wkly, XLI, No. 23.

Reddy, V. R. 2006.Getting the Implementation Right: Can the Proposed Watershed Guidelines Help? Econ. Polit. Wkly, XLI, No. 40.

Reddy, V.R., Syme, G.J., 2015. Integrated Assessment of Scale Impacts of Watershed Interventions: Assessing Hydro-Geological and Bio-Physical Influences on Livelihoods. Elsevier INC.

Reddy, V.R., Reddy, M.G., Galab, S., Springate-Baginski, O., 2001. Watershed Development and Livelihood Security: An Assessment of Linkages and Impact in Andhra Pradesh, India. CESS/University of Leeds, Draft Report.

Reddy, V.R., Reddy, M.G., Galab, S., Soussan, J., Springate-Briganski, O., 2004. Participatory watershed development in India: can it sustain rural livelihoods? Dev. Chang. 35 (2), 297–326.

Reddy, V.R., Reddy, M.G., Reddy, Y.V.M., Soussan, J., 2008. Sustaining rural livelihoods in fragile environments: resource endowments and policy interventions—a study in the context of participatory watershed development in Andhra Pradesh. Indian J. Agric. Econ. 63 (2), 169–187.

Reddy, V.R., Reddy, M.G., Soussan, J., 2010a. Political Economy of Watershed management: Policies, Institutions, Implementation and Livelihoods. Rawat Publishers, Jaipur.

Reddy, V.R., Rout, S.K., Chiranjeevi, T., 2010b. The Impact Assessment Study of Watershed Development Projects in Rajasthan. Project Report, Livelihoods and Natural resource Management Institute, Hyderabad.

Renner, M., 2010. Troubled Waters: Central and South Asia Exemplify some of the Planet's Looming Water Shortages, May 1. World Watch Institute.

Rout, S.K., 2013. Collective Action and Natural Resource Management: A Study of Watershed Development in India. Unpublished Ph. D Dissertation, Centre for Economic and Social Studies (CESS), BRAOU, Hyderabad.

Shah, A., 1998. Watershed Development Programmes in India: Emerging Issues for Environment Development Perspectives. Econ. Polit. Wkly, 32(26).

Shah, H., 2008. An Assessment of Participatory Integrated Watershed Management in the Hilkot Watershed, Mansehra, Pakistan. Thesis submitted to the University Putra Malaysia, Malaysia, in Fulfilment of the Requirement for the Degree of Master of Science, October.

Shah, A., 2009. Equity in the impact of watershed development: class, gender and regions. In: Shah, A., Wani, S.P., Sreedevi, T.K. (Eds.), Impact of Watershed Management on Women and Vulnerable Groups. Proceedings of the Workshop on Comprehensive Assessment of Watershed Programs in India. ICRISAT, Patancheru, Andhra Pradesh, India.

Sheikh, M.l., n.d. Approaches in Watershed Management in Areas Affected by Overgrazing and Misuse of Range Land Resources. Forestry Department, Pakistan, http://www.fao.org/docrep/006/ad085e/ad085e10.htm [09-07-2015 17:04:04].

Singh, K., 1991. Determinants of people's participation in watershed development and management: an exploratory case study. Indian J. Agric. Econ. 46 (3), 278–286.

Singh, N., Karki, S., 2004. Discourse, Legislative Framework and Practice on Integrated Water Resources Management in Bhutan. IUCN, Nepal Country Office.

Singh, S.L., Kharel, B.P., Joshi, M.D., Mathema, P., 2004. Watershed Management Case Study: Nepal-Review and Assessment of Watershed Management Strategies and Approaches. Food and Agricultural Organisation (FAO). Rome.

Sri Lanka Upper Watershed Management Project, http://www.rrojasdatabank.info/envplansrilanka.pdf.

Sthapit, K. M., n.d. Watershed Management in Nepal. ICIMOD, Kathmandu.

Thapa, R.S., n.d. Contribution of Watershed Management Program to the livelihoods of the local people: a case study from Kathekhola Subwatershed, Baglung, Nepal. Masters Thesis, Land Management and Land Tenure in Urban and Rural Areas; Technical University of Munchen, Germany.

Tiwari, K.R., Bajracharya, R.M., Sitaula, B.K., 2008. Natural resource and watershed management in South Asia: a comparative evaluation with special references to Nepal. J. Agric. Environ. 9, 72–89.

Tsering, K., 2011. A Roadmap for Watershed Management in Bhutan: From Policy to Practice. Watershed Management Division, Department of Forests and Park Services, Ministry of Agriculture and Forests, Thimphu, Bhutan.

Virgo, K.J., Aslami, M.H., Ahmed, B., 2006. Participatory watershed management: examples from Herat, western Afghanistan. P. World Assoc. Soil Water Conserv. 1, 65–81.

Wimalasinghe, H.M., Wijayaratne, C.M., 1998. Participatory watershed management: present status and future prospects. In: Paper Presented at the Conference on The Status and Future Direction of Water Research, S Colombo, Sri Lanka, 4–6 November. http://publications.iwmi.org/pdf/H023525.pdf.

World Bank, 2008. World Development Report 2008. World Bank, Washington, DC.

Further reading

Kumar, M.D., Reddy, V.R., Narayanamoorthy, A., Bassi, N., James, A.J., 2017. Rainfed areas: poor definition and flawed solutions. Int. J. Water Resour. Dev. 34, 278–291. https://doi.org/10.1080/07900627.2017.1278680.

CHAPTER 4

From watershed to sub-basins: Analyzing the scale effects of watershed development in Narmada river basin, central India

Jayesh Talati[a], **M. Dinesh Kumar**[b]
[a]Formerly with IWMI-TATA, International Water Management Institute, Anand, Gujarat, India
[b]Institute for Resource Analysis and Policy, Hyderabad, Telangana, India

4.1 Introduction

Narmada basin is one of the water-rich basins of India with an estimated renewable surface water resource of 27 Million Acre Feet (MAF) as per a 1993 estimate (GOI, 1999), though the latest assessment carried out by Central Water Commission (2017) showed a total renewable water availability of 45.24 BCM (CWC, 2017). The Narmada River originates from East of Madhya Pradesh, enters into Gujarat, and ends up in the Gulf of Cambay. Approximately, 90%of the basin is falling in Madhya Pradesh (MP) (Fig. 4.1). Since the mid-1990s, MP government has made concerted efforts in watershed development and water harvesting through decentralized administration and governance under Rajeev Gandhi Watershed Mission and Rajeev Gandhi Rural Drinking Water Supply (RWSS) mission. Subsequently, intensive watershed development work was taken up in 1694 watersheds of the basin between 1995–95 and 2001–02. Besides this, many donor agencies have supported local NGOs to independently implement watershed development projects.

The intensive rainwater harvesting activities, and groundwater development, which occur as an aftermath of intensive watershed development program, have the potential to cause major socioeconomic and hydrological changes at various levels in the basin from village to watershed to sub-basin to the entire basin. Limited field observations from intervention areas have made many NGOs argue that local groundwater recharge improved

Current Directions in Water Scarcity Research, Volume 1
https://doi.org/10.1016/B978-0-12-814851-8.00004-5
© 2019 Elsevier Inc.
All rights reserved.

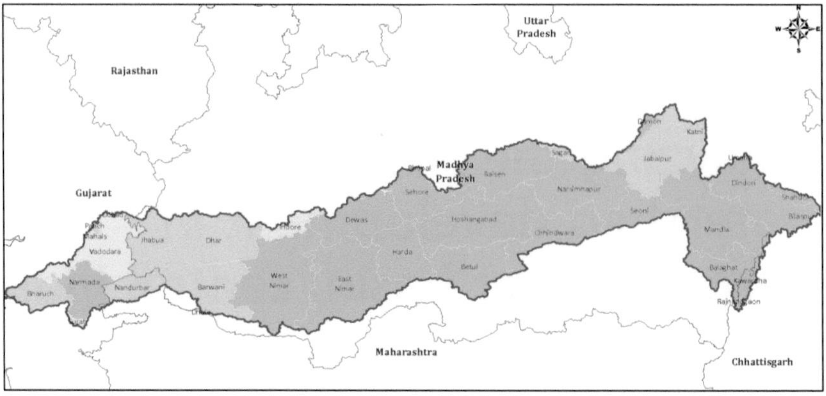

Fig. 4.1 District falling in Narmada basin.

as a result of watershed/water harvesting interventions, and the improved hydrological regime conditions not only help people improve their socioeconomic conditions through enhanced water uses, but also increases the lean season flows in downstream areas and increases the local groundwater buffer for drought years. Notwithstanding the limitation induced by the neglect of the change in local water demands due to induced supplies, the unit of such analysis is never extended beyond the village or the watershed selected for the project.

The effects of watershed-based catchment treatments and groundwater use on the basin hydrology were not considered when water from the river was allocated by the Narmada Water Disputes Tribunal allocated among the four party states and when Narmada valley development projects were originally planned (Kumar, 2010, 2014; Kumar et al., 2005). Watershed development was rather viewed as a complementary work for sustainable basin management, often limited to a few hilly watersheds.

But, as some researchers observe, since the days of conceptualization of the Narmada Master Plan, the water management concerns and priorities in the basin have changed with changing socioeconomic conditions and political climate. During the period from 1993 to 2003, the focus of the Congress-led government in the state has been entirely on watershed development, decentralized water harvesting, and groundwater use. This according to the researchers had effects on the hydroecology in the form of reduced runoff generation, reduced stream flows, and faster groundwater drawdown. They use the recent hydrological records that show much lower basin yields to support their argument (see Kumar, 2010; Kumar et al., 2005).

4.2 Objective, methodology, and scope

The study had two major objectives: analyze the socioeconomic and hydrological impact of watershed development and water harvesting activities at the local level; and assess the effect of watershed development and water harvesting activities on the availability of flows at sub-basin level. As mentioned earlier, approximately, 90% basin area is falling in MP. Geographically, 22 districts of Madhya Pradesh (MP) are situated in Narmada basin either fully or partly. The basin covers 100% area of three districts, 60%–90% area of five districts, 30%–60% area of seven districts, and 3%–30% area of another seven districts.

Narmada River has 41 tributaries, of which 39 tributaries are draining from MP. Data on tributary-wise area of treated microwatersheds are not available. Data collected from Rajiv Gandhi Watershed Mission show that intensive watershed development work was carried out in Khandwa, Khargone, and Jhabua. Among these four districts, which are leading in watershed development, stream flow-gauging stations are available in Kundi sub-basin of Khargone district and Hathni sub-basin of Jhabua district. Therefore, these two districts were purposively selected for the study. Since Khargone district is falling in Kundi sub-basin, we have used district-level data as representing the sub-basin. In case of Jhabua district, which occupies only 66% of Hathni sub-basin area, we have culled out sub-basin level data from the district-level data. From each district, one microwatershed, which is intensively treated, was chosen for rigorous study involving primary data collection. These microwatersheds are those which have more than 90% their geographical area treated under employment assurance scheme (EAS) during 1995–96 to 1999–2000. From both the districts, a nontreated microwatershed, which is contiguous to the treated microwatershed, was selected for comparison.

4.2.1 Lines of analysis

⇒ Use of satellite imagery: to assess the changes in land use pattern and cropped area at sub-basin, watershed, and microwatershed levels due to the watershed interventions.

⇒ Geographic Information system (GIS) to capture the temporal and spatial variations in the land use at sub-basin and watershed levels using primary survey data as well as satellite imagery.
⇒ Longitudinal and cross-sectional analysis: comparison of pre- and post-watershed intervention periods for the same watershed as well as treated and nontreated watersheds.
⇒ Mapping: changes in land use, particularly cropping pattern and cropping system, were mapped on village revenue maps.
⇒ Estimation of water balance: by considering the groundwater recharge and extraction figures for the year 2004 for one of the treated microwatersheds using water-level fluctuation approach.
⇒ Regression between runoff and rise in well water level (dependent variables) and rainfall (independent variable) carried out for analyzing hydrological changes at the macrolevel.

4.2.2 Tools and techniques

The study involved: (1) analysis of data collected from complete household census of the selected microwatersheds by administering structured questionnaire to collect socioeconomic data; (2) analysis of data on water-level fluctuations in wells in the treated and control watersheds during pre- and postmonsoon periods for the year 2004; (3) analysis of stream flow data at 10-day intervals for one of the selected sub-basins collected from Narmada Control Authority, Indore, MP; (4) analysis of rainfall and land use data collected from the corresponding district land record offices of the selected sub-basins, and water level fluctuation data and specific yield values for the aquifer formations of the sub-basins from the state geohydrology survey unit of the sub-basins; (5) analysis of agricultural statistics obtained from district agriculture office for the districts representing the two sub-basins; (6) analysis of data on watershed development programs collected from district level CEO, Panchayat Office for the two sub-basins.

4.3 Results and discussion

4.3.1 Socioeconomic changes

Analysis of socioeconomic data showed that there was a remarkable growth in population at the sub-basin level during the study period in both Kundi and Hathni sub-basins, with annual compounded growth rate of 2.4% and 3.2%, respectively. But, the treated watersheds show much lower growth in population as compared to the control (untreated) watersheds in both the basins—3.45% against 4.5% in Kundi watersheds and 0.39% and 4.49% in Hathni watersheds, though their growth rates are higher than the aggregate sub-basin level growth rates. A marked increase in livestock population during the study period was also observed at Kundi sub-basin, with much lower growth in the microwatershed. But, the growth in the treated watershed (0.56%) was slightly higher than that in the untreated watershed (0.17%).

Cross-sectional analysis brings out sharper differences. The livestock population per household is 4.1 and 6.29 animals in treated watersheds, in Kundi and Hathni sub-basins, respectively, compared to 2.56 and 5.56 animals per household in nontreated watersheds of the respective basins. This indicates more intensive livestock-rearing activities in the treated watersheds. The rapid rise in livestock holding at micro- and macrolevels would lead to increased demand for water for domestic uses, cattle drinking, and animal feeding.

Our analysis showed not so significant increase in number of people engaged in different rural occupations such as agriculture, animal husbandry, and local employment, for the watershed in Kundi sub-basin. Differences are sharp when compared with the situation in nontreated watersheds. People in treated watersheds, viz., Bada Bhawata and Khedi, remained engaged in agriculture for an average of 5.4 and 8.4 months compared to 4.1 and 6.6 months in nontreated watersheds, viz., Bahedwa and Jaljyoti. In Khedi watershed, there was marginal rise in agricultural employment after the intervention from 8.3 to 8.4 months. A higher percentage of the people are engaged in animal husbandry in treated watersheds (11.75% and 24.4% in Bada Bhawata and Khedi, respectively) as compared to untreated watersheds (9.8% and 21.35% in Bahedwa and Jaljyoti). This may be explained by the larger livestock holdings in the treated watersheds. This can be attributed to better availability of fodder.

It is a truism that better management of natural and environmental resources in rural areas leads to reduced distress migration. There has been sharp reduction in the number of family members engaged in migration in both treated and untreated watershed, the reduction being higher in the untreated watersheds (1.75% against 2.3%) of Kundi sub-basin, for which data are available. One could argue that the watershed interventions have helped contain the impact of consecutive drought. Of the 244 respondents from the treated watersheds in Hathni sub-basin, 115 reported that getting temporary employment during implementation of WDP and reduction of migration period are two major impacts they realized so far.

However, informal discussion with stakeholders of WDP and interaction with watershed community during field work revealed that temporary nature of employment generation did not significantly affect the overall migration in general except a little change in the period and pattern of migration.

The study did not show any marked difference in the local employment generation potential in both the basins. Increased agricultural production, improvement in soil condition, and introduction of income generation activity through self-help group are other benefits reported by 8, 4, and 21 respondents, respectively.

4.3.2 Changes in land use

As per the data from agricultural census, 1989–90 and 1995–2000, land use shows declining trend between 1989–90 and 1999–2000 in Kundi sub-basin. Data collected from district land records office, Jhabua, also corroborated this for Hathni basin. Total reported area came down from 1,349,000 to 1019,000 hain Kundi basin and from 271,000 to 269,000 hain Hathni basin. Forest area has drastically shrunk from 34.85% to 9.75%in Kundi, while it reduced from 3.22% to 2.75% in Hathni. Land that is not available for cultivation increased from 15.3% to 22%in Kundi sub-basin and from 21.3% to 23.6%in Hathni. Cultivable waste land increased from 2.3% to 3.6% and 2.3% to 2.6% in Kundi and Hathni, respectively. Net sown area increased from 46.7% to 63.2% (of the cultivable area) in Kundi and did not change much in Hathni basin. Gross cropped area increased from 51.6% to 73.6% in Kundi, but marginally reduced from 70.90% to 68.8% in Hathni.

Kharif and winter cropped area were found to be showing declining trend in Kundi basin. This is mainly because of the reduction in reporting area. Therefore, it is important to look at the changes in cropping intensity. The cropping intensity (expressed as a ratio of gross cropped area and net sown area) increased after the watershed interventions from 110% to 116%. The analyses of agricultural statistics for the 5-year periods 1992–96 (preintervention) and 1996–2003 (postintervention) showed that the average production and productivity of many of the crops also declined with decline in cropped area. While the reduction in crop production could be attributed to the reduction in aggregate area under each crop, reduction in yield of some crops would have contributed to this. It was interesting to note that the yield of crops such as wheat and cotton had increased after the watershed interventions, while that of many rain-fed crops had reduced.

The reduction in yield of rain-fed crops could be the overall impact of the consecutive poor monsoons in many years after the watershed interventions.

Analysis of data collected from primary survey carried out in the micro-watersheds, presented in Table 4.1, indicated that cultivated area, irrigated area, Kharif cropped area, and winter cropped area increased by 8.5%, 45.8%, 14.2%, and 53.2% in treated watershed and by 21.5%, 46%, 57.6%, and 25.6% in nontreated watershed of Kundi basin postwatershed intervention period. One can argue that while increase in winter cropped area was by virtue of improved water availability in treated watersheds, in the untreated watersheds the farmers are expanding the area under rain-fed crops to meet the increased food and fodder requirements. Cropping intensity increased by 5% in the treated watershed but decreased by 3.25% in the untreated watershed during the study period (Table 4.1).

In case of Hathni sub-basin, data pertaining to prewatershed intervention were not available for any of the watersheds. Postwatershed intervention, the percentage of irrigated area and winter cropped area to the geographical area in the treated watershed was higher than that of untreated watershed by 3.9% and 9.45%. The percentage of Kharif cropped area to geographical area remained the same for both the microwatersheds. This indicates higher cropping intensity and area under irrigation after watershed intervention in the sample watershed. In the absence of baseline data, one cannot say with a high level of confidence that the higher cropping intensity in the treated watershed of Hathni sub-basin is because of watershed intervention.

Source wise area under irrigation increased in Kundi basin. Maximum expansion in irrigated area occurred under sources other than tanks. The small water harvesting structures (WHSs) may be functioning as the alternative irrigation sources in the treated watersheds. Drastic reduction in tank-irrigated area was observed in the basin, which declined from 11,200 to 1200 ha. Area under well and tube well irrigation increased by 36%. Canal-irrigated area had reasonably increased. It clearly shows that groundwater development took place at a greater speed in the basin during the last decade. While surface irrigation shows mix trend because canal irrigation is progressing, tank irrigation is declining.

4.3.3 Hydrological changes

Data on annual rainfall available for 44 years in case of Kundi sub-basin area and 28 years in case of Hathni sub-basin area were analyzed. Five-year moving average of these rainfalls were analyzed and presented in Fig. 4.2.

Table 4.1 Change in cropped area at microwatershed level (area in acres).

		Kundi basin		Hathni basin	
	Cropped area	**Treated-Khedi**	**Nontreated-Jaljyoti**	**Treated-Bada Bhawata**	**Nontreated-Bahedwa**
Period	**Geographical area**	**1092.00**	**1331.00**	**1411.00**	**971.00**
Prewatershed	Reported area	872.25	528.82	—	—
	Cultivated area	855.25	492.75	—	—
	Irrigated area	355.00	196.75	—	—
	Kharif cropped area	751.25	344.50	—	—
	Rabi cropped area	110.00	65.25	—	—
	Cropping intensity	114.64	118.94	—	—
Postwatershed	Reported area	938.75	662.82	—	—
	Cultivated area	927.75	598.75	585.50	417.50
	Irrigated area	517.50	287.25	140.50	59.00
	Kharif cropped area	858.25	543.00	896.50	626.50
	Rabi cropped area	168.50	81.90	169.00	24.50
	Cropping intensity	119.63	115.08	118.85	103.91

Source: Primary survey, 2004.

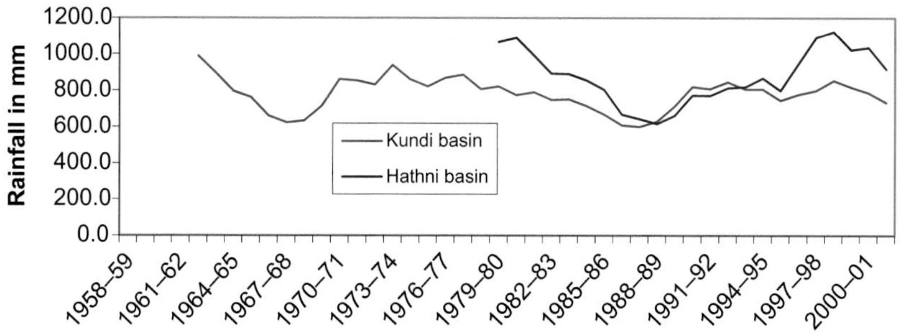

Fig. 4.2 Moving average of annual rainfall in two sub-basins.

In the case of Kundi sub-basin, the moving averages indicate cycles of approximately 20 years with steep rise reflecting consecutive wet years, followed by steady decline reflecting consecutive dry years. The year corresponding to lowest moving average value is same for both the basins. In the case of Hathni, steep decline and rise in moving average was found indicating a sharper difference between rainfall magnitudes of wet years and dry years. With a mean annual rainfall of 880 mm, Hathni sub-basin receives higher rainfall as compared to Kundi, which has a mean annual rainfall of 784 mm (source: authors' own analysis based on data from Office of Land Records, Khargone and Jhabua, MP).

Both the sub-basins experienced remarkable strides in groundwater development. Number of WHS created under WDP directly affects the local-level water availability. Data on pre- and postmonsoon well water levels of 39 observation wells in Kundi sub-basin and 51 observation wells in Hathni sub-basin available from Geo hydrological Survey units of the respective sub-basins were analyzed to estimate the average annual rise/fall in water levels at the sub-basin level. But apart from watershed treatment, annual rainfall also will have an effect on monsoon-water-level fluctuations. To factor out the effect of yearly variations in rainfall on groundwater recharge, we worked out recharge fraction, as a ratio of *Rise in water level* and *Average annual rainfall.* However, this method of estimating the "recharge fraction" has its own limitations, as it ignores the likely nonresponsiveness of water levels in shallow aquifers to excessively high rainfall due to the phenomenon of "rejected recharge" appearing as base flow. When the rainfall becomes excessively high, presence of recharge structures may not result in differential recharge as the shallow aquifer may not have adequate storage space to store the infiltrating water from the small reservoirs created by the recharge structures.

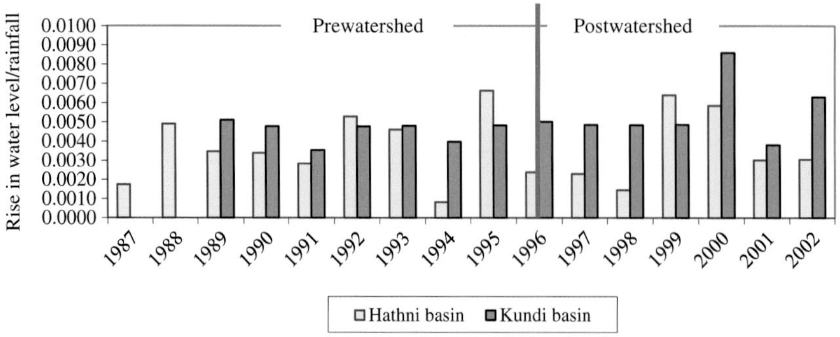

Fig. 4.3 Effect of watershed interventions on recharge fraction. *(Source: Geo-hydrological survey unit, Khargone and Jhabua, MP.)*

The estimated values of recharge fraction for the two sub-basins indicate positive change after introduction of WDP at the sub-basin level (Fig. 4.3), meaning higher rate of percolation of water for the same quantum of rainfall, which is indicative of the positive effect of watershed interventions in terms of increasing opportunity time for runoff water to infiltrate. The average recharge fraction for Kundi sub-basin was 0.0055 post-WSD (for the period from 1997 to 2002), against 0.0046 for the pre-WSD (1988 to 1996). The difference in "recharge fraction" was very marginal (0.0037 post-WSD against 0.0036 pre-WSD) in the case of Hathni sub-basin. Based on these results, one can argue that land treatment and WHS created under WDP might have contributed to augmenting ground recharge at the local level. But, alternatively, one could also argue that the higher percentage of rainfall getting underground is due to the effect of varying magnitudes of rainfall on the recharge process (that higher rainfall may not result in higher recharge rates but higher runoff) and not due to any effect of land treatment on the recharge rates. Therefore, it is important to compare the recharge behavior for treated and untreated watersheds for the same magnitude of rainfall.

Detailed analysis was done for the treated watersheds and the control watersheds in Hathni sub-basin, which enabled comparison of groundwater recharge for the same quantum of rainfall in different years. Like in the case of the sub-basins, the recharge fraction had substantially increased, what is striking is that the difference in recharge fraction between treated watershed and untreated watershed is larger at higher magnitudes of rainfall (Fig. 4.4). At lower magnitudes of rainfall, the portion of the rainwater that percolates down the soil to the shallow aquifer is more or less same for the treated and untreated watersheds.

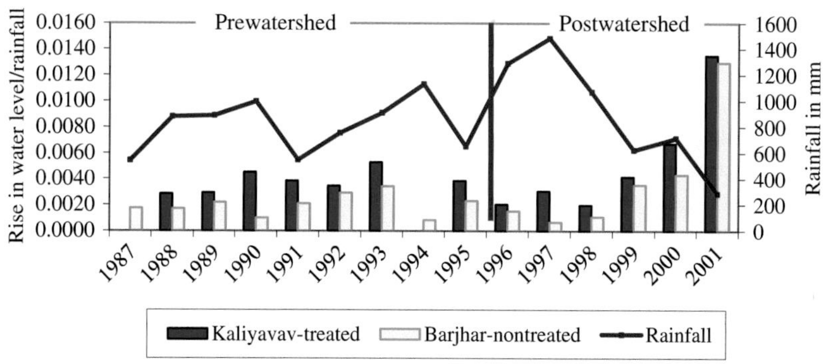

Fig. 4.4 Effect of watershed interventions on recharge fraction in microwatersheds, Hathni sub-basin. (Source: *Geo-hydrological survey unit, Jhabua.)*

4.3.3.1 Water balance at microwatershed level

Microwatershed level water balance was estimated on the basis of groundwater recharge and groundwater extraction for the year 2004 in Kundi basin. The groundwater recharge and groundwater extraction were estimated using the estimated values of average water-level fluctuations in wells during monsoon ($\boldsymbol{FL_{WL}}$), the specific yield of the aquifer ($\boldsymbol{S_y}$), and the geographical area of the watershed ($\boldsymbol{A}$) to arrive at the net recharge ($\boldsymbol{RECH_{NET}}$). To the net recharge figures, the estimated groundwater draft during kharif ($\boldsymbol{PUMP_{WELL}}$) was added to arrive at the actual monsoon recharge.

$$\text{Net recharge during monsoon} = RECH_{NET} = FL_{WL} \times S_y \times A$$

$$\text{Actual monsoon recharge} = RECH_{ACTUAL} = F_{WL} \times S_y \times A + PUMP_{WELL}$$

Estimates of water balance shown in Table 4.2 clearly indicate that not only the aggregate recharge is higher in the treated watershed as compared to the untreated one (102,300 against 59,500 m^3), but the recharge rate is also higher (0.027 MCM/km^2 against 0.022 MCM/km^2). On the other hand, the groundwater development reached a plateau in the treated watershed in spite of high recharge, owing to high rates of pumping. Though number of wells and bore wells and the well yields are nearly same for both the watersheds, annual pumping hours are remarkably higher in treated watersheds, giving warning signals against future problems of groundwater overdevelopment. The balance recharge in the treated watershed was only 2000 m^3. Over and above, the abstraction figures include only that for irrigation. If we consider groundwater use in other sectors, especially, livestock,

Table 4.2 Water balance at microwatershed level in Kundi basin.

Details	Unit	Treated watershed	Nontreated watershed
No. of wells and borewells	No.	74	71
Average annual pumping hours per well	h	423	229
Average well discharge	m^3	32	33
Reported area	ha	376	265
Rise in well water level during monsoon	m	3.65	3.63
Specific yield of the basalt aquifer		0.02	0.02
Total groundwater recharge	ha m	102.30	59.50
Groundwater abstraction	ha m	100.17	53.65
Recharge rate (MCM/km^2)		0.027	0.022
Groundwater balance	ha m	2.11	5.89

Source: Author's calculation based on Primary survey, 2004.

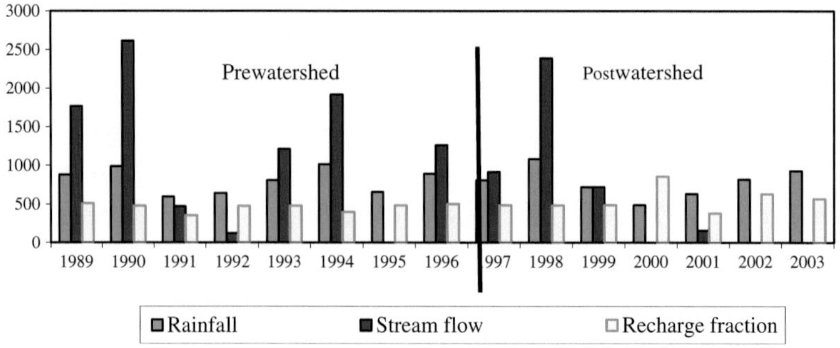

Fig. 4.5 Effect of watershed interventions on streamflows in Kundi sub-basin.

domestic purpose, there would be hardly any water left as buffer. It means that the extra water made available at the local level through WDP interventions was being utilized to its fullest potential by watershed community. Therefore, unless measures are taken to increase groundwater recharge rates on a sustainable basis, the communities should exercise precautions in further appropriation of the resources in areas where the abstraction levels are touching or even exceeding the recharge rates.

4.3.3.2 Impact of WDP on stream flow at sub-basin level

Fig. 4.5 presents the annual rainfall, annual stream flows (MCM), and recharge fraction (multiplied by 10^5) for Kundi sub-basin. Though "recharge fraction" is showing slightly increasing trend, stream flow of Kundi sub-basin was found to be showing declining trend over a period of time. There are two possibilities: first: the land treatment reduces runoff rates; and second: the WHS capture a portion of the runoff. Alternatively, one could argue that ideally the stream flows would be lower at lower magnitudes of rainfall. Such an argument gets strengthened when one looks at the annual rainfall values during the post WDP period.

Therefore, to measure the real impact of watershed treatment on stream flow generation and the effect of water harvesting on downstream flows, we ran a regression between rainfall and stream flow for two time periods: pre-water harvesting period (15 years) and time period including post-WDP period (20 years). The results of the regression analysis are presented in Table 4.3. The beta coefficient of the regression equation is slightly smaller (3.97) for the period encompassing the post-WDP period, meaning the runoff corresponding to a unit rainfall (runoff coefficient) reduced after the watershed treatments. Also, the value of the *Y* intercept was smaller (−1914) than the pre-WSD situation (−1884). Hence, it confirms the preliminary findings discussed in the beginning of the paragraph that runoff rate is reduced in post-WDP period.

Regression was also run between annual rainfall and average rise in water levels (pre-post monsoon) in wells for the same time periods (i.e., 1989–95 and 1989–2003). The regression equation for the first time period (i.e., 1989–95) suggested drop in groundwater levels during the monsoon period for zero annual rainfall, whereas for the latter, the groundwater-level change was positive during monsoon (0.76 m), even for zero annual precipitation. However, the beta coefficient for the latter was lower (0.004) than that of the former (0.0047). A 1000-mm rainfall raised the water level by 4.75 m in the case of the latter, whereas the estimated average water level rise for the former was 4.55 m. The nature of regression equations suggests that the recharge would be more for higher magnitudes of annual rainfall if watersheds are treated.

Table 4.3 Regression between average annual rainfall vs rise in well water level and annual stream flow.

Variables	Period	"*a*"	*b*-Coefficients	*t* Value	R^2
Rise in well water level	Pre (1989–1995)	−0.145	0.0047	4.730	0.789
	Post (1989–2003)	0.759	0.0040	3.611	0.501
Stream flow	Pre (1980–1995)	−1883.727	3.988	8.427	0.845
	Post (1980–2001)	−1913.545	3.974	10.358	0.863

Source: Authors' analysis.

4.4 Conclusion and policy implications

The study provided sufficient evidence of positive socioeconomic impact of watershed development at the local level. The study also threw pointers to the fact that WDP led to increased land use with expanded winter cropping, enabled by greater access to irrigation sources. Also, the watershed communities increased the intensity of livestock keeping with greater access to water and biomass. Intensive treatment of land and extensive WHSs led to increased rate of rainfall recharge into the shallow aquifers at the sub-basin level, indicated by high recharge fraction for the same runoff during the postintervention period. The treated watershed showed higher recharge fraction than the untreated watershed, with higher effect in high rainfall years. However, water balance estimates done for the treated microwatershed indicates that groundwater recharge and extraction reached equilibrium, meaning almost all the recharged water gets used up within the watershed.

Analyses using limited hydrological data show that the availability of downstream runoff had reduced after the watershed/water harvesting interventions, manifested by lower estimated runoff coefficients. This can be attributed to the increased local pumping for irrigated agriculture, which followed recharge augmentation. Hence, there is no evidence of improved sustainability of groundwater use or ecological sustainability due to watershed development. Therefore, we need to consider such socioeconomic and hydrological changes that are likely to occur as a result of improved water availability locally, before planning and implementing any new water resource development program to ensure sustainable use of the resource. Also, it is desirable to plan and design WDP and water harvesting projects keeping in view the committed flows for downstream development to reduce the unintended effects. Such planning should start from the basins or sub-basins and not from the microwatersheds unlike as being done currently, so that there can be greater clarity on the constraints induced by the downstream development needs.

References

Central Water Commission, 2017. Reassessment of Water Availability in India Using Space Inputs, Basin Planning and Management Organization. Central Water Commission, New Delhi.

Government of India, 1999. Report of the National Commission for Integrated Water Resources Development. vol. I. Ministry of Water Resources, Government of India, New Delhi.

Kumar, M.D., 2010. Managing Water in River Basins: Hydrology, Economics, and Institutions. Oxford University Press, New Delhi.

Kumar, M.D., 2014. The hydro-institutional challenge of managing water economies of federal rivers: a case study of Narmada river basin, India. In: Garrick, D., Anderson, G.R.M., Connel, D., Pittock, J. (Eds.), Federal Rivers: Managing Water in Multi Layered Political Systems. Edward Elgar Publishing, Cheltenham, UK and Northampton, MA, USA, pp. 229–242.

Kumar, M.D., Ghosh, S., Singh, O.P., Ranade, R., Ravindranath, R., 2005. Changes in Groundwater Ecology and Its Implication for Surface Flows: Studies From Narmada River Basin, Madhya Pradesh, India, Water Policy Research Highlight #25. IWMI-Tata Water Policy Research Program, Anand, Gujarat, India.

CHAPTER 5

Catchment assessment and planning for improving effectiveness of watershed development interventions: A case study from South Gujarat

A.J. James[a], James Batchelor[b], Nitin Bassi[c], Grant Milne[d]
[a]Environmental & Natural Resource Economist, New Delhi, India
[b]Digital Specialist, Atkins, London, United Kingdom
[c]Institute for Resource Analysis and Policy, Delhi Liaison Office, New Delhi, India
[d]Senior Natural Resource Management Specialist, World Bank, Washington, DC, United States

5.1 Introduction

From 2013 to 2015, the World Bank supported a Catchment Assessment and Management Planning (CAMP) Study, aiming to improve the understanding of hydrology issues in watershed management in India.[a] The study examined the utility of hydrological modeling as a tool to investigate the potential multiscalar impacts of watershed development and to support planning at different institutional levels. It looked at the potential of integrating hydrological models into the planning methodology used by the flagship national watershed development scheme of the Government of India (GOI)—the Integrated Watershed Management Program (IWMP). The study's findings were expected to not only contribute to the design of new World Bank-supported watershed programs in India, the IWMP, and the

[a] This work (James et al., 2015a,b) was funded by the Program on Forests (PROFOR), a multidonor partnership managed by a core team at the World Bank. PROFOR finances forest-related analysis and processes that support the following goals: improving people's livelihoods through better management of forests and trees; enhancing forest governance and law enforcement; financing sustainable forest management; and coordinating forest policy across sectors. In 2013, PROFOR's donors included the European Commission, Finland, Germany, Italy, Japan, the Netherlands, Switzerland, the United Kingdom, and the World Bank. Learn more at www.profor.info.

Current Directions in Water Scarcity Research, Volume 1
https://doi.org/10.1016/B978-0-12-814851-8.00005-7
© 2019 Elsevier Inc.
All rights reserved.

newly announced PMKSY, but to also guide hydrological assessments in watershed program development in other regions. The main motivations of the study were to understand how a greater focus on hydrology in planning and implementation can offset some of the unintended negative impacts of watershed development projects (Batchelor et al., 2002; Calder et al., 2008; Kerr et al., 2006; Kumar et al., 2006; World Bank, 2006).

While watershed development can and often does deliver significant beneficial impacts, it can also result in negative externalities that can affect downstream water users. The construction of rainwater harvesting structures to catch rainfall and subsequent surface runoff can enhance groundwater recharge and increase local water availability for irrigation and other uses. The increased water availability often encourages farmers to not just expand irrigated area but also to use irrigation more intensively (e.g., for double or triple cropping). This will normally lead to greater crop yields and higher farmer incomes, thereby achieving key social and economic objectives of national watershed programs.

At the same time however, enhanced water availability is not a guaranteed outcome of watershed development. Poor field designs, placement and/or maintenance of water-harvesting structures, unfavorable groundwater characteristics, and many other factors may reduce the effectiveness of these investments. Watershed development can also *increase* local water loss—both as a result of evapotranspiration from higher intensity of irrigation and due to evaporation directly from the water storage structures themselves—leading to reductions in downstream water flows. Any resulting reduction in downstream flows may not be a problem in areas of high rainfall or in basins that are still hydrologically "open." However, the severity of water scarcity is growing across India and many basins are now "closed," with all available water fully allocated to, or appropriated by, different water users and water requirements. The situation is likely to be exacerbated by factors such as climate change and an increasing demand for water. In addition, as watershed development is intensified and expanded to new areas, the downstream impacts are likely to accumulate and become more acute.

The Government of India has funded watershed development in selected watersheds across the country for several decades. National watershed guidelines were last revised in 2011 (MoRD, 2011). Under the current guidelines, states are now responsible for both the selection of watersheds to be "treated" and the field implementation of watershed programs. Watersheds are selected using multicriteria analysis that combines 13 different parameters and uses Geographic Information Systems (GIS) to identify those watersheds most in need of treatment. Although the guidelines implicitly recognize the importance of hydrology, the criteria for the selection of watersheds are heavily weighted toward social and economic goals, with little account taken of the potential downstream externalities of, for example, intensification of agricultural water use. This is partly due to the lack of easily accessible hydrological data that can be used as inputs into the selection of watershed and the planning of watershed development, especially in the headwater areas where most watershed development is focused.

Assessing and monitoring the impact of watershed development at a catchment scale is challenging and time consuming, partly because recharge rates from individual rain water harvesting structures can be highly variable in both space and time. Consequently, there have been relatively few cases in which watershed development planning has directly accounted for the impact of watershed development at the catchment or landscape scale. One method that can be used to assess its impact on water resources is hydrological modeling. Hydrological models have been used widely in research studies for decades. An emerging global trend is to now use hydrological models in practical applications to support watershed development, for example in site selection, assessing downstream impacts of investments in soil and water conservation interventions, etc. This trend is driven by the falling costs of computing power, the increasing availability and utility of open-source modeling software, and remotely sensed digital data that can be downloaded from the web and used at various scales.

5.2 The issue

Although national watershed development programs have been implemented by various Ministries of the Government of India since the early 1970s, even the latest Common Guidelines for Watershed Development Projects (MoRD, 2011) do not define a "watershed" in the hydrological sense. Instead of defining watersheds on the basis of topography and drainage lines, contiguous plots of land measuring around 1000–5000 ha are designated a "watershed" and are intensively treated to control and capture the water in that area.[b] The Guidelines advocate a "multitier ridge-to-valley sequenced approach" where the uppermost part of the catchment (usually

[b] This used to be 500 ha till 2008, when Common Guidelines were formulated by the DoLR with the Planning Commission of India, recommending a "cluster approach" of "geo-hydrological units normally of average size from 1000 to 5000 ha comprising clusters of microwatersheds" (MoRD, 2008, p. 7), which was applied to all watershed projects taken up from April 1, 2008.

hilly and forested, is treated first) followed by the "middle tier" (intermediate slopes just above agricultural lands) followed by the lowest tier, which are the "plains and flat areas." In India, as also in other parts of the world, such an approach, however, can have three key water-related consequences:

- **Reduced water flows to the lower parts of the catchment** and more importantly, the larger rivers that flow in the valleys—because of greater capture of water and increased evapotranspiration from resultant expansion in both the area that is cropped and the area that is irrigated in the upper parts of the catchment (where there are often farmers cultivating terraced fields).
- **Overabstraction of groundwater**, in the absence of decision making with a view to incorporating water use priorities and improvements to the productivity and sustainability of water use, as farmers benefiting from the additional water captured in new and existing Water Harvesting Structures (WHSs)—even in the upper and middle "tiers" of catchments—invest in more numbers of bore wells, deepen bore well depth, and increase pumping capacity to expand cultivation into previously unirrigated areas.
- **Less water in streams and aquifers**, because of increased cultivation, groundwater extraction, and increased evapotranspiration (from irrigated and rainfed farming systems), which, in some cases, can leave villages worse off than prior to the treatment—because (1) most options for augmenting water supply have already been exhausted and (2) that area is barred from benefiting from another IWMP project.[c]

Following sustained critical analysis (e.g., Batchelor et al., 2002, 2003, 2014; Calder et al., 2008; Kumar et al., 2006), which were reiterated during the preparations for the World Bank-assisted Neeranchal Watershed Management Project of the GOI, the then Joint Secretary of the Department of Land Resources, Ministry of Rural Development of the GOI, asked what the ideal size of a watershed should be, and how could it be operationalized as part of a revamped IWMP. Accordingly, the three key objectives of the CAMP study were to:

- Derive a methodology to incorporate hydrological assessment into catchment management planning in the Indian context.
- Demonstrate this methodology in one subcatchment.
- Create practical tools to apply (relevant parts of this methodology) in the IWMP.

[c] A village once selected for an IMWP project cannot be selected again for another IWMP project.

5.3 Study planning

5.3.1 Catchment choice

Following discussions with DoLR and Water Resources specialists at the World Bank, it was decided to carry out the pilot in the western Indian state of Gujarat, for the following reasons:

- The relatively better availability of data, given that a State Water Data Centre had been set up by the state Department of Water Resources, combining surface and groundwater data, as part of the Hydrology Project Phase 2 (HP 2), which was supported by the World Bank.
- The presence of a Bank-supported Integrated Coastal Zone Management project emphasizing a single-database approach to hydrology.
- The interest of the World Bank team in preparing HP3, which is aiming to take a more decentralized look at planning, right down to watershed level—an issue that was not addressed in HP1 and HP2.
- The willingness of the State-Level Nodal Agency (SLNA) for implementing the IWMP in Gujarat to support and assist the pilot.

Within Gujarat, it was decided to select the catchment of the Sukhi river, which lies within the Orsang sub-basin of the Narmada river basin, and where the IWMP was being implemented (Figs. 5.1, 5.2, and Box 5.1).

5.3.2 Choice of IWMP watersheds

A field visit to the Sukhi catchment confirmed ease of access and that data were available: the bridge over the Orsang river at Bodeli had a river-gauging station and there was a well-equipped weather station in Tejgadh near Bodeli. Also, if the area above the Sukhi reservoir was selected, there would be no dams, canals, or other structures in the catchment selected.

During the field visit, the DWDU office in Chhota Udeypur provided a map showing the locations and start dates of IWMP projects in Chhota Udeypur district. Since a softcopy of the data was not available, this information could not be overlaid onto the DEM and SWAT sub-basins. But by using the district boundary as a guide (a common feature to both maps), a rough assessment could be made of the location of IWMP watersheds in the upper Orsang catchment (Fig. 5.3).

Using SWAT and Google Earth allowed a detailed examination of each sub-basin and the identification of villages within them. By matching these with the villages in the IWMP map, it was possible to identify which

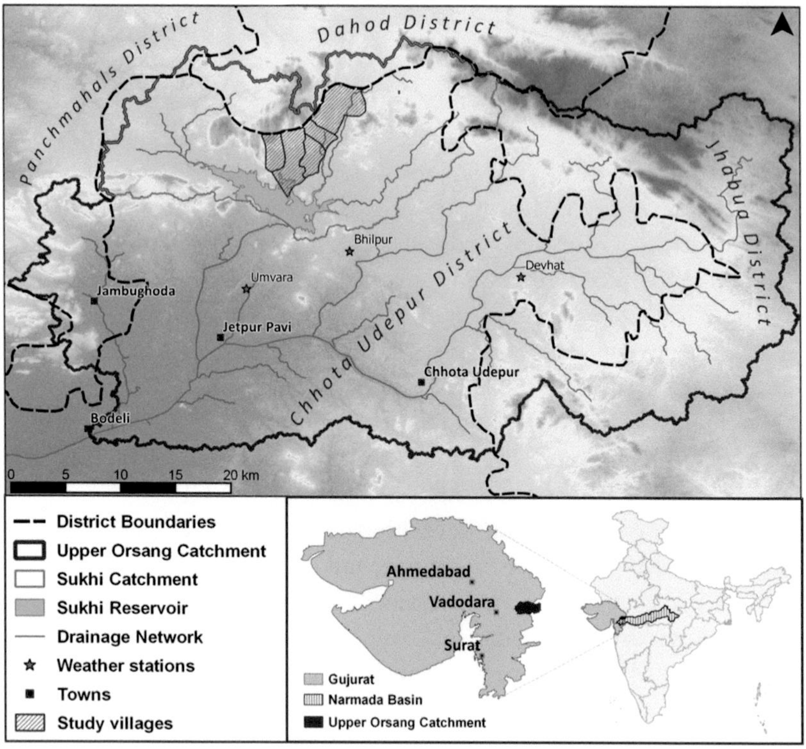

Fig. 5.1 Location of the Sukhi catchment.

sub-basins contained IWMP watersheds. This analysis showed that the area above the Sukhi reservoir had the most promising group of IWMP watersheds for detailed study. Further investigation (using QGIS and Google Earth) identified five IWMP Batch 1 (started in 2009–10) project villages (Kevdi, Dholisimel, Dungarbhint, Kundal, and Ghata) that looked to meet a key criterion (sloping from "ridge to valley") as locations for the choice of catchment.

With sub-basins delineated at 10 km, these five villages cover the large majority of two sub-basins with Kevdi, Dholisimel, and Dungarbhint forming an upstream-downstream continuum within one 21.5-km^2 sub-basin (Fig. 5.3). Another advantage of these villages is that the sub-basins lie almost completely within Chhota Udeypur district (Fig. 5.1) with the small area lying in Panchmahal district covered almost entirely by forests. Hence, secondary data (on land use, irrigated and cropped area, etc.) needed to be procured only from one district.

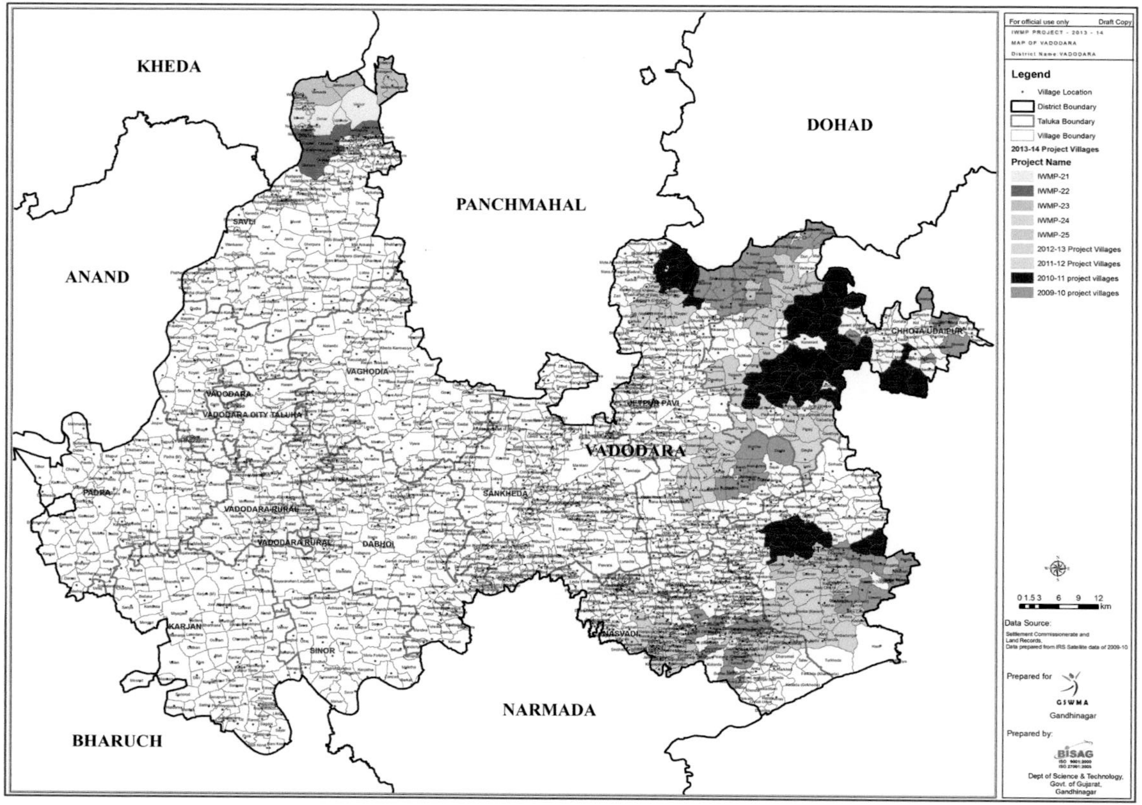

Fig. 5.2 Location of IWMP watersheds in the Sukhi catchment.

BOX 5.1 Key features of the Sukhi catchment

- **Location**: The major part of the catchment is in eastern Gujarat, with 79% of its area in Chhota Udeypur district, 15% in Dahod district, 1% in Panchmahal district, and the remaining 5% in Jhabua district in the adjoining state of Madhya Pradesh.
- **Area**: The catchment has an area of 393 km^2, which is mostly agricultural and forest land, and a significant part of the catchment falls within the Ratanmahal wildlife sanctuary.
- **Water infrastructure**: It contains a reservoir created by the Sukhi dam, which ought to have inflow data, and there is also a gauging site at Bodeli downstream of the Sukhi dam.
- **IWMP**: Around 50 villages in the part of the catchment lying in Chhota Udeypur district have been selected by IWMP to undergo watershed development.

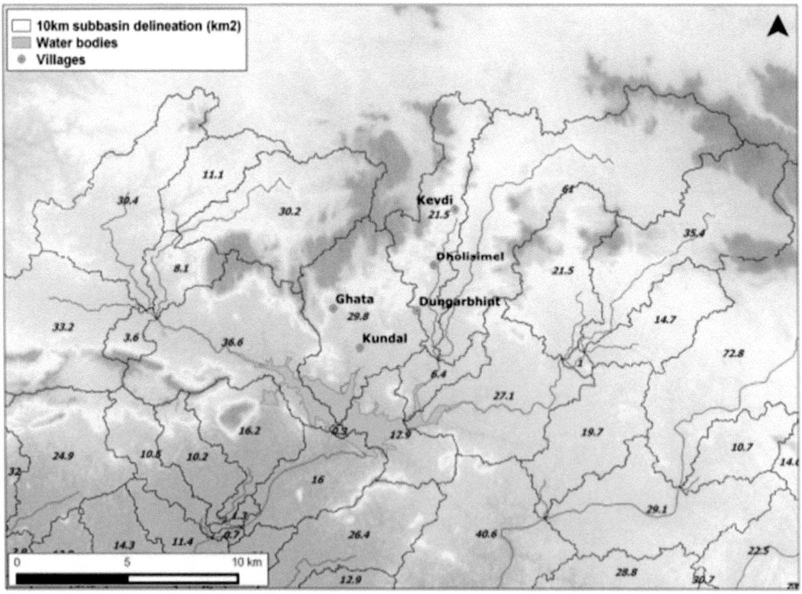

Fig. 5.3 IWMP watersheds above Sukhi reservoir (SWAT delineation).

5.3.3 Hydrological assessment

Two broad and mutually supportive methods can be used for hydrological assessments of catchments: a one-time assessment for a given hydrological unit and for a given time frame based on hydrological principles for aggregate-level assessment of catchments, using *conceptual* (or perceptual) models; and, understanding and simulating the hydrological processes in the catchment on a spatial and different time scales, using *hydrological* models. Developing perceptual models is an important step in a modeling process, but it is not an alternative for a calibrated and validated hydrological model.

The fundamental difference between perceptual and hydrological models—that seek to simulate catchment hydrology (also called "simulation models")—is that the former is static, not spatially and temporally distributed, rarely calibrated and only deals with the outcomes of hydrological processes. Simulation models, in contrast, are calibrated and validated to deal with the processes themselves (which are both time and space dependent); and can be run on various time scales (daily, weekly, monthly, and annually) and for different scales (small watersheds, sub-basins, and the whole basin).

Insights from the perceptual model, which are based on field visits and a basic analysis of available data, are invaluable inputs into the hydrological model and hence, a good starting point for the modeling process.

5.3.4 Choice of model

The ease of setting up and using SWAT for selecting the catchment, as illustrated here, for rapid assessment of catchment characteristics (using freely available global datasets) was one of the factors for considering SWAT as a serious option for the hydrological assessment. A review of the literature and discussion with experts confirmed several reasons why SWAT was a suitable model to study the impact of watershed development in India (see also Box 5.2)[d]:

- A large user base that has grown substantially over the last decade leading to frequent model improvements and updates.
- A large number of successful applications both in India and worldwide that give the model good credibility as a useful source of information
- An active online community that can provide useful support regarding all aspects of model use.
- It is a semidistributed model and can therefore account for spatial variation in important catchment characteristics such as land use.
- It is adept at using remote-sensing data, both as inputs and for calibration and validation, making it a good choice of model in locations where other sources of data are scarce.
- The delineation of subwatersheds during model setup is flexible allowing SWAT to be applied at any scale from the plot level up to the continental.

[d] See, inter alia, Mishra et al. (2007) and Glendenning et al. (2012).

BOX 5.2 Key features of the Soil and Water Assessment Tool (SWAT)

- SWAT is a semidistributed hydrological model that operates on a daily time step and is designed to predict the impact of management decisions on the water balance components and the sediment and agricultural chemical yields of a specified domain.
- There are two primary interfaces available for SWAT: ArcSWAT and MWSWAT. Both options run the same version of the model but use different GIS interfaces: ArcSWAT uses ESRI's ArcGIS (including spatial analysis extension); MWSWAT uses the open-source MapWindow GIS (although it does require Microsoft Access).
- ArcSWAT is the more widely used version, but the ArcGIS software needed is expensive. An advantage is the extensive online support and tutorials available for ArcSWAT. The extensive literature database is also a valuable resource.
- The major advantage of MWSWAT is that the software needed is much cheaper than the ArcSWAT version. Although less supporting material is available for MWSWAT compared to ArcSWAT, useful online resources available can be accessed at http://www.waterbase.org/documents.html and https://groups.google.com/forum/#!forum/waterbase.
- In most cases, the data needed for running SWAT at default settings can be found online, but the resolution/accuracy of the data is not always sufficient, especially for smaller catchments.
- SWAT can be combined with other models such as MODFLOW and WEAP.

- It is open source, which allows users to access the source code, both to see how the model works and to make changes to the model so that it better suits their needs.
- It can be easily linked with other models including the US Geological Survey Modular Finite Difference Flow model (MODFLOW), and the Water Evaluation and Planning model (WEAP).
- It is computationally efficient, which allows detailed modeling of large catchments.
- Simulation of crop growth and yields means that it can be used to examine the economic outcomes of different scenarios.
- Additional software, such as SWAT-CUP and SWAT-Check, help the modeling process and improve model outputs.
- There are several interfaces available for different GIS including ArcGIS, MapWindow, and QGIS, which allows users to choose one with which they are familiar.

Given these advantages, and despite the fact that other models had been suggested by some other experts consulted (e.g., WEAP by the National Institute of Hydrology, Roorkee), the Team decided to use SWAT for the hydrological modeling.

5.4 Setting up and using the model

5.4.1 Secondary data

Data inputs into SWAT were summarized and their various sources identified (Table 5.1), based on discussions with Government Department officials in Gujarat (both in Gandhinagar, the capital, as well as on site in Bodeli, Sukhi, and Chhota Udeypur). Secondary data requirements are summarized in Table 5.1.

The secondary data, however, could not be used in the form in which they were obtained, and hence had to be cleaned and processed for input into SWAT. Gaps in the rainfall data were filled by considering the observed measurement in the neighboring gauging station, while solar radiation was estimated using a complex equation, which takes sunshine hours as an input among other parameters. The modeling, however, had started by using global datasets, the strategy being to replace these rather coarse datasets with more accurate datasets from Indian sources as and when they became available.

5.4.2 Primary data

Five villages on three different streams discharging into the Sukhi reservoir were selected to get an understanding of the extent of watershed interventions undertaken in the watershed and the impact of these interventions on resource conservation and on livelihoods (including crop yields and returns). These five villages comprised seven different subwatersheds as delineated by the SWAT model using global data sets.

The initial field visit identified numerous structures across the five villages including many non-IWMP structures and revealed a varied terrain ranging from flat areas close to the Sukhi reservoir, with many structures and irrigated areas, to steeper forested areas upstream. Discussions with local villagers and staff of the District Watershed Development Unit (DWDU) revealed that the Gram Panchayats wanted more large check dams (seeing the benefits from the ones already constructed) but since there was little room left for more (large) structures in the flatter areas, they were planning

Table 5.1 Data requirements for SWAT and their sources

Data type	Parameters	Central government sources	Location	State government sources	Location
Satellite and map data	DEM (Raster data) Land use (Raster data or shape file) Soil type (Raster or shape file) Drainage network (shape file) Village Cadastral Maps (shape files) Shape files of village boundary and watersheds Canal network (shape file)	India WARIS website, MoWR NRSC, Hyderabad NBSSLUP, Bangalore	Orsang sub-basin	Gujarat SWMA BISWAG CSWCRTI, Vasad, Gujarat	Orsang sub-basin
Meteorological data	Daily minimum and maximum temperature Daily rainfall Daily sun-shine hours or solar radiation Daily average relative humidity Wind speed Daily pan evaporation	IMD	# Jetpur # Jambugam # Bhilpur # Chhota Udeypur # Bodeli # Jambughoda	Gujarat SWDC Gujarat SWMA	# Dhandhoda # Bhilpur # Jetpur Pavi # Devhat # Bodeli # Umarva
Hydrological and ground-water data	Daily river discharge Sediment Water quality	CWC	# Bodeli # Chandwada (last site before Orsang meets Narmada)	SWDC	# Bodeli # Chhota Udeypur

	Pre and post monsoon depth to groundwater level	CGWB	Observation sites at Pavi, Bodeli, Chhota Udeypur, and Kevdi	GSWMA SWDC	Observation sites in Chhota Udeypur, Pavi Jetpur, and Sankheda blocks
	Estimated groundwater recharge	CGWB	Different blocks in Vadodara district	GSWMA SWDC	Different blocks in Vadodara district
Reservoir data	Water inflows Water release in canals and spillway during different hydrological years Gross and live storage (including storage at the beginning and end of the hydrological year) Evaporation from the reservoir	CDO Gandhinagar		Gujarat SWRD	Sukhi dam

BISWAG, Bhaskar-Acharya Institute for Space Applications and Geo-Informatics; *CDO*, Central Design Organization; *CGWB*, Central Ground Water Board; *CSWCRTI*, Central Soil and Water Conservation Research and Training Institute; *CWC*, Central Water Commission; *IMD*, Indian Meteorological Department; *MoWR*, Ministry of Water Resources, Government of India; *NBSSLUP*, National Bureau of Soil Survey and Land Use Planning; *NRSC*, National Remote Sensing Centre; *SWDC*, State Water Data Centre; *SWMA*, State Watershed Management Agency; *SWRD*, State Water Resources Department; *WARIS*, Water Resources Information System (India).

Table 5.2 Field formats for primary data collection.

Format	Type	Purpose
1	Well survey	Mapping of all wells (with GPS coordinates) and to assess the impacts of watershed interventions on water availability in the wells, a range of attributes were collected including type, date of construction, condition, funding source, and cost
2	Survey of water harvesting structures	Mapping of all water harvesting structures (with GPS coordinates) and to assess their impacts in terms of resource conservation and increased water availability
3	Information from village records	Understanding village social dynamics and agriculture (including irrigation) pattern
4	Household survey of socioeconomic, agricultural, and irrigation details	Water productivity analysis and assessing crop economics at the household level
5	Focus group discussions with the village community	Time series analysis on the condition water resources, forest, cropland, livestock, and returns from agriculture Capturing the impact of droughts and wet years on the natural resources and livelihoods Assessing effectiveness of local institutions (such as GPs, WCs, SHGs, UGs) in resource management

to use funds under available government programs, including the IWMP, to construct additional structures in upstream areas as well. To collect primary data more systematically from these villages, however, five different questionnaires were prepared (Table 5.2).

A dedicated field team member organized the primary data collection, which was completed in parallel with the secondary data collection, in two months (mid-September to mid-November 2014). The primary data collection from the five villages in a part of the Upper Sukhi Watershed was greatly facilitated by the fact that the IWMP projects were still ongoing in these villages. This not only gave access to the information collected by the IWMP baseline survey but also facilitated entry-point discussion and rapport building with the villagers.

5.5 Understanding the catchment and its problems

The collected data were used not just as inputs into the modeling process but were also analyzed separately. The main findings are summarized below.[e]

5.5.1 Watershed characteristics

- **Rainfall varies sharply between wet and dry years**: Average annual rainfall for the study period (1999–2013) was 1062 mm with a standard deviation of 386. Nearly all the rainfall falls during the monsoon during the months of June, July, August, and September. The majority rainfall occurs on days with more than 25 mm of rain. On average, 62% of rain occurs on days with rainfall of more than 25 mm, while only 14% of rainfall occurs on days with between 0.1 and 10 mm of rain. This has large consequences for runoff, erosion, and the effectiveness of rainwater harvesting (RWH) structures. The rainfall in the watershed also varies sharply across dry and wet years, as shown by the rainfall statistics:
  - *Total rainfall*: Thus, total annual rainfall was 1724 mm in 2006, the wettest year in the period from 1999 to 2012, while it was 358 mm in 2009, the driest year—which is just about 20% of the 2006 figure.
  - *Daily rainfall*: The highest amount of rainfall recorded in a day was 290 mm in 2006 and just 80 mm in the driest year in the period from 1999 to 2012.
- **Annual inflows into Sukhi reservoir also vary sharply with rainfall**: This variation impacts inflows into the Sukhi reservoir and, by extension, the water released for irrigation in downstream villages. The Sukhi dam, completed in 1987, has a reservoir with an effective storage capacity of 178.47 million cubic meters (MCM) and a surface area of 29.04 km^2 when full. Built for irrigation, two canals flow downstream with a total command area of 31,532 ha. Inflow data for the Sukhi reservoir showed that while the annual inflow into the reservoir for the period 1990–91 to 2012–13 was around 175 million cubic meters (MCM), it ranged from a high of around 450 MCM (in 2006) to a low of around 30 MCM (in 2009).

5.5.2 Village-level information

- **All five villages have large elevation ranges and steep topography**: The five villages surveyed, Kevdi, Dholisimel, Dungarbhint, Kundal, and Ghata, are situated toward the center of the Sukhi watershed and are contiguous, with Kevdi being the uppermost and Kundal and Ghata edging

[e] Further details are in the Technical Note on SWAT Modeling (James et al., 2015b).

the Sukhi reservoir (Fig. 5.3). The villages are characterized by large elevation ranges and steep topography. All five lie on the boundary of the wildlife reserve, and can be divided into flatter areas (e.g., in the villages of Kundal and Ghata near the Sukhi reservoir) where agriculture is the dominate land use and steeper areas (e.g., in Kevdi) that are mainly covered by forest.

- **Forest and degraded forest are the dominant land cover in all the villages**: Kundal has the highest proportion of forest, at over 40% of the village area, while in Dungarbhint the area of degraded forest is larger than the area of forest. The areas of both forest and degraded forest stay relatively stable through the study period.
- **There has been a large shift from kharif only to double cropping**: The main crops include cotton, rice, and maize during the monsoon (*kharif*) season and maize during winter (*rabi*) season. The major change in cropping pattern over three periods, 2004–05, 2008–09, and 2012–13 (for which land use data were procured from National Remote Sensing Centre (NRSC), Government of India) is a large shift from *kharif* only to double crop agriculture—most evident for the two villages closest to the reservoir, Ghata and Kundal. While in 2004–05 the majority of agricultural land is *kharif* only for all villages (apart from Ghata where the areas of double crop and *kharif* only agriculture are relatively similar), in 2008–09 and 2012–13 all the villages see a major shift to double cropping.
- **Wells doubled in number since 2000 and are heavily concentrated along drainage channels**: There are 402 wells in total, of which 70% are dug wells and 30% bore wells, giving a well density of roughly 10 per km^2 for the village areas. These wells, however, are heavily concentrated along the drainage channels and very few are located in the forested areas. The highest concentrations are in the villages of Dholisimel and Dungarbhint. Areas closest to the Sukhi reservoir have the highest proportion of wells with year-round availability, a probable reason why the proportion of agriculture that is double cropped is higher in that area than further upstream. A noteworthy feature about the wells in these villages is their growth over time, which more than doubled from 2000 to 2014. The 77 wells of 1990 doubled to 178 wells in 2000, while the 265 wells in 2007 increased to 402 wells by 2014. The primary survey data and discussions in the villages showed that while there are very few dry or defunct wells and more than half (54%) of wells have water round the year, a large number (around 40%) of wells are seasonal, running dry after March (the festival of *Holi*). The "best wells" are found in the villages of Ghata and Kundal near the Sukhi reservoir, which are also the villages with the highest proportion of wells with year-round availability.

- **Water Harvesting Structures increased prior to the IWMP**: In total, the survey mapped 251 RWH structures in the five villages, of which 77 were funded IWMP. Only 9 check dams, from a total of 72, were funded by IWMP, as the most optimal locations had already been used by the time the project started in 2009. No structures have been built within the forest itself due to opposition from the Forest Department. Many of the larger check dams were funded by the Irrigation Department. The density of structures is similar in all the villages apart from Dungarbhint where only a few structures have been constructed by IWMP. Many of the structures are located along, or close to, the edge of the forest. For example, in Ghata, nearly all the streams and gullies that flow from the forest have been blocked by check walls or gully plugs. The majority of RWH structures funded by IWMP in the five villages were built in 2012 and 2013. Prior to this, the number of structures built on an annual basis was relatively low in comparison. However, the average number of structures built from 2007 to 2011 was around 10 per year compared to an average of around 5 per year from 2001 to 2006. This increase in the number of structures built is clearly related to the parallel increases seen in double cropped area and the number of wells and indicates that watershed development was occurring in the area prior to the investments by the IWMP. In total 55% of the structures are in good condition, 23% are damaged but still working, and 22% are defunct.

  The survey could not capture structures that were built earlier in the period, but which were subsequently completely destroyed by floods: Kevdi village has the highest proportion of damaged and destroyed structures that can still be observed and currently there is no check dam still standing on the main stream. The local community identified high runoff from steep slopes during large rainfall events as the cause, as this brought branches and boulders down from the upstream forest and along the main stream (the number of large rainfall events and the steep topography of the villages support this observation.)
- **There is significant interannual variability in ground water levels**: The depth to water levels during summer is generally very low after an abnormally good monsoon year and very high after a very low rainfall year. This suggested that not all the water that goes into the hard rock aquifers underlying the watershed gets discharged after the monsoon, in spite of the steep terrain.[f]

[f] The option in SWAT to use expert knowledge to specify some parameters was used to incorporate these insights.

- **Water-level fluctuation is greater for very low rainfall than for abnormally high rainfall**: The four observation wells selected for detailed analysis (Ambala, Chhota Udeypur, Ferkuwa, and Kevdi) showed the highest water level fluctuation was in a drought year (1987, when the values varied from 3.05 to 0.47 m), and was much less in an abnormally wet year (1988, when the values varied from 6.37 to 5.25 m).
- **Following years of drought (e.g., 1985–87) a good monsoon (in 1988) results in excessively high groundwater recharge**, owing to the emptying of the aquifer in the previous years, which helps store extra water in the formation from the deep percolation.
- **The postmonsoon depth to water level in a high rainfall year is lower than in very low rainfall years**. Along with the high fluctuation between pre- and postmonsoon water levels, this suggests high rates of recharge.
- **There is neither long-term decline nor long-term rise in water levels overall**, in spite of increase in number of wells and bore wells in the area and presumably increase in groundwater draft. This could be because natural groundwater discharge also plays a significant role in altering groundwater balance along with groundwater draft through wells. The increase in groundwater draft is limited by the insignificant amount of static groundwater in the area, which is underlain by hard rock formations.

  Village-level discussions and cropping data from local government sources confirmed the survey findings and also detailed three key trends:
- **Increase in cropped area** in both monsoon and winter seasons
- **Increase in irrigated area** and more so for winter crops such as maize although the area under irrigated maize and rice during monsoon season has also increased. It is to be noted that the entire cropped area during winter and summer seasons is irrigated.
- **Expansion to irrigated crops**: There is a growing trend of cultivating irrigated long-duration cotton varieties, which grows over both the *kharif* and *rabi* seasons, replacing single-season cotton. Also, groundnut is a minor summer crop introduced in 2006–07 in Ghata and Kundal, the two villages nearest to the Sukhi reservoir, but cultivated only by farmers who have access to wells with year-round water availability.

5.5.3 Villagers' perception of water problems

The population of these five villages is largely tribal and poor.[g] Most households (78%) have agriculture as their main occupation, while around 15%

[g] According to GoI classification, they are Scheduled Tribes (ST) and Below-Poverty Line (BPL) households.

work as farm labor. Around 60% of farmers have marginal land holdings (less than 1 ha), while 24% have small holdings (1–2 ha). While farmers generally see their future prosperity in shifting to long-duration cotton, only a few farmers with the necessary resources can afford the shift. To supplement their agricultural incomes, many migrate from these villages, especially after the *kharif* cropping season (August—November), to fairly distant locations such as Kachchh at the western tip of Gujarat and also to Mumbai.

Since rainfall is relatively high, and there are many open wells, water is not generally perceived to be a problem for drinking or livestock, but there are two cases where villagers reported problems:

- **Water scarcity in summer months and during rabi**: Many wells run dry after the *holi* festival in March, creating both drinking water problems and for providing the last two irrigations of the winter *rabi* crop. The problem increases as one goes further upstream from Kundal up to Kevdi. Even in these villages, drinking water is not a problem faced by all farmers but only by those who do not have wells with all-round water supply. The scarcity of water for the *rabi* crop, however, is more widespread although, as noted earlier, some farmers in the lowest villages if Kundal and Ghata are able to grow a summer groundnut crop.
- **Low rainfall years bring acute scarcities**: Water scarcities are acutely felt by a large number of villagers in all five villages during low rainfall years (as in 2009). Consecutive years of drought—a possible consequence of increasing climate variability—would create serious water problems in this area.

5.5.4 The challenge for watershed management

The hilly and rocky Upper Sukhi watershed has relatively high rainfall, occurring in a few months in a year resulting in high-velocity flows that do not allow water harvesting structures to remain intact and functional along the main streams. More wells and RWH structures built in the last decade has supported a growth in irrigated agriculture in (micro) watershed villages.

Although the local community is tribal and poor, mostly farming small and marginal land holdings during the single crop monsoon season, some farmers have wells with year-round water to support irrigated agriculture. Cropped and irrigated areas have increased in the last decade, with a preference for longer-duration cotton, a cash crop. All the area under winter crops (mostly maize) and the small area under the summer groundnut crop are irrigated.

Wells also provide water for drinking and livestock although there is a shortage in summer months even in normal rainfall years, which affects the last irrigations for the winter (*rabi*) crop. Low rainfall years and droughts worsen the situation considerably. Many in this agriculture-dependent community migrate for work after the monsoon *kharif* crop to supplement their livelihood.

Watershed management in this must take account of the following:

- The relative lack of treatment of the uppermost parts of the watershed which are forested, uninhabited, and directly under the control of the state Forest Department
- A large number of structures already built on the (smaller) drainage lines, implying that there are few suitable sites left to build more numbers of large RWH structures, such as check dams
- Most structures built on the upper reaches of the main streams flowing into the reservoir have been broken by the monsoon rainwater carrying branches and boulders, and there is not much space in villages to build additional RWH structures.
- There are large variations in rainfall, evapotranspiration, and runoff across dry and wet years that affect inflows into the reservoir, which, in turn, affect canal releases to downstream communities.

This basic understanding of the characteristics of the watershed is an essential first step to modeling the watershed, based on which the impact of watershed interventions on the water resources in the local villages and in downstream villages can be assessed.

5.6 Modeling the catchment characteristics

The aim of the hydrological assessment was not just to understand the key hydrological features of the watershed but also to create "what if" and "what's best" scenarios of potential impacts of watershed management interventions, especially in downstream communities. Only when a model captures the dynamics of the watershed well can it be used to generate scenarios of potential impacts.

Since experience suggested that data collection would take time, the team decided to start the modeling using freely available global data sets simultaneously with the collection of national, state, and local data, i.e., from mid-September. As data became available, the rather coarse global datasets would be replaced, thereby improving the results. The results, in this case,

BOX 5.3 Basic steps in modeling

- Calibrate and validate for current conditions.
- Identify which hydrological processes are impacted by the management interventions.
- Select SWAT parameters that represent these processes
- Change sensitive parameters to represent management practices and different scenarios (different combinations, locations, and magnitudes of interventions). It's important that any parameter changes are realistic.
- Assess the reasonableness of results.
- Look at impacts (local and downstream) in wet, dry, and normal years.

Validate against local data.

were how well the values predicted by the model match observed values of inflows into the Sukhi reservoir (see Box 5.3).[h]

5.6.1 Model run with global datasets

The model set up in SWAT with global datasets (see Box 5.4) were checked against measured (or calculated) inflows into the Sukhi reservoir.

The global data sets are coarse-grained, as seen from the depiction of the Upper Sukhi catchment (Fig. 5.4).

Compared to actual inflows into the Sukhi reservoir, the SWAT run with global data was a reasonable but not very good fit to the data, either for monthly or annual inflows into the reservoir (Figs. 5.5 and 5.6). (Simplistically, this is seen by the fact that the bars are not of the same height, whereas in a "perfect fit" they would be the same.)

5.6.2 Model run with improved datasets

With the inclusion of some national and state data (32-m CartoDEM from NRSC, rainfall, and temperature data from Bhilpur and Dhandhoda stations and the reservoir dimensions from the State Water Resources Department), the performance of the SWAT model improved (Figs. 5.7 and 5.8).

The "goodness of fit" of the model simulated values to the actual values as also seen in the simple linear regression between the simulated and the actual inflows, both monthly and annual, also improved. The improvement in the "fit" of the model is seen in the fact that the R^2 values are much better

[h] The modeling approach was discussed with Prof. Ashwin Gosain of IIT Delhi and Dr. Durga Rao of NRSC in Hyderabad, and suggestions incorporated.

BOX 5.4 Global data sets used

- Aster GDEM (30 m)—Global Digital Elevation Model produced by NASA and the Japanese Ministry of Economy, Trade and Industry downloaded from http://gdex.cr.usgs.gov/gdex/
- CFSR weather data (http://globalweather.tamu.edu/)—Global weather dataset that includes rainfall, minimum and maximum temperatures, wind speed, solar radiation, and relative humidity, for the period 1979 to present. The Climate Forecast System (CFSR) uses a high-resolution, coupled atmosphere-ocean-land model to resample conventional weather data to produce a global dataset with a 38 km resolution.
- FAO digital soil map of the world data at 10 km resolution downloaded from http://www.waterbase.org/download_data.html
- MODIS land cover (from US Geological Service)—500-m resolution global land use dataset developed using 10 years of data from the MODIS instrument aboard NASA's Terra satellite. Downloaded from http://landcover.usgs.gov/global_climatology.php
- Reservoir areas delineated from satellite image using QGIS and capacity details acquired from the India WARIS website

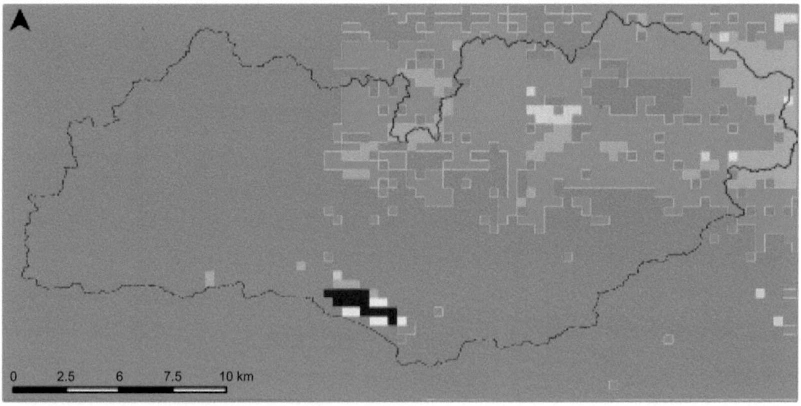

Fig. 5.4 Upper Sukhi catchment from global data.

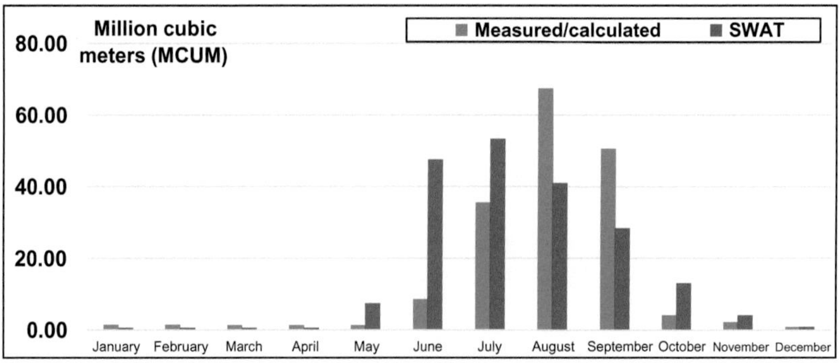

Fig. 5.5 Monthly inflows into the Sukhi reservoir, 1991–2012, global data.

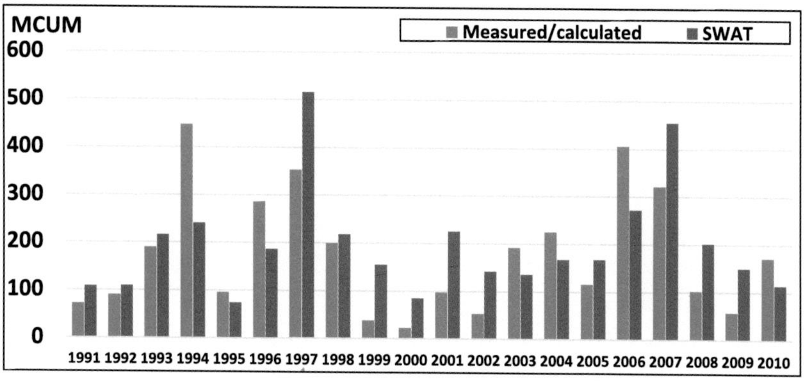

Fig. 5.6 Annual inflows into the Sukhi reservoir, 1991–2012, global data.

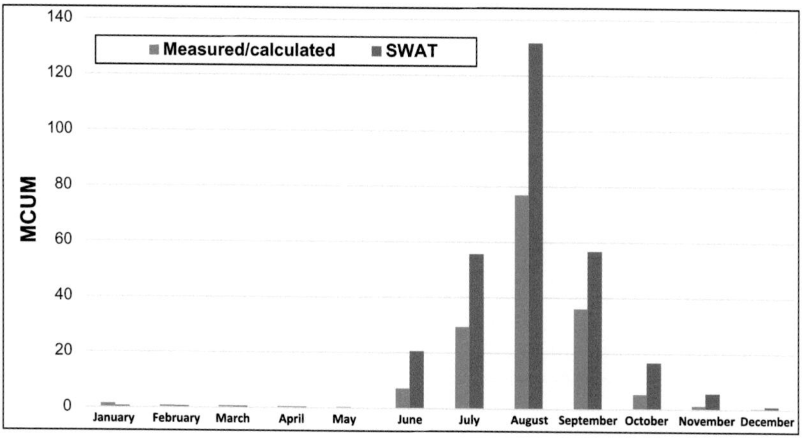

Fig. 5.7 Monthly inflows into the Sukhi reservoir, 1991–2012, improved data.

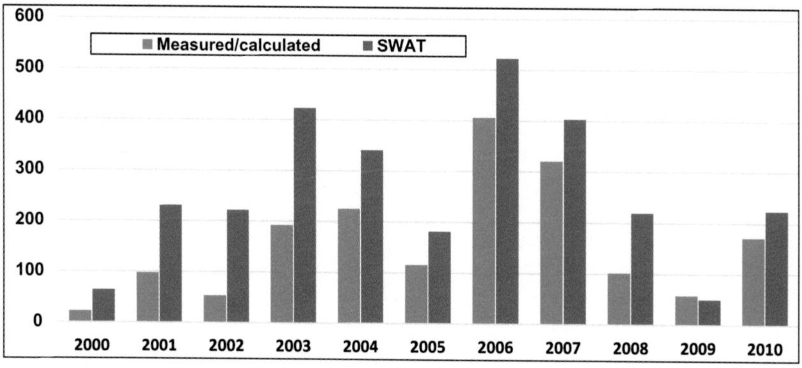

Fig. 5.8 Annual inflows into the Sukhi reservoir, 1991–2012, improved data.

for the SWAT model run with (some) local data compared to global data: for annual inflows into the reservoir, the R^2 improved from 0.45 to 0.81, while for monthly inflows it improved from 0.24 to 0.87 (Figs. 5.9 and 5.10).

5.6.3 Final model run

After replacing all the global data with national, state, and local data, including those from the primary survey (see Box 5.5), the outputs from the final run of the model were much better.

The final results showed that the model simulation of monthly inflows into the Sukhi reservoir was a good match to the actual data (Fig. 5.11).

The main results showed that the model captures the dynamics of the watershed quite well—as indicated both by excellent NSE and R^2 values and visual comparison of the observed and simulated reservoir inflows.

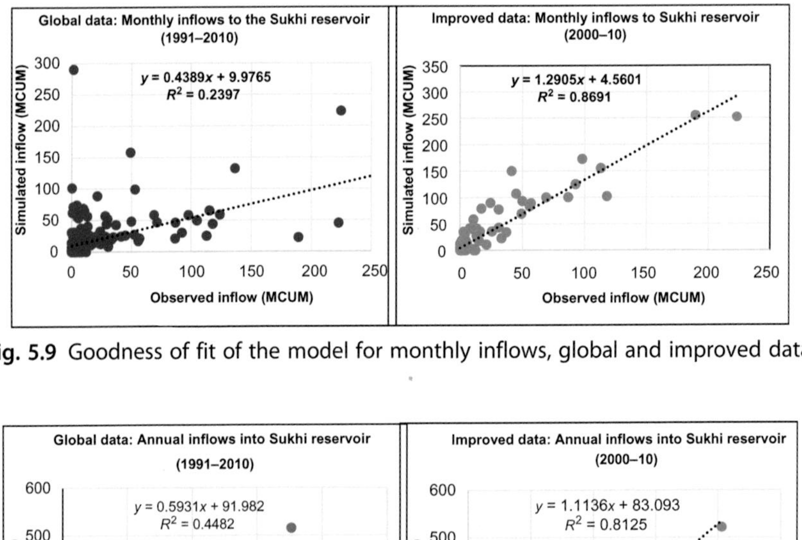

Fig. 5.9 Goodness of fit of the model for monthly inflows, global and improved data.

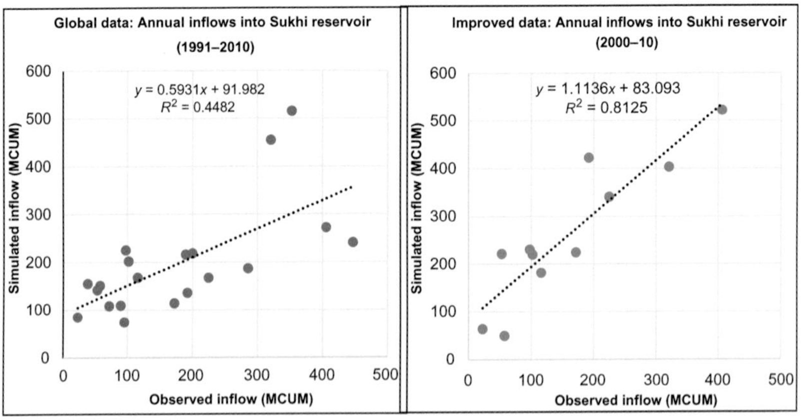

Fig. 5.10 Goodness of fit of the model for annual inflows, global and improved data.

BOX 5.5 Main data used in the final model run

Digital Elevation Model (DEM): CartoDEM from NRSC with 30-m resolution.
Land Use Land Cover (LULC): 250k LULC for 2004–05, 2008–09 and 2012–13 from NRSC.
Soil data: NBSLUPP Soil Handbook for Gujarat.
Climate data: Rainfall data from the three local weather stations nearest to the catchment and minimum and maximum temperatures, wind speed, solar radiation, and relative humidity from one station (Dhandhodi).
Discharge: Monthly data for the Sukhi reservoir from the State Water Resources Department.
Cropping patterns: State Agricultural Department and primary survey in five villages.
Irrigation data: Primary survey in five villages.

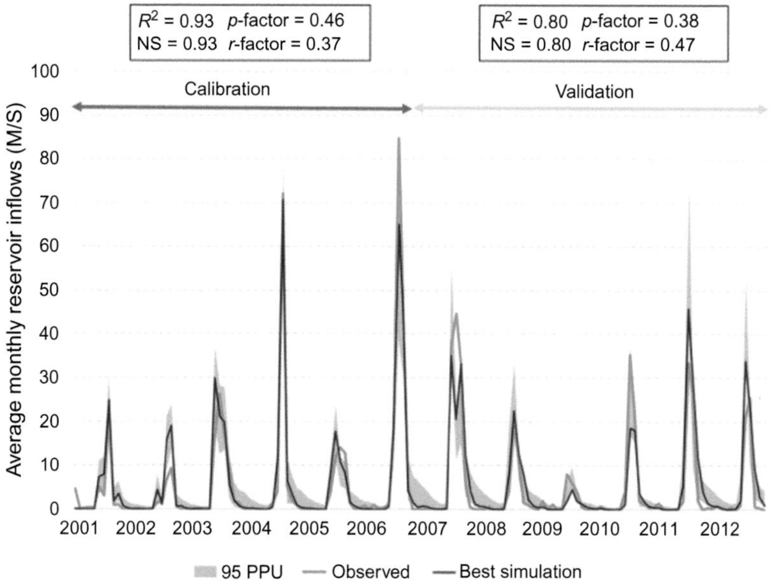

Fig. 5.11 Final model simulations and actual monthly inflows into Sukhi, 2000–12.

There was some uncertainty in the model outputs. The two obvious sources of uncertainty are: (1) the rainfall data (a common source of error in many hydrological models due to the difficulty of extrapolating point measures at rain gauges to areal estimates of rainfall) as they come from gauging sites outside of the watershed and at lower elevations; (2) the observed reservoir inflow data used for calibration, which were not measured directly but calculated (by the Water Resources Department staff at the dam site) using a water balance equation and are therefore likely to be less accurate than discharge data recorded at a control structure. Outflows from the Sukhi reservoir occur not just over the spillway of the dam, but also through two irrigation canals, which makes the accurate measurement of total outflows more difficult. Calculation of evaporation from the reservoir could also introduce errors into the water balance equation especially because of the need to account for large and rapid changes in the surface area of the reservoir. Finally, neither the seepage from the reservoir nor pumping of water from the reservoir by the surrounding villages was accounted for in the water balance equation.

The model, nevertheless, was adjudged ready for simulations.

5.7 Scenarios of impact

The development of model scenarios enables an analysis of the potential impacts that investments in water infrastructure and other factors (e.g., changes in climatic variables) may have on the hydrology and water resources of the area of interest. Such analysis makes it possible to explore tradeoffs involved in decision making, for example, at what level of watershed developed does the reduction in downstream flows starts affecting downstream benefits such as irrigation and drinking water supplies—and become politically, economically, socially, or environmentally unacceptable?

Scenario development is a stage at which all stakeholders can provide their input based on their vision of the future of the watershed (see Box 5.6).

BOX 5.6 Useful characteristics of scenarios

- Must be a realistic vision of the future conditions of the study area, as far as possible.
- Focuses on two or three key factors (such as proposed water and land management interventions, environmental change, and economic development) that are *important* (i.e., they can have relatively large impacts) but also relatively *uncertain* (i.e., their potential impacts are not entirely predictable in either their effects or magnitude).
- Relative accuracy is often greater than that of the underlying model is compared to observed values. This means that model scenarios can still provide useful outputs for watershed planning even if there is some uncertainty regarding baseline model performance.

Source: *Kauffman et al. (2014)*

The development of scenarios for the Sukhi watershed focused on two main factors:

- **Changes in the levels of in situ and ex situ rain water harvesting** (RWH) in the watershed and
- **Changes in cropping intensity**, in particular, the expansion in double-cropped area

The study team felt that model scenarios based on these two issues could provide useful information for planning watershed development in the Sukhi watershed, and consider the important and uncertain impacts it may have downstream, in view of the criteria outlined in Box 5.6

- *Importance*: These two factors are likely to have a big impact on the watershed in the future given the investments being made in watershed development by IWMP and other programs and the expansion in the double cropped area driven by population growth and the microeconomic needs of the agriculturally dependent communities.
- *Uncertainty:* Though the expected impact of increased RWH and expansion of double cropped area would be greater water use within the watershed, there is significant uncertainty on the magnitude of this increase and the impact on downstream flows.

Two versions of the scenarios were developed, one for the entire catchment and the second for the five villages where detailed analysis was undertaken.

5.7.1 Catchment-level scenarios

5.7.1.1 Scenario description

Two main scenarios were developed to look at the impacts of land use change and watershed development on the hydrology of the Sukhi watershed:

- **Scenario 1: The watershed prior to the modeling period**: The watershed in the absence of the watershed development and the intensification of agriculture that has occurred over the last two decades, when most of the agricultural land was *kharif* only and there were very few large RWH structures. Although it is difficult to envisage a return to the pre-1999 situation in the watershed, scenario 1 helps to illustrate the impact that agriculture and watershed development has already had on the hydrology of the watershed.
- **Scenario 2: A significant intensification of agriculture and watershed development from current levels**. This is the most plausible future for the watershed given the investment by IWMP and other programs in watershed development and the increasing demand placed on agriculture to produce more food and improve local livelihoods.

Both scenarios 1 and 2 are disaggregated into three parts:

- Part (a) models the combined impacts of changes in RWH and land use
- Part (b) models the impacts of changes in RWH alone
- Part (c) models the impacts of changes in land use alone

The best simulation from the model calibration was used as the baseline and the scenarios evaluated for the period 2007–12 (Table 5.3).[i]

Table 5.3 Description of model scenarios.

Scenarios		Description
1	**A.** No development	This scenario models the watershed as if no watershed development or increase in the area of double cropped agriculture had occurred. This is represented by the removal of all reservoirs, apart from Sukhi, Jamli, and Jogpura, and by using the 2004–05 land use dataset for the entire modeling period
	B. No watershed development	Removal of all RWH structures (represented as reservoirs in SWAT) apart from Sukhi, Jamli, and Jogpura
	C. No land use change	Use of the 2004–05 LULC for the entire modeling period
2	**A.** Intensification of agriculture and RWH	This scenario represents the more likely future for the watershed: intensification of agriculture and RWH, represented by (1) an increase in RWH structure capacity to 40 m^3/ha and implementation of reservoirs in forested watershed areas; and (2) an increase in double crop area from 30% to 40% of watershed area but no change in reservoir capacity
	B. Intensification of RWH	Increase in RWH structure capacity to 40 m^3/ha and implementation of reservoirs in forested watershed areas
	C. Intensification of agriculture	Increase in double crop area from 30% to 40% of watershed area but no change in reservoir capacity

[i] This period was chosen because it witnessed significant land use change that was incorporated into the model and therefore provides a stable baseline from which to evaluate the impacts of the scenarios. If the whole modeling period had been used, land use change would have had to be considered when evaluating scenario outputs.

5.7.1.2 Scenario impacts

The overall impacts of the scenarios are as expected:

- **Inflows into Sukhi reservoir**
  - Scenario 1 (A, B, and C) increases inflows.
  - Scenario 2 (A, B, and C) decreases inflows.
  - Scenario 1A has largest increase in reservoir inflows
  - Scenario 2A has the largest decrease in reservoir inflows
- **Impact on downstream flows**
  - Watershed development has the largest impact on downstream flows dry years (i.e., the relative impacts of the scenarios are greatest in 2009, the driest year)
  - Scenario 1A increases downstream flows by 24%
  - Scenario 2A decreases downstream flows by 19%.

Whether it is a dry year or a wet year also affects various water balance components—evapotranspiration (ET), percolation, artificial recharge from RWH structures, and reservoir inflows. The changes in these components for the baseline and other scenarios, as a percentage of rainfall, show the following for 2009, the driest year; and 2011, the wettest year:

- **Changes in the double cropped area have the largest impact on the amount of water lost as ET from the watershed with the impacts largest in the driest year (2009).** In this year, ET for scenarios 2a and 2c, in which the double cropped area is increased to 40% of the watershed area, is close to 100% of rainfall, around 10% more than for the baseline. This is because a large amount of water is removed from the shallow aquifer for irrigation and is subsequently lost as ET. The water stored in the shallow aquifer is nearly exhausted in 2009, so two or more consecutive low rainfall years could have been calamitous for the local community. For scenarios 1a and 1c, in which the double cropped area is reduced to 9% of the watershed area, ET in 2009 is reduced to 77.4% of rainfall, more than 10% lower than the baseline figure.
- **The impact of the scenarios on ET in the wettest year (2011) is far less as a percentage of rainfall**, although the absolute difference is similar.
- **Recharge from RWH structures in the driest year (2009) is double the average recharge as a percentage of rainfall.** Differences in percolation rates between the scenarios are mainly a result of recharge from RWH structures, which more than doubles from an average of 0.4% of rainfall for the baseline, to 1.1% for scenario 1a. This is in line with the increase in RWH structure capacity from 15 to 40 m^3/ha for scenario 1a and an expansion of structures into the mainly forested subwatersheds.

5.7.2 Village-level scenarios

5.7.2.1 Extracting village areas

While simulation models can create scenarios that enable insightful inferences to be drawn about the behavior of the watershed in response to watershed development interventions, it is the impacts of interventions in the five villages that are of direct interest. The choice of the watershed was, however, largely decided by the data needs of the model, viz., the availability of reliable data from the Sukhi reservoir. This situation, where the availability of data for calibration and validation requires the modeling of a watershed much larger than the area of interest, is likely to found for other areas where IWMP watersheds are located, as many are found in headwater areas, a long way upstream of any gauging point that could provide reliable data for calibration and validation.

One of the advantages of SWAT is that the subwatersheds can be delineated at a user-specified resolution, which allows results to be analyzed at local levels, even for large watersheds. A tradeoff is that models of large watersheds with many subwatersheds can substantially increase model runtime and make processing model outputs a time-consuming process. One solution is to vary the resolution of subwatersheds across the watershed so that areas of interest are defined in more detail (Notter et al. 2012). *This process is of particular interest as it allows the selection of a larger catchment for modeling and simulations, and a focus on a smaller part within the catchment for detailed analysis.*

The process of extracting model outputs for areas of interest from within a larger model is demonstrated here by analyzing model outputs for the five survey villages. During model setup, the subwatersheds were defined so that they were of similar size to the village areas used by IWMP for watershed development to allow model outputs to be more closely associated with each village. It was not possible, however, to match the boundaries exactly as neither the village boundaries nor the boundaries of the government-delineated microwatersheds matched the hydrological boundaries as defined by the DEM. Although watershed development under IWMP is meant to be planned for the government-delineated microwatersheds, it appears that in the Sukhi watershed most of the planning was done using village areas. The SWAT subwatersheds overlain onto the areas of the five survey villages shows that the village areas cover parts of a number of different subwatersheds, but the majority of the areas are covered by nine (numbered) subwatersheds (Fig. 5.12).

The land use of each subwatershed has a large impact on its hydrology and therefore on the impact of the scenarios. The nine subwatersheds shown in Fig. 5.12 can be divided into three distinct groups based on their dominant land uses:

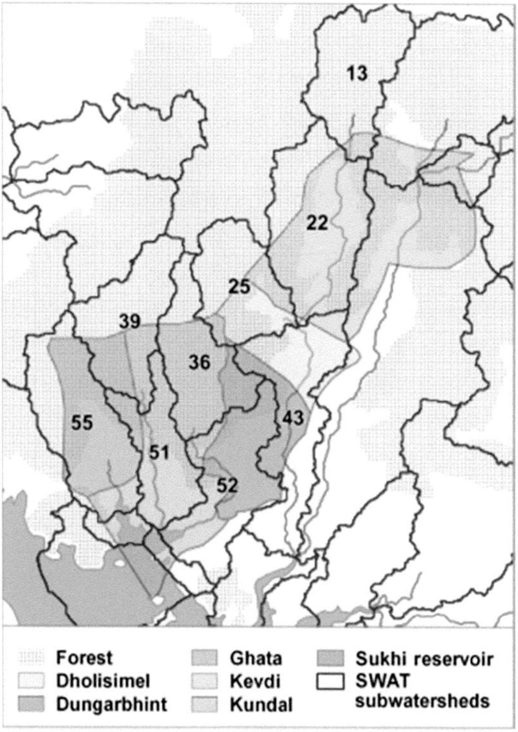

Fig. 5.12 SWAT subcatchments covering the five survey villages.

1. **High degree of forest cover**: Subwatersheds 13, 36, and 39
2. **Mix of forest and agricultural land**: Subwatersheds 22, 25, 51, and 55
3. **Mix of agricultural land, grassland, and degraded forest**: Subwatersheds 43 and 53

These nine subwatersheds can be divided into four subcatchments, all of which drain directly into the Sukhi reservoir, making them a useful unit of analysis (Fig. 5.13).[j]

A detailed (water infrastructure) plan was developed for Kevdi village for RWH structures based on preliminary model findings, analysis of satellite images, and consultation with the local community. The main aim of the plan was to ensure that sufficient water was available for the whole *rabi* irrigation, as villagers reported that on a majority of farms away from the river,

[j] The only large area of the five villages that falls outside of the nine subwatersheds is the eastern half of Kevdi village, which covers a small part of a much larger subcatchment. This illustrates the difficulties of using administrative units, such as village areas, when analyzing and planning hydrological interventions.

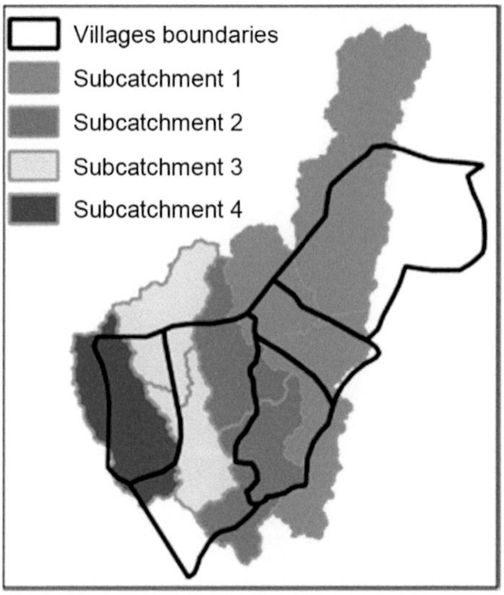

Fig. 5.13 Microcatchments covering the five surveyed village.

wells run dry before the end of the *rabi* season even during normal rainfall years, and to minimize reductions in downstream surface water flows. This plan was extrapolated to the other survey villages by calculating the amount of extra capacity needed for each hectare of agricultural land (Table 5.4). The split between on- and off-channel storage was 45:55, based on the number of structures of different types recommended in the plan for Kevdi village. (The off-channel storage is higher due to the large number of farm ponds recommended.)

5.7.2.2 Scenario description

The 40 m^3/ha figure applied in scenario 2A earlier (to represent intensification of RWH) is slightly lower than the average of 42 m^3/ha capacity of RWH structures calculated as a function of village area (Table 5.4). The capacity calculated is also higher than the figure of 40 m^3/ha for all the villages other than Kevdi. In the next round of village-level scenario generation, therefore, these revised figures for additional RWH structures were applied to the nine subwatersheds covering the village areas (rather than for the village areas per se). Using information from the village plans, the generic scenarios applied to the whole watershed were adapted to investigate the impact of these plans at the local level (Table 5.5).

Table 5.4 Extrapolation of the Kevdi Village Demo Plan to the other survey villages.

	Ghata	Kundal	Dungarbhint	Dholisimel	Kevdi
Village area (ha)	533	1052	517	373	1287
Total agricultural area (ha)	121	231	104	62	179
Rabi area (ha)	114	228	95	38	127
Baseline RWH capacity (m^3)	7995	15,780	7755	5595	19,305
Additional off-channel storage	14,911	28,791	13,341	8490	26,164
Additional on-channel storage	8602	16,455	7425	4429	12,719
Total extra storage (m^3)	19,116	36,567	16,500	9842	28,265
Total RWH capacity (m^3)	27,111	52,347	24,255	15,437	47,570
Total RWH capacity (m^3/ha)	51	50	47	41	37

Table 5.5 Description of village scenarios.

Scenarios		Description
V1	No development	Identical to scenario 1A and models the watershed as if no watershed development or increase in the area of double cropped agriculture had occurred. This is represented by the removal of all reservoirs and by using the 2004–05 land use dataset for the entire modeling period
V2	Village plan	This scenario represents an increase in RWH capacity based on a watershed plan developed for Kevdi village. The plan is extrapolated to all the subwatersheds covering the village areas based on the RWH capacity needed for each hectare of agricultural land. The split between on and off channel storage for Kevdi village is also applied to all subwatersheds. Reservoirs are also implemented in forested subwatersheds 13, 25, 36, and 39, the size of which are calculated using a 15-m^3/ha figure
V3	Intensification	This scenario represents a 25% increase in double cropped area within each subwatershed and an associated increase in RWH capacity, calculated using the per hectare figure from the Kevdi village plan. The capacity of reservoirs in subwatershed 13, 25, 36, and 29 is doubled compared to scenario V2

Note that for scenarios V2 and V3, reservoirs are also implemented in the forested subwatersheds upstream of the villages (i.e., in subwatersheds 13, 25, 36, and 39).[k]

5.7.2.3 Scenario results

Water balance components: Extracting data on the water balance components for the nine subwatersheds reveals differences to the Sukhi watershed as a whole for the baseline, due to the differences in the proportions of the different land uses, topography, and soil characteristics:

- **Evapotranspiration (ET) is lower than for the watershed as a whole**: 30.8% compared to 40%. This is due to the large areas of forest and degraded forest in the nine subwatersheds, which have lower ET than irrigated double cropped areas, which make up a higher proportion of the rest of the watershed.
- **Variations in reservoir inflows the driest and wettest years are significant**: While baseline reservoir inflows for the nine subwatersheds are like those of the whole watershed—the average over the period is 40.5% compared to 40%—reservoir inflow in 2009 under the baseline scenario is only 15% of rainfall compared to 19% for the whole watershed.
- **Percolation and (artificial) recharge from RWH are like those for the entire watershed**.
- **Removing RWH structures and decreasing double cropped area, sees only small increase in average reservoir inflows**: While this change is only from 40.5% to 41.2% of rainfall, Scenario 1A for the whole watershed saw an increase from 40% to 43% of rainfall. This highlights the fact that the study villages are located in an area of the watershed that so far has seen relatively less intensive agricultural and hydrological development.
- **Reservoir inflows decrease in scenarios V2 and V3, as expected, but the decrease is larger for scenario V3**. This is due to increase in double cropped area and the larger increase in RWH capacity.

[k] Although no structures are allowed to be constructed in the forest, many new structures constructed by IWMP are concentrated at the edge of the forest where gullies and streams emerge. As reservoirs are conceptually placed at the outlet of sub-watersheds in the model, adding reservoirs to these subwatersheds is an effective way of representing the concentration of structures at the forest borders.

Scenario V2, which models the plan developed for Kevdi village, sees a large increase in recharge from RWH structures compared to the baseline, especially in 2009, the driest year. The increase in ET is not large due to the fact that there is sufficient groundwater in the baseline scenario for full irrigation of the *rabi* crop. Therefore, in scenario v2, the additional groundwater is not needed for irrigation and is instead stored in the shallow aquifer with a proportion contributing to stream flow. This is a key area in which the model does not accurately reflect reality at a local level, as the local community has reported that in most years there is insufficient groundwater available to irrigate the entire *rabi* crop. Groundwater in the area is very shallow and drains quickly into the reservoir following the end of the monsoon.[1]

Impact on yields: SWAT also models average crop yields and those in the nine subwatersheds, for the period 2007–12, compared well with those reported in the primary village survey, with the exception of *kharif* maize (Table 5.6).

Table 5.6 SWAT estimated and farmer-reported crop yields, five villages, 2007–12.

Crop	Season	SWAT baseline scenario estimate	Average farmer estimate
Rice	Kharif	2112	2193
Maize	Kharif	3796	2087
Maize	Rabi	1886	2181

[1] The problem is also highlighted by the fact that reservoir inflows are lower in 2009 for scenario V2 than they are for V3, due to increased runoff that results from irrigation of a larger double cropped area. In reality it is unlikely in 2009, even with intensive RWH, that enough groundwater would be available such that runoff resulting from irrigation would increase reservoir inflows. Limitations in the modeling of groundwater affect the ability to use the model to assess the beneficial impacts of watershed development and highlight the need for further model development. Groundwater representation is acknowledged as a weakness of the SWAT model, which is why in many studies, where groundwater plays an important role, it is coupled with MODFLOW, a groundwater model, which allows groundwater to be represented in far greater detail (Kim et al., 2008). This would be an option of the Sukhi catchment but would obviously increase the amount of data and time needed to complete the modeling.

There is little variation in crop yields between the scenarios because in all of them there is deemed to be sufficient water available for irrigation in nearly all the subwatersheds. Only in 2009 does the water in the shallow aquifer come close to being completely depleted, which limits irrigation for some HRUs in the following year.

5.7.3 Discussion

Overall, the model helps to provide a solid hydrological foundation to the behavior of water in the watershed, putting concrete numbers on hydrological phenomena such as evaporation, transpiration, groundwater recharge, and percolation, and thus explaining the observed inflows into the Sukhi reservoir downstream of the selected watershed. Such information is vital in planning the management of the (larger) watershed and also in the prioritization of (smaller) watersheds to be taken up for treatment.

The simulation of watershed hydrology by the model enables the creation of scenarios to assess the impacts of watershed management interventions both locally, in villages within that particular watershed, and downstream of it. Supplemented by multistakeholder interactions, the main observations made possible by the use of this approach are the following:

- **Watershed inflows exceed outflows**: With "watershed inflows" referring to rainfall falling over the watershed (the basic source of the water) and "outflows" referring to the water that reaches the Sukhi reservoir (as "inflows into the reservoir"), this observation is borne out by the fact that there are inflows into the Sukhi reservoir even in very low rainfall years. While good rainfall years swell the Sukhi reservoir, increasing its ability to provide water to downstream farmers, dry years see a huge shrinkage, even compared to the average rainfall years.
- **Water demand exceeds supply**: The demand for water for domestic uses, livestock, and irrigation is more or less met in the five villages studied in detail, although villagers reported summer scarcities and problems even in a single low-rainfall year. Consecutive years of low-rainfall or droughts could severely constrain drinking water demand for human and livestock populations. The demand for water for irrigation has been growing over the last decade or so, illustrated by the large increase in wells and RWH structures, and is likely to grow further. While population increases could further increase water demand even in average rainfall years.
- **Need alternative rain water harvesting options**: The substantial rainwater harvesting prior to the IWMP projects of 2009–10 had resulted in

check dams and gully plugs being built in the best possible locations, leaving few options for such interventions in the IWMP. There are, however, still several options for decentralized storage and recharge structures, such as subsurface tanks and dykes and farm ponds, which can provide the farmers the extra *rabi* irrigation water that they mentioned as a major issue. All these, however, have to be considered in the specific local context as what may work well in one ecological, sociopolitical, and cultural context may not perform similarly in another context. Careful evidence-based analysis will be necessary before deciding on such options.

- **Need increased water productivity**: A shift to crops that use the available water but give higher profits per unit of water could improve agricultural livelihoods. This is not easy to realize in practice, however, as international experience has shown. Even strategies to reduce crop water use, such as shifting to less-water intensive crops, microirrigation (drip and sprinklers) and (plastic or organic) mulching to reduce evaporation from crop lands (and thus reduce the need to irrigate), appear sensible and practical (as do options to raise agricultural profits, by reducing cultivation costs and increasing output prices), but these can only be achieved by careful formulation of these strategies, based not only on strong local evidence of impact but also on the local (social, political, regulatory, legal, and cultural) contexts, besides effective implementation. This apart, whether it is considered necessary to improve livelihoods in upstream areas at the cost of improvements with higher social value in downstream areas, is an issue for multicriteria policy analysis and decision.
- **Nonwater-based livelihoods must be encouraged**: The unmet demand for irrigation water is reflected in the inability of all farmers to grow a *rabi* crop, causing many to migrate for work after the *kharif* season. Providing support for feasible nonwater-based livelihoods that are at least as profitable as migration could help to stem such stress migration. This again is a policy decision to be taken at the level of the watershed as a whole, where local perceptions and priorities will have to be considered along with government interests (e.g., for long-term development of the region as a whole and for future generations).

5.8 Lessons for hydrology-based watershed management

5.8.1 Approach and methodology

The basic lesson is the fundamental need for sound hydrological analysis and sustained and well-facilitated stakeholder consultations to underpin decisions on watershed management strategies. This has, however, to be undertaken within an integrated policy and institutional framework based on (economic, environmental, social, and cultural) development priorities for water allocation across large watersheds. As in countries like Australia, this will require political cohesion and broad agreements but the actual frameworks for action will have to be context-specific.

The six key steps in the methodology developed for the hydrological assessment are the following:

- Field visits, stakeholder dialogue, transect walks, and model selection
- Model set up and modeling using data from open-access global datasets
- Acquiring, quality controlling, and using secondary data from (nonopen access) national, state, and other databases
- Participatory collection of biophysical primary data (e.g., well and RWH structure surveys)
- Participatory collection of societal primary data (e.g., development timelines of water and other natural resources)
- Modeling and scenario analysis to assess and map the scale of current and future watershed management benefits and externalities

The more innovative aspects of the methodology piloted in this study include the following:

- **Using data from global and regional databases as a starting point**. Being readily available, this could enable the setting up the model very quickly—which is very different to normal modeling practice
- **Iterative incorporation of official and primary data over a period of time**. Refining the model as better-quality (official) data became available.
- **Using only open source software**. While projects and organizations are still willing to spend on software, this is not necessary currently—and thus the use of GIS and modeling software should no longer be constrained by the cost of licenses.
- **Savings in the cost of modeling work**. This is due not only to the availability of freely available open source software, but also because models can be set up fairly quickly with available global data and refined as better data are made available.

The planning of management interventions, however, requires a longer process of sustained and well-facilitated engagement with the key stakeholders, within clear policy and institutional frameworks of water allocation and water use.

5.8.2 Application: Watershed management projects

The basic steps for applying the methodology in a stand-alone watershed management project are the following:

(1) the selection and prioritization of watersheds

(2) the hydrological assessment (with stakeholder interaction) and

(3) intervention planning.

Steps (2) and (3) are interactive and iterative, with inputs from primary surveys and discussions feeding into model set up, preliminary model outputs informing discussion on intervention planning, and monitoring and evaluation of project impacts being used to rerun the model to track changes in the watershed.

5.8.2.1 Selection of project area

- **The ridge-to-valley approach is a good starting point**, as in the case of the IWMP, but multiscalar biophysical analysis is now possible, with the expansion of data availability, and can be used to decide the starting point. River basin maps and topology provide an overview of the various smaller watersheds in the basin-level watershed, and the subwatersheds of each watershed. In the case of preselected project sites, placing the selected site within its larger hydrological unit would give a better context to project planning and management (than treating it as a standalone site with no hydrological connections either upstream or downstream).
- **Accessibility issues may affect site selection**: Even if selected, the hilly and forested upstream areas of watersheds may not be accessible for a variety of reasons. For instance, in many places, including India, these areas are under the jurisdiction of the Forest Department and no project can be implemented there without their supervision and control. Some areas have military installations (e.g., wireless and microwave repeater stations) under the control of the Ministry of Defence and are therefore inaccessible. The project area to be selected for interventions thus depends as much on the access to the site as on budget and development priorities (e.g., the criteria for selection of IWMP projects).
- **The size of the watershed for the hydrological simulation modeling depends more on the need for accuracy of model findings**. Freely available global data sets and improved "open source" software means that modeling can be done at any convenient scale, but generally speaking, the larger the watershed, the easier it is to find stream gauging points or reservoirs, and better the (simulation) model at simulating

watershed hydrology. As demonstrated in Section 5.3, the area selected for project interventions can lie within this larger watershed selected for modeling.

- **Hydrological unit vs administrative boundary is an "old chestnut"** (e.g., Farrington et al., 1999) and while it would be helpful for model predictions if the project area were determined by hydrological criteria (e.g., the subwatersheds delineated by SWAT), these rarely coincide—and most projects prefer administrative boundaries for ease of planning and implementation. However, rather than planning project interventions without any reference to hydrological dynamics (as in the case of the IWMP), it is possible to carry out the modeling for the larger area and extract outputs for the project area (as demonstrated for the five villages in the Upper Sukhi watershed) and thus use available hydrological information in project planning.
- **Overall, however, the best starting point depends on the context**. Given a choice, however, it is best to start in the upper reaches of the watershed in the project area (to reduce the velocity of rain water and thus prevent adverse impacts such as soil erosion and increase groundwater infiltration) even though the entire watershed is not being taken up for treatment under the project. The modeling can provide watershed-level information and scenarios that can be used for planning interventions.

5.8.2.2 Hydrological assessment

- **A review of earlier hydrological assessments is a necessary first step to a new or updated hydrological assessment**. This essentially involves understanding the watershed, its dynamics and complexities, usually from experience with other similar watersheds elsewhere. Collecting secondary information, trawling and mining the internet, and undertaking a drive-through or a fly-over the watershed are useful parts of this activity, and can be undertaken in any order that is convenient, but discussions with key stakeholders are a must, not only to add details of interest and to assimilate local information but also to create support for the proposed management activities.
- **The choice of model is a key step in the hydrological assessments**. There are a large number of suitable models and the final choice has to be based on a number of issues including the skills and experience of the modeler, the availability of suitable data, and the time and resources budgeted for the modeling exercise. As explained earlier, SWAT was a good choice for the Indian context.

- **Model set up involves deciding values for a large number of technical parameters in these simulation models**, using available data, the literature (for approaches used by other researchers), and expert opinion (of specialists who have worked in the area). These will have to be adjusted as better information emerges, but some parameters (like subwatershed size in SWAT) have to be fixed at the beginning, keeping in mind the need to extract information later about the smaller project area. As mentioned earlier, the watershed size will be decided not only by data availability and accessibility to the site for the field visits but also by the nature of "what if" and "what's best" questions to be answered by the model.
- **Collecting primary and secondary data depends on the model chosen.** But listing the data requirements of the model and identifying potential sources of each dataset is the first step. There are three basic sources of data for any modeling exercise: (1) global datasets (DEM, soils, weather, etc.); (2) country-specific data sets (on reservoirs and canal networks, soils, crop yields, weather, etc.); and (3) local information (on cropping seasons, crop durations, irrigation applications, the number (georeferenced) location, and development timelines of wells and RWH structures, etc.). Thus, while the model can be set up quite easily using global data sets, additional information can be added to improve the quality of the model outputs. The primary and secondary surveys can start simultaneously, not only to reduce time but also to ensure that the primary information is available to input into the model.
- **Model calibration and validation are essential to checking how well the model is able to represent watershed hydrology**. Here it is important to note that not only is the "goodness of fit" to be checked but also the extent of uncertainty inherent in the model predictions.[m] The assumptions that went into detailing the model specifications as well as the data need to be reviewed and revised till the "goodness of fit" is as close to ideal as possible and uncertainty is minimized. Also, incorporating land use changes into the model set up and scenarios is essential in areas where land use has changed substantially during the period taken for modeling. It would also be helpful to have an expert or group of experts review the modeling process at regular intervals during the modeling process, as done in the demonstration detailed in Section 5.3, to assess the technical details of the modeling (e.g., the separation of the data set for use in warm up, calibration and validation, the setting of model parameters, and the use

[m] In the SWAT, these are represented by the *p* and *r* statistics.

of manual functions instead of automatic functions as in the case of SWAT).

- **Main model outputs are the basic water balance components of the watershed** (including key features such as evapotranspiration, groundwater recharge, and runoff outside the watershed), which can be used to classify the watersheds along with information from stakeholder interactions about demand and supply
- **Understanding future watershed behavior requires the creation of a baseline and other scenarios that work in relation to the baseline**. Interpreting the scenarios in relation to the baseline produces more reliable results than the absolute value of the scenario outputs. While watershed-wide baselines and scenarios are useful and necessary, the model can also be used to create scenarios for the project area in detail.

5.8.2.3 Intervention planning

- Although discussed under separate headings, the hydrological assessment and the stakeholder-led planning are part of the same mutually supportive process and need to be undertaken synchronously and iteratively. Building cohesion and trust among the key stakeholders is vital as is the creation of an appropriate institutional platform and conducive conditions for effective multistakeholder discussions.
- **Management planning can be initiated as a part of the process of stakeholder consultations.** Using information from the earlier stakeholder interactions and field visits and supplemented with model scenarios, this needs to be viewed as iterative and adaptive. Participatory village planning can be done more easily by involving villagers in discussions of local watershed characteristics and problems using maps based on remotely sensed data. While Google Earth is a relatively cheap and simple tool to create such interactive map-based discussions, it can also be done with more sophisticated maps and tools. Detailed planning, however, will involve visiting each farmer's field to discuss specific options to improve water productivity (e.g., leveling, bunding, terracing, crop choice, cultivation techniques such as mulching, irrigation options such as drips and sprinklers and farm ponds), visiting village water bodies (streams, ponds, wells, and RWH structures) to assess what can be done to augment water supply (and protect and prioritize drinking water for humans and livestock) using traditional knowledge and skills.
- **Assessing plan implications using the model is an iterative next step**. This will introduce the key features of the watershed plan (including

the total additional capacity of wells and RWH structures planned and options to reduce water demand and improve water productivity) into the model and creating a separate scenario to assess potential hydrological impacts in terms of the water balance components.

- **Addressing downstream impacts will become important as planned interventions are likely to reduce surface and or groundwater outflows**. At the minimum, discussions will have to be held with downstream stakeholders to see if some part of the plan can be scaled down (to reduce such adverse impacts) or if adverse impacts can be minimized. Water productivity enhancements can be suggested and introduced, so that even reduced quantities of water may produce the same level of profit per unit area. It would be ideal if the community benefiting directly from the project undertakes to guide the downstream community in the adoption of these practices, based on their experiences through project interventions. But this is a much larger issue if there are interbasin transfers, dams, urban areas, and ecological requirements in the selected watershed. These will require locally relevant and effective sociopolitical institutions and mechanisms, as well as policy frameworks, to be set up and operate to resolve the tradeoffs that are inevitable when downstream flows reduce.
- **Finalizing the management plan will require additional considerations**: These include understanding the support required to promote nonwater-based livelihoods and assess the strength and capability of local organizations and local government officials in order to plan for the capacity building of individuals and groups and the strengthening of local institutions. The Watershed Plan will also have to include arrangements for the participation and contribution of local villagers to construction, implementation, monitoring, and management of these planned interventions. Provided the budget is sufficient to implement the plan, all these options can be aggregated into a Village Plan, to be discussed and finalized with the villagers and their representatives, for implementation. Implementing, managing, and monitoring the plan will require continuous hand-holding by project staff to ensure that these activities proceed as per the agreed Village Plan.

5.9 Postscript

To answer the question posed by the Joint Secretary in the Ministry of Rural Development GOI: No, there is no "ideal" size of a watershed, but yes, it is possible to use a hydrological model at a suitably large scale (e.g., 400 km^2, so as to get data to run the model), get a good fit to the observed data, and then "zoom in" to any set of contiguous villages within the basin to guide the planning of watershed management interventions therein—but within an iterative and participatory planning process. Applied hydrology can indeed improve watershed management!

References

Batchelor, C.H., Singh, A., Rao, M.S.R., Butterworth, J., 2002. Mitigating the unintended impacts of water harvesting. In: Paper Presented at the IWRA International Regional Symposium on 'Water for Human Survival', 26–29 November 2002, Hotel Taj Palace, New Delhi.

Batchelor, C.H., Rao, M.S.R., Rao, S.M., 2003. Watershed development: a solution to water shortages in semi-arid India or part of the problem? Land Use Water Resour. Res. 3 (2003), 1–10.

Batchelor, C.H., Reddy, V.R., Linstead, C., Dhar, M., Roy, S., May, R., 2014. Do water-saving technologies improve environmental flows? J. Hydrol. 518A (Pt. A), 140–149.

Calder, I., Gosain, A.K., Rao, M.S.R., Batchelor, C.H., Garratt, J., Bishop, E., 2008. Watershed development in India. 2. New approaches for managing externalities and meeting sustainability requirements. Environ. Dev. Sustain. 10 (4), 427–440.

Farrington, J., Turton, C., James, A.J., 1999. Participatory Watershed Development: Challenges for the Twenty-First Century. Oxford University Press, New Delhi.

Glendenning, C.J., van Ogtrop, F.F., Mishra, A.K., Vervoort, R.W., 2012. Balancing watershed and local scale impacts of rainwater harvesting in India—a review. Agric. Water Manag. 107, 1–13.

James, A.J., Kumar, M.D., Batchelor, J., Batchelor, C., Bassi, N., Choudhary, N., Gandhi, D., Syme, G., Milne, G., Kumar, P., 2015a. Catchment Assessment and Management Planning for Watershed Management, Volume I: Main Report. World Bank, Washington, DC.

James, A.J., Kumar, M.D., Batchelor, J., Batchelor, C., Bassi, N., Choudhary, N., Gandhi, D., Syme, G., Milne, G., Kumar, P., 2015b. Catchment Assessment and Management Planning for Watershed Management, Volume II: Annexes. World Bank, Washington, DC.

Kauffman, S., Droogers, P., Hunink, J., Mwaniki, B., Muchena, F., Gicheru, P., Brindraban, P., Onduru, D., Cleveringa, R., Bouma, J., 2014. Green water credits—exploring potential to enhance ecosystem services by reducing soil erosion in the upper Tana Basin, Kenya. Int. J. Biodivers. Sci. Manag., Ecosyst. Serv. Manag. 10 (2), 133–143.

Kerr, J., Milne, G., Chhotray, V., Baumann, P., James, A.J., 2006. Managing Watershed Externalities in India: Theory and Practice. Available at http://www2.gsu.edu/~wwwcec/docs/doc%20updates/New%20Folder%20%282%29/published%20version%20of%20Kerr%20et%20al.%20Externalities%20article%20in%20EDS%20%28jkerr%20v1%29.pdf.

Kim, N.W., Chung, I.M., Won, Y.S., Arnold, J.G., 2008. Development and application of the integrated SWAT-MODFLOW model. J. Hydrol. 356, 1–16.

Kumar, M.D., Ghosh, Shantanu, Patel, Ankit, Singh, Prakash, O., Ravindranath, R., 2006. Rainwater harvesting in India: some critical issues for basin planning and research. Land Use Water Resour. Res. 2, 1–17.

Mishra, A., Froebrich, J., Gassman, P.W., 2007. Evaluation of the SWAT model for assessing sediment control structures in a small watershed in India. Trans. Am. Soc. Agric. Biol. Eng. 50 (2), 469–477.

MoRD, 2008. Common Guidelines for Watershed Development Projects. Department of Land Resources, Ministry of Rural Development, Government of India, New Delhi. Available at http://dolr.nic.in/dolr/downloads/pdfs/CommonGuidelines2008.pdf.

MoRD, 2011. Common Guidelines for Watershed Development. Projects 2008 Revised Edition 2011. Department of Land Resources, Ministry of Rural Development, Government of India, New Delhi. Available at http://dolr.nic.in/dolr/downloads/pdfs/Common%20Guidelines_2011[1].pdf.

Notter, B., Hurni, H., Wiesmann, U., Abbaspour, K.C., 2012. Modelling water provision as an ecosystem service in a large East African river basin. Hydrol. Earth Syst. Sci. 16, 69–86.

World Bank, 2006. Managing Watershed Externalities in India. World Bank, Washington, DC. Agriculture and Rural Development Sector Unit, South Asia Region, Discussion Paper Series, Report No. 1. Available at http://documents.worldbank.org/curated/en/2006/04/7047191/managing-watershed-externalities-india.

CHAPTER 6

Water accounting for understanding water tenures: A case study of water rich Warna sub-basin of Krishna river basin, Maharashtra

M. Dinesh Kumar[a], K. Siva Ram Kishan[a], A.J. James[b], Nitin Bassi[c]
[a]Institute for Resource Analysis and Policy, Hyderabad, Telangana, India
[b]Environmental & Natural Resource Economist, New Delhi, India
[c]Institute for Resource Analysis and Policy, Delhi Liaison Office, New Delhi, India

6.1 Introduction

Water tenure is a legally or customarily defined relationship among people, as groups or individuals, with respect to water resources. A water tenure approach to understanding people's relationship with water encourages a comprehensive, rigorous, and interdisciplinary analysis of the rights and uses, rather than understanding the laws and policies in relation to ownership or usufruct rights in water. The proper use of a "water tenure analysis" into designing water governance framework would help bring in transparency, fairness, and accountability in decision making. For water tenure analysis to be effective, it should be supported by proper water accounting data, upon which consensus has to be built among competing users of water (Lopez-Gunn, 2013).

Water accounting studies, based on an analysis of data related to various components of the hydrological system (e.g., rainfall, stream flows, groundwater recharge, evaporation from water bodies and swamps, storage changes in reservoirs, and land use and water use) at appropriate hydrological units (e.g., basin, catchment, or watershed level), can give a range of useful information about potential water-related extreme events, how much of water in the basin is used in different sectors (drinking and domestic uses, municipal use, industrial use, livestock use, power generation, crop consumptive uses,

Current Directions in Water Scarcity Research, Volume 1
https://doi.org/10.1016/B978-0-12-814851-8.00006-9
© 2019 Elsevier Inc.
All rights reserved.

and various in-stream uses), how much of the basin's water goes uncaptured into natural sinks (e.g., saline formations, seas), and how much is lost through nonbeneficial uses. It could also help the water administration be active in the basin to plan actions to mitigate the adverse impacts of hydrological stresses on the socioeconomic system, such as drinking water shortages for human and livestock populations in affected areas, crop losses, and reduced income and livelihoods of affected populations, and damage to water infrastructure.

From the point of view of water governance and management, water accounting studies are useful in the sense that they can provide insights into: (1) optimum level of investments in water augmenting measures (e.g., watershed development programs, water harvesting systems); (2) possible measures to (a) increase effective water availability (e.g., reducing evaporation from surface reservoirs and preventing soil moisture depletion from rain-fed crop land), (b) reduce water demand (e.g., reducing nonbeneficial [both consumptive and nonconsumptive] use of water in irrigated fields) and (c) convert other nonbeneficial uses of water in the basin (e.g., evaporation from barren land) into productive uses; and (3) better intersectoral water allocation, including from lower-valued to higher-valued uses.

In this chapter, we discuss a methodology for water accounting and apply the methodology to one sub-basin of a river in western Maharashtra for one hydrological year. The objective is to assess how much of the renewable water resources in the basin are utilized for various consumptive uses, and how much are available for further exploitation to support future uses. The hydrology, socioeconomic characteristics, water resource systems, water balance of surface and groundwater, and water accounts of the sub-basin are presented and discussed. Based on the findings, some inferences are drawn with regard to water tenure for the sub-basin.

6.2 Hydrology and geohydrology of Warna river basin

6.2.1 Topography and geography

Warna is a sub-basin of the Krishna river basin situated in the south-west part of State of Maharashtra (Fig. 6.1). Krishna river basin is an interstate river basin having a total drainage area of 258,948 km^2, falling in four Indian states, viz., Maharashtra, Karnataka, Telangana, and Andhra Pradesh. The total sub-basin area is about 2095 km^2, with elevation ranging from 540 to 1100 m above MSL. Warna river arises in Sahyadri near Patherpunj in Patan block of Satara district of Maharashtra at an altitude of 914 m above MSL.

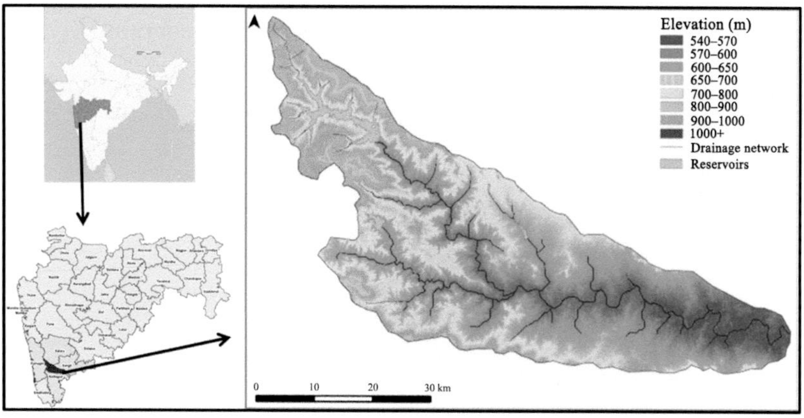

Fig. 6.1 Location of Warna river basin along with its elevation gradient.

The total length of the river is around 150 km. It joins Krishna near Haripur village in Sangli district at an altitude of 548 m. The sub-basin falls in the rain-shadow zone of the Western Ghats and has three distinct seasons, viz., monsoon, winter, and summer. It is spread over nine talukas of four districts, mainly Sangli and Kolhapur, and some parts of Satara and Ratnagiri districts. The talukas it covers are: Shirala, Walwa, and Miraj talukas of Sangli district; Shahuwadi, Panhala, Hatkanangale, and Shirol talukas of Kolhapur district; Patan taluka of Satara district; and Sangameshwar Taluka of Ratnagiri district of Maharashtra.

The sub-basin has a complex topography and bears a transitional character between the Konkan on the west and the Deccan plateau on the east. The eastern boundary of the sub-basin is marked by the flood plain of the river Krishna. The general trend of the sub-basin is from west to east. The western part of the sub-basin comprises a series of elongated hillocks and undulating hilly and rugged terrain. In these areas, the slope is more than 60°. The eastern half of the sub-basin has mostly flat terrain.

6.2.2 Rainfall

There is very high spatial variation in the rainfall within the sub-basin (Fig. 6.2). Rainfall is very high in the western parts of the basin adjoining Konkan and comprising the western ghat portion of the basin. The rainfall falls rapidly from west to east. It varies from 4000 mm in upper western parts around Sahayadries and Konkan portion and less than 700 mm in the lower eastern portions. About 85% of the rainfall occurs during the south-west monsoon, i.e., in the months of June to September.

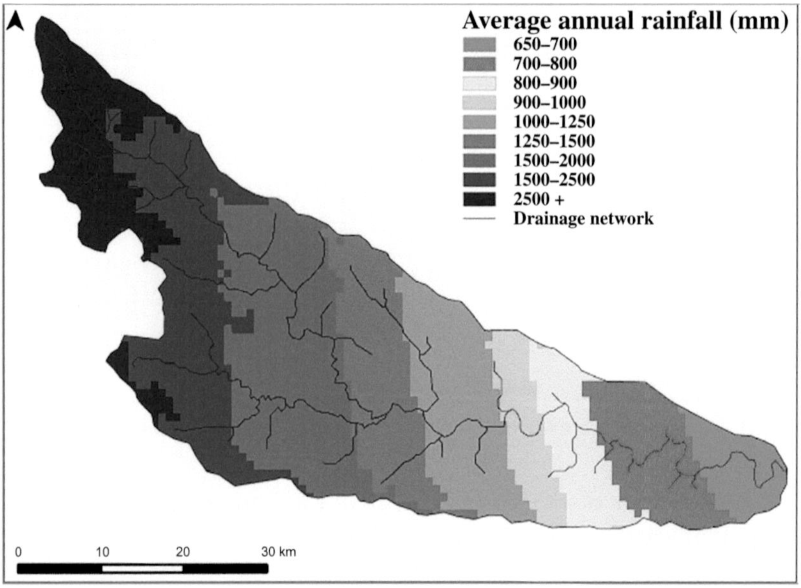

Fig. 6.2 Spatial variation in average annual rainfall in Warna river basin.

Table 6.1 Spatial and temporal variation in rainfall in Warna river basin (1998–2012).

		Rainfall parameters			
District	**Taluka**	**Min (mm)**	**Max (mm)**	**Mean (mm)**	**CV (%)**
Sangli	Miraj	321.0	908.9	588.7	29.3
	Walwa	291.0	1412.6	740.2	37.2
	Shirala	481.8	1794.0	1002.6	33.0
Kolhapur	Shirol	272.0	1050.0	643.4	34.2
	Hatkanangale	364.1	1609.6	843.5	38.3
	Panhala	1290.0	2735.0	1760.7	28.2
	Shahuwadi	1521.5	3450.5	2054.5	26.7

(*Source*: Authors' own analysis based on information available from Department of Agriculture, Government of Maharashtra.)

In Sangli district, between 1998–99 and 2008–09, highest annual rainfall (around 1750 mm) was recorded during 2005–06 in Shirala block and lowest (about 250 mm) in Walwa block during 2003–04. During the same period, the highest recorded rainfall (around 3400 mm) in Kolhapur district was in Shahuwadi block and lowest (about 250 mm) in Shirol block. The spatial varation in rainfall in six talukas of two districts falling partly in Warna sub-basin is analyzed and presented in Table 6.1. However, these figures need to be interpreted carefully, as the gradient of rainfall is very steep, with reducing magnitude of rainfall from west to east (Fig. 6.2). A small shift in location can change the quantum of annual rainfall significantly.

6.2.3 Climate and agroecology

Warna sub-basin of Krishna lies in the temperate climatic zone. Like most other parts of the country, three distinct seasons are observed in the basin: hot summer (from March to May), rainy season (from June to October), and winter (from October to February). Temperature goes up to 45°C in summer, but generally it varies between 20°C and 30°C. The western upper part of the basin is humid, whereas the eastern part exhibits semiarid conditions.

The reference evapotranspiration (PET) gradually increases from western to eastern part of the river basin—from 1300 to 1600 mm in upstream areas, 1600 to 1700 in midstream, and 1700 to 1800 mm in downstream areas of the basin.

The sub-basin falls in two different agroecological zones, viz., south western Maharashtra and north Karnataka plateau (hot dry semiarid ESR with shallow and medium loamy black soils, medium to high available water capacity and length of growing period 90–120 days); and north Sahyadris and western Karnataka plateau (hot dry subhumid ESR).

6.2.4 Hydrology

On the basis of the spatial pattern in occurrence of precipitation in the catchment, the sub-basin can roughly be divided into three parts, i.e., upstream (from the origin up to Chandoli/Warna dam), midstream (from Chandoli dam to Chavare village), and downstream (from Chavare to Haripur village) (Mohite and Samant, 2013). Warna sub-basin is mainly drained by river Warna along with its numerous tributaries. The main river originates in the Sahyadri near Patherpunj in Patan taluka of Satara district. Major tributaries, which join Warna from the northern side, are Morna and Meni. Other major streams are Kadavi and Kansa, both of these flows west to east and join the southern bank of the river.

There are a number of striking features of the river courses of the Warna and its tributaries. Morana river, an important left bank tributary of Warna, takes a "V" bend at Danoli & Samdoli and "U" bend at Khochi. All these bends are mainly controlled by joint structure of the bedrock. In lower reaches, the river flows with prominent loops at Kokarud, between Ambole and Sonarli near Shirla. The length of the loop formed by the river channel is about 10 km. The neck distance is 1.5 km.

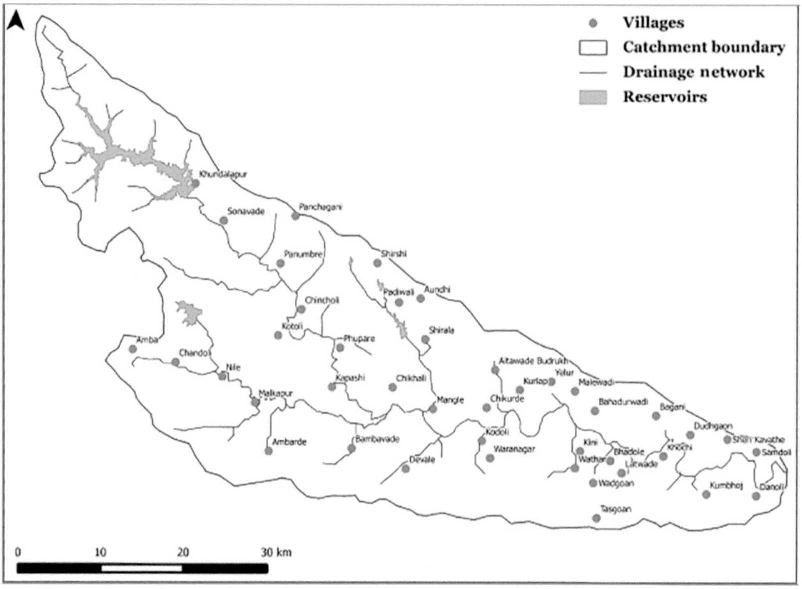

Fig. 6.3 Drainage pattern in the Warna sub-basin of Krishna river basin.

Many drainage patterns of the sub-basin are revealed by first-, second-, third-, and fourth-order streams in the basin (Fig. 6.3). The most common drainage pattern are dendritic, radial, paralleled, and rectangular. However, the drainage is not uniform over the basin. Being less affected by tectonic activities, the drainage pattern is mainly controlled by erosion activities and lithology. The basin's individual drainage pattern is mostly dendritic. Except a few streams in western regions, all major and minor streams are seasonal and are fed by monsoon rains only. Therefore, the period of recharge does not exceed more than 6 months in a year (July to October).

The major water bodies in the basin include 4 waterfalls, 5 reservoirs, 13 streams, 14 minor tanks and ponds, and 21 springs. Due to hilly undulating terrain, local geomorphology, and climatic gradient, the upper catchment represents most diverse riparian habitats, whereas the lower catchment has lowest diversity of riparian habitat. Among the five reservoirs, Warna reservoir is a large one.

6.2.5 Geology and geohydrology

The sub-basin area is mostly underlain by Deccan Traps formations (mainly Basalt). Laterites are also found in some parts of the basin. In Deccan Trap there is no well-defined, uniformly distributed, homogeneous aquifer system. The availability and productivity of groundwater in these formations is entirely dependent in its inherent physical property such as the size and distribution of vesicles, number and spacing of interconnected joints and fractures, and degree of weathering.

Groundwater in the basaltic aquifers occurs under phreatic and semiconfined conditions. The massive lava flows and thick red-bole layers tend to inhibit vertical movement of groundwater and thus act as confining aquiclude. The productive aquifers, when favorably situated, receive recharge and groundwater moves down the slopes till it is withdrawn by groundwater abstraction structures such as dug wells or discharges naturally through springs and wetlands. The average depth of wells varies from 9 to 15 m and diameter varies from 4 to 8 m. The range of water level varies from 3 to 7 m and the yield ranges from 75 to 100 m^3/day. The upper part of the basin is mainly covered with red-brown soil and at places by the lateritic soils, while the lower most portions is known for black cotton soil. The middle part of the basin mostly comprises coarse shallow soils and alluvium.

6.3 Water resource system of Warna sub-basin and water supplies

6.3.1 Surface reservoirs and irrigation projects

There are three main irrigation projects in the sub-basin, Warna being the major and Kadavi and Morna being the medium irrigation projects. Table 6.2 provides the salient features of the dams. The total water availability through these diversion structures is around 1.08 BCM. Out of these, water availability from Warna reservoir is about 0.97 BCM, from Kadavi dam reservoir is 0.07 BCM, and from Morna dam reservoir is 0.03 BCM.

Annually, about 60% (i.e., 0.64 BCM) of the available water in Warna reservoir is consumed for irrigation and industrial use. Whereas, all the available water from Kadavi and Morna dam is diverted for irrigation and industrial use. The gross irrigated area of Warna irrigation project, which includes a lift irrigation scheme (LIS), is about 111,823 ha and that of Kadavi and Morna irrigation projects are about 1160 and 110 ha, respectively.

6.3.2 Groundwater

As per the assessment carried out jointly by Central Ground Water Board and Groundwater Survey and Development Agency of Maharashtra, in 2008–09, the net annual groundwater availability in the basin is around 49 MCM, i.e., 47.4 MCM in noncommand areas and 1.6 MCM in command areas.

Table 6.2 Salient features of the reservoirs in Warna river basin.

Particulars	Reservoirs		
	Warna	**Kadavi**	**Morna**
River	Warna	Kadavi	Morna
Purpose	Irrigation, hydropower	Irrigation	Irrigation
Catchment area (Th ha)	30.10	2.33	8.55
Full reservoir level (m)	626.90	601.25	595.50
Gross storage capacity (MCM)	974.19	71.24	21.16
Live storage capacity (MCM)	779.35	70.56	15.15
Design flood (m^3/s)	2135.00	744.35	1075.00

(*Source*: Water Resources Information system of India.)

This works out to be roughly only around 1.3% of the total precipitation in the basin, or 23.5 mm. The investigation carried out by National Geophysical Research Institute in Deccan trap formations (in Kukadi basin and Godavari–Purna basin) shows that the recharge from rainfall could be in the range of 7.5%–8.6%, with the mean recharge values of 46 mm in Kukadi and 56 mm in Godavari–Purna, respectively (Athavale et al., 1983). However, two points need to be kept in mind, while using the "percentage rainfall" as the basis for arriving at recharge estimates. In hard rock areas with limited groundwater storage potential, the recharge cannot increase with quantum of rainfall. Second: the extent of recharge from rainfall is also a function of the terrain conditions in the hard rock areas. As noted by Raju et al. (1979) based on field studies in South India, hard rock areas with slope more than 5% do not hold surface water for sufficient time for percolation. In the case of Warna sub-basin, nearly 80% of the basin areas have steep slopes, and therefore are not conducive to natural recharge from rainfall. Hence, these low values of recharge estimated the sub-basin can be reconciled.

The gross groundwater draft for irrigation was 0.5 MCM in command areas and 25.3 MCM in noncommand areas. The total groundwater draft for domestic and industrial groundwater use was only 0.3 MCM in command areas and 1.5 MCM in noncommand areas (GSDA and CGWB, 2011). With a stage of groundwater development of 56.3%, the basin can be categorized as safe. However, in some talukas such as Hatkanangale, Shirol, Miraj, and Walwa, groundwater development is already in semicritical or critical stage. Nevertheless, as pointed out by Kumar et al. (2012), these estimates of groundwater balance need to be looked at very carefully, as the assessment of natural recharge from rainfall does not take into account groundwater outflows, which is quite significant in regions with steep topography and groundwater flow gradients, with the result that the utilizable groundwater gets overestimated.

6.4 Socioeconomic system of Warna sub-basin

Based on isohyets, Warna sub-basin is divided into three different zones (Fig. 6.4) (Joy and Paranjape, 2011). Isohyets were chosen in a way that they closely matches with the characterization and zoning of the sub-basin based on slopes, soils, and socioeconomic conditions of the sub-basin.

Zone 1 is the western part of the basin and receives an average annual rainfall of above 1600 mm. This zone covers 42.5% of the total basin area. This zone is mostly hilly and comprises western part of the Shirala and Shahuwadi taluka. It includes the Chandoli National Park and the catchments of two large dams. Out of the total area of 89,000 ha, only 15.5%

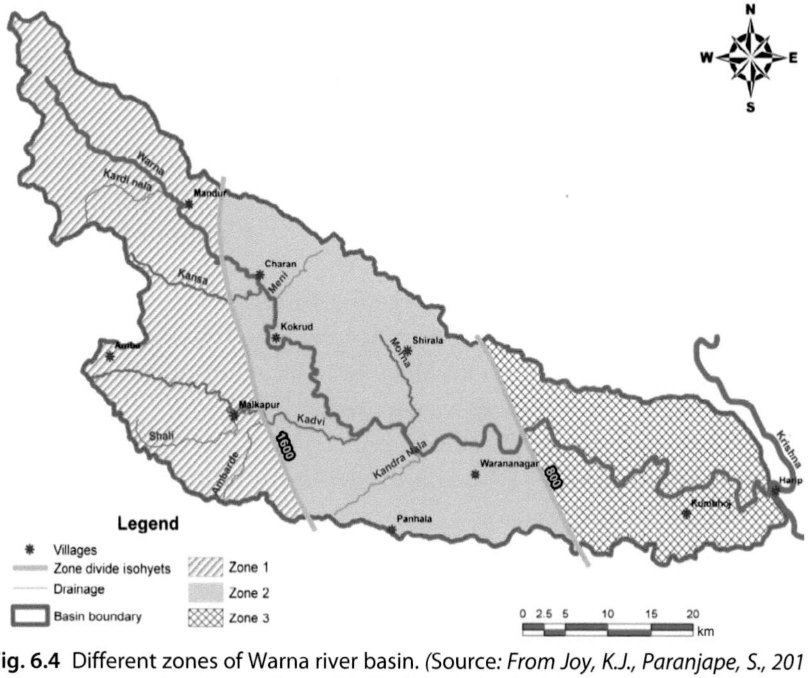

Fig. 6.4 Different zones of Warna river basin. (Source: *From Joy, K.J., Paranjape, S., 2011. Water resources planning and people's livelihoods in the Northern part of the Western Ghats: the case of Warna basin. Paper Presented at the Consultation Meeting on Water Resources of the Western Ghats by WGEEP. KFRI, Trichur.*)

is under agriculture. Almost 64% of the area is under reservoir submergence and covered by the National Park. The area under wasteland is 16%. Because of the presence of large amount of forest land and common land, the arable land availability is very low in this part of the sub-basin. Farming is not a major economic activity in this sparsely populated and remote region. With excessively high rainfall and humid climate, water demand for agriculture is also negligible.

The zone has about 19,600 households comprising a total population of 93,500. Total cattle population is around 37,700 U. Majority of the farmers belong to small land holding category. Almost the entire cropping is rain-fed and only single crop is taken by the farmers. Major crops include rice and finger millet. Only a small proportion of cultivated area is irrigated through wells and bore wells. Thus, this zone is characterized by subsistence form of agricultural practice. Large-scale migration to towns and cities has also been reported from this region.

Zone 2 is the central part of the basin and receives an average annual rainfall between 800 and 1600 mm. It is the considered to be the transition zone between the western and eastern part of the basin. It is spread over 35.4% of the total basin area. Out of the total area of 74,200 ha, about 55% is under agriculture and 44% comprise of wasteland.

The zone has about 94,200 households comprising a total population of 4.4 lac people. Total cattle population is around 1.05 lac. Though most of the agricultural area is rain-fed, irrigation is performed in about 30% of the flat terrain of the region. Major crops include rice, finger millet, sugarcane, and groundnut.

Zone 3 comprises the portion of the sub-basin, which receives an average annual rainfall of less than 800 mm. It comprises the eastern most part of the basin, which is in a low-lying plain area and has fertile alluvial soil. It has characteristics of a semiarid region with high variability in annual rainfall.

The zone is spread over 22.1% of the total sub-basin area. Out of the total area of 46,300 ha, almost 91% is under agriculture and only 8.8% comprises wasteland. The zone has about 52,300 households comprising a total population of 2.6 lac people. Total cattle population is around 72,600. Most of the agricultural land is irrigated as the zone lies in the command of the upstream dams and is also served by a number of lift irrigation schemes on the river itself. Sugarcane is the major cultivated crop in this zone. In some parts, soybean and potato are also grown.

6.5 Methodology and analytical framework for water accounting study

In water accounting for blue water, we can look at: (a) the renewable water resources (annual surface water flows and groundwater replenishment) as the "total inflow" into the basin during a hydrological year; (b) how much is being used up in various consumptive uses during the same year (various outflow); and (c) the "balance," which is in the form of unutilized water at the drainage outlet of the sub-basin and the changes in groundwater and surface storage occurring during the hydrological year.

The amount of water that is being used up in various consumptive uses during the year consists of evaporation from open water bodies, swamps and ET from crop land, nonrecoverable deep percolation, and the "net" of water used by cities and rural areas for domestic and industrial uses minus the "return flows" to the natural system in the form of wastewater. Here, the outflows from cropland would NOT consider the water directly used by the cropland from rainfall (effective precipitation or the use of water naturally available in the soil profile by the crops) and would only consider the consumptive use from irrigation of the crops grown during the three seasons.

The runoff as part of the total inflow ("virgin flows") can be estimated by adding up the "observed flows" and the "effective diversion" by the major reservoirs, other storages, and diversion points in the basin. The effective diversion would be the total water diverted from the rivers and tributaries for various purposes minus the estimated return flows to the stream. The return flows can be from irrigation commands and urban centers. The "total water diverted from rivers and tributaries" can be estimated using data on reservoir releases, river lifting, reservoir evaporation, and the net storage change during the hydrological year. There are 13 major and medium reservoirs in Luni river basin.

The water accounts for the basin can be estimated as:

$$\begin{aligned}\mathrm{INFLOW_{TOTAL}} = {} & \mathrm{CU_{IRRIGATION}} + \mathrm{CU_{RURAL-DOMESTIC}} + \mathrm{CU_{URBAN}} \\ & + \mathrm{CU_{LIVESTOCK}} + \mathrm{CU_{INDUSTRY}} \\ & + \mathrm{EVAP_{RESERVOIR}} + \mathrm{OUTFLOW_{STREAM}} \\ & + \mathrm{GWS_{CHANGE}} + \mathrm{SC_{RESERVOIR}} \quad (6.1)\end{aligned}$$

$$\mathrm{INFLOW_{TOTAL}} = \mathrm{VFLOW_{STREAM}} + \mathrm{GWR_{RENEW}} \qquad (6.2)$$

But, in this case, the estimation of renewable recharge should not consider the recharge from command area, as this would lead to double counting. The virgin flow ($VFLOW_{STREAM}$) can be estimated as:

$$VFLOW_{STREAM} = OUTFLOW_{STREAM} + EWD \tag{6.3}$$

If we assume that the return flows from irrigated fields only contribute to groundwater recharge in the command area, return flows from irrigation schemes can be treated as zero. However, it can be considered that the wastewater outflows from urban areas return to the streams.

In such situations, EWD can be estimated as the sum of total water released from reservoirs, water lifted from diversion points along the stream/river through the KT weirs and other river lifting points, evaporation from the reservoirs and their annual (+ve) storage change minus the wastewater return flows from urban areas:

$$EWD = RELEASE_{CANAL} + LIFT_{RIVER} + EVAP_{RESERVOIR} + SC_{RESERVOIR} - RF_{WASTEWATER} \tag{6.4}$$

The consumptive use of water in urban area (CU_{URBAN}) can be treated as 80% of the total water supplied to meet the municipal water needs, whereas all the water supplied to meet the rural domestic water needs can be considered as the $CU_{RURAL-DOMESTIC}$.

Irrigation includes four components, viz., beneficial evapotranspiration by crops (ET); nonbeneficial evaporation from the soil (both from the soil not covered by canopy and the barren soil in the field after crop harvest); nonrecoverable deep percolation (also the water flows into saline formations); and return flows to streams or groundwater system, which can be recovered for reuse. How much of the water applied in the field would be available for these components would be determined by the technical efficiency with which water is applied (Allen et al., 1998; Kumar and van Dam, 2013). Therefore, consumptive water use from irrigation includes three major components,ET, nonbeneficial evaporation, and nonrecoverable deep percolation:

$$\text{Irrigation water consumed in crop production } (CU_{IRRIGATION}) = AX\left[\Delta_{IRRIGATION} - \{(\Delta_{APPLIED} - ET) \times F\}\right] \tag{6.5}$$

Here, $\Delta_{APPLIED}$ is the sum of irrigation dosage ($\Delta_{IRRIGATION}$) and total soil moisture available from rainfall (also known as effective rainfall). For purely irrigated crops of winter and summer, the effective rainfall (P_{EFF})

can be assumed to be zero. A is the cropped area. The factor "F" is introduced to take account for the fraction of the total water applied in excess of crop water requirement, which is available for reuse.

In regions with subhumid climatic conditions, the nonbeneficial evaporation from the barren soil will be insignificant. Though a small part of the basin has semiarid climate, we ignore the effect of the same for simplifying the analysis. Again, since the area has shallow water table conditions, the deep percolation from the irrigated field will be recoverable, meaning water applied in excess of ET would return to groundwater.

Hence, the value can be assumed as 1 (one). In the latter case, real water saving from the use of microirrigation systems will be negligible. Given the fact that a large part of the basin has arid climatic conditions, and groundwater table is deep, the value of F can be considered as zero, which means the excess water applied in the field would not contribute to groundwater recharge. In other words, the irrigation in excess of the irrigation requirements would be lost in soil evaporation and nonrecoverable deep percolation. For using this procedure, the value of depth of irrigation should be known from primary survey.

Alternatively, the irrigation application $\Delta_{IRRIGATION}$ can be estimated as the difference between ET and effective rainfall of the crop (using FAO CROPWAT model), plus the extra water required to take care of field application efficiency.

$$\Delta_{IRRIGATION} = (ET - P_{EFF})/IE_{APPLICATION} \tag{6.6}$$

Since we do not have the estimates of irrigation water application, the estimate of irrigation water requirement based on FAO CROPWAT was treated as the consumptive water use from irrigation (without factoring in water application efficiency). It was assumed that the allowance in irrigation dosage provided for taking care of inefficient water application in the field would be recoverable from the shallow aquifer, and therefore would not change the final values of consumptive use of irrigation water applied.

6.6 Data used for the analysis and sources

Data on groundwater resources (block-wise) in the basin were obtained from Groundwater Survey and Development Agency (GSDA) of Maharashtra, which are the estimates provided by GSDA, which were validated by Central Ground Water Board. Data on point rainfall for different locations in the basin for the period from 1998 to 2012 were obtained from http://mahaagri.gov.in/rainfall/hlevelSelect1.asp (*Source*: Department of Agriculture, Government of Maharashtra). Data on stream-flows for 1986–87 to 2009–10 were obtained from CWC's published hydrological data books. Time series data on reservoir inflows, live storage, volumetric water utilization for various uses, and reservoir evaporation for Warna irrigation scheme for 5 consecutive years (i.e., 2006–07 to 2007–08, 2008–09, 2009–10, and 2010–11) were obtained from the water audit reports of major and medium irrigation projects of Maharashtra (GoM, 2007, 2008, 2009, 2010, 2011). Data on area under different irrigated crops (for the year 2011–12) were obtained from Directorate of Agriculture, Eco-state division, Government of Maharashtra. Blockwise data on population for the blocks/districts falling in the basin were obtained from Census 2011 report. In addition to these, maps showing isohyets and variation in reference evapotranspiration (ET_0) were obtained from various online sources and report of the study "Sustainable Livelihoods and Biodiversity in Developing Countries," Deliverables 5.1 (Phuong and Xuan, 2011). In addition, some scientific papers available on topics such as rainfall-runoff modeling and groundwater recharge (Athavale et al., 1983; Krishnaswamy et al., 2012; NIH, 1998–1999; Raju et al., 1979) were used.

6.7 Analysis of water accounts of Warna sub-basin

6.7.1 Basin outflows, reservoir storage and water utilization pattern

The data on annual streamflow recorded at Samdoli gauging station, which is near the confluence point of Warna river with the Krishna river, are available for the period from 1986–87 to 2009–10. They are analyzed and presented graphically in Fig. 6.5. The highest streamflow during the 24-year period (5839 MCM) occurred during 2005–06, when all the six rain-gauging stations in the basin experienced excessively high rainfall.[a] The runoff rate for the year was 2.79 m. These stations recorded very high rainfall during the successive year (i.e., 2006–07) also, which was reflected in the second highest streamflow (5650 MCM) observed during 2006–07. But, the streamflow declined sharply in the subsequent years, as the rainfall was much lower as compared to 2005–06. The lowest runoff observed in the basin during the 24-year period was in 2002–03 (1361 MCM)

[a] The rain-gauging stations in the basin are Shahuwadi, Panhala, Hatkangle, and Shirol in Kolhapur and Miraj, Valva, and Shirale in Sangli district.

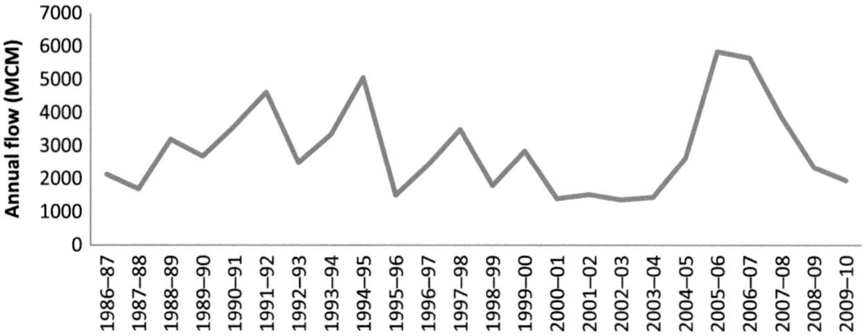

Fig. 6.5 Streamflows in Warna sub-basin, at Samdoli (1986–87 to 2009–10). (Source: *Authors' own analysis based on CWC (2006, 2007, 2009, 2012).)*

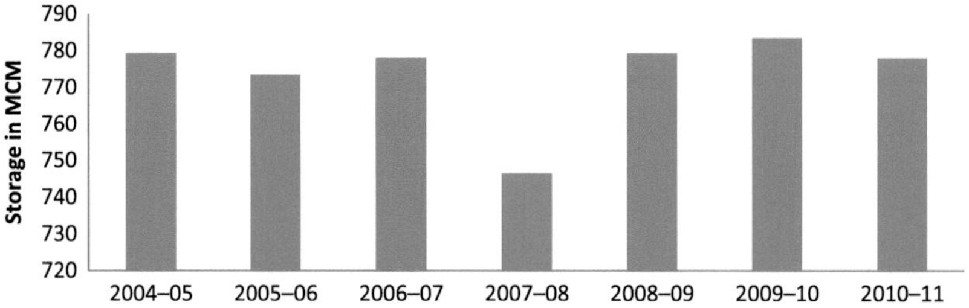

Fig. 6.6 Live storage of Warna reservoir (2004–05 to 2010–11).

(a drought year), in which most parts of the basin experienced very low rainfall. This corresponds to a runoff rate of 0.66 m.

But, these figures do not provide clear indications of the total runoff generated in the basin. The reason is the flows are altered by a major reservoir (Warna reservoir), which diverts water for irrigation through canals, in addition to several small water-lifting schemes along the river. Since the command area of this major irrigation scheme is not fully developed, the potential utilization is very low. However, there are several lift irrigation schemes, which utilize the water released from the reservoir irrigating a total area of around 1.18 lac ha using river lifting and canal lifting. We assume that the total amount of water "effectively diverted" by the reservoir is equal to the gross storage capacity of the dam.

Inflows in Warna reservoir for the period from 2004–05 to 2010–11 are presented in a graphical form in Fig. 6.6. The 7-year data do not show any sharp variations in live storage of the reservoir. The lowest live storage was 746.5 MCM, while the highest was 783.4 MCM (during 2009–10).

One reason for this is the carryover storage from the previous year. The carryover storage is significant in most years, ranging from a lowest of 149.5 MCM in 2008–09 to a highest of 357.44 MCM in 2010–11.

In order to understand the actual utilization of water from Warna reservoir, it is important to look at the water audit report of the reservoir (which is available for 5 consecutive years). It provides data on the following: the total live storage on October 15 of every year soon after the monsoon inflows; total amount of water utilized for the river lift irrigation schemes of the government, the water released for canal irrigation during winter and summer, water released for nonagricultural uses, evaporation from the reservoir; the inflow received from the upper catchment during the lean season; and the total amount of storage in the reservoir at the end of the hydrological year (i.e., June 30). However, the data on water utilization do not confine to the geographical area of the basin alone. It also includes areas outside the basin that are being served by the reservoir and lift irrigation schemes.

Since there is no irrigation water use in the basin during kharif season, the outflows from the reservoir during the monsoon season would go up-captured and would get recorded in the last drainage point in the basin. The irrigation demand is during the winter and summer. Since there is no outflow at the last drainage point of the basin in any of the years, it can be assumed that the water releases from the reservoir through canals and river for gravity irrigation, river lift irrigation (by both farmers and the LISs), and nonirrigation uses are fully utilized and there is no effective return flow to the main trunk of the river. Hence, by applying the simple continuity equation, the total water utilized from the reservoir can be estimated. The estimates are provided in Table 6.3 (Column 9). The annual water utilization varies from a highest of 710.9 MCM in 2006–07 to 420.55 MCM in 2010–11. The table provides us the estimates of volumetric consumption of water released from the reservoir by private lift irrigators (Column 10).

In order to estimate the effective water diversion from Warna river during 2009–10 due to the presence of the reservoir, we consider the storage change in Warna reservoir during 2008–09 and 2009–10, which is 265.55–149.52, i.e., 116.03 MCM. Hence, going by Eq. (6.4) defined in the methodology section, the effective water diversion from the river is 692.03 MCM.

Since the command area development of Warna irrigation project is not yet completed, a large share of the water from the reservoir is released in the river for the lift irrigation schemes of the government. They utilize a significant share of the water from the reservoir scheme. The volumetric water use by the LISs ranges from 250.69 MCM in 2010–11 to a highest

Table 6.3 Effective water utilization for various sectors and storage changes in Warna reservoir.

Hydrological year	Live storage as on Oct. 15 (MCM)	Inflow during lean season (MCM)	Balance storage on June 30 (MCM)	Reservoir evaporation (MCM)	Irrigation through		Nonirrigation use (MCM)	Total water utilization (MCM)	Water use for river lift
					Canals (MCM)	LIS (MCM)			
2006–07	778.03	190.02	257.15	22.85	7.66	343.43	7.50	710.90	329.45
2007–08	746.52	89.83	258.99	20.80	9.54	264.01	5.46	577.36	277.55
2008–09	782.06	0.0	149.52	22.41	8.35	287.80	5.35	632.54	308.63
2009–10	783.41	58.24	265.65	22.96	7.00	348.78	8.57	576.00	188.69
2010–11	777.99	0.0	357.44	21.99	6.81	250.69	9.89	420.55	131.18

(*Source*: Authors' own estimates based on Reports on water auditing of irrigation projects of Maharashtra state for 5 years from 2006–07 to 2010–11.)

of 348.78 MCM in 2009–10. Over and above this, there is large amount of water, which farmers directly lift from the river for irrigation. This water available for lifting is also from the reservoir releases, as the river doesn't have flows during times of the year when there is demand for irrigation. The volumetric diversion from the river ranges from 329 MCM in 2006–07 to 131.18 MCM in 2010–11. Not all that water might get used up in the basin. Substantial amount of water is taken out of the basin for irrigation purpose, as the basin does not have extra arable land, which lies uncultivated and which can absorb the water.

Under lift irrigation, water application is more controlled than under gravity irrigation and therefore watering of the fields does not lead to excessive surface and underground drainage. Since slopes in the area receiving water from lifting are mild, field runoff does not expect to return to the streams. However, since the irrigation service provided through the LIS is highly subsidized, water intensity is unlikely to be the main consideration for farmers in selecting their cropping pattern.

Farmers receiving water by gravity tend to grow highly water intensive crops such as paddy, which are also less sensitive to water availability and stress. This is due to the comparatively poor reliability of water supply under gravity irrigation and inadequate control over water application. Further, since the farmers are not confronted with marginal cost of using irrigation, there is less incentive to adopt water-efficient crops and efficient irrigation methods. Since irrigation dosage is likely to be excessive, the chances for return flows from irrigated fields are high. Such return flows would be particularly high for paddy. But, because of plain topography of the areas receiving canal water, the chances of surface runoff would be less, and the excess irrigation dosage is likely to end up in the shallow aquifers. The alluvial nature of the aquifer further results in steady replenishment of the wells in the command area, benefiting both the farmers in the designated command area and farmers dependent on only wells.

This is evident from Table 6.3, which provides estimates of groundwater resource availability in Warna sub-basin. As Table 6.4 indicates, the non-command areas have higher degree of exploitation of groundwater. This could be because of two reasons. First: in the noncommand areas, which are mostly in the hilly and hard rock terrain (basalt and laterite), the utilizable recharge figures generally tend to get overestimated. Second: in the command areas, which are located in the alluvial plains, good replenishment of groundwater also encouraged abstraction. Table 6.4 also indicates that the overall groundwater balance is positive, with a net change in storage of 2662 ha m (i.e., $5806-3144=2642$), i.e., 26.44 MCM.

Table 6.4 Ground water availability and draft in Warna river basin (ha m per year).

		Renewable groundwater recharge			Gross groundwater draft		
District	**Block/taluka**	**Noncommand**	**Command area**	**Total**	**Noncommand**	**Command area**	**Total**
Kolhapur	Hatkanangale	908.53		908.53	784.76		784.76
Kolhapur	Shirol	123.93		123.93	108.08		108.08
Kolhapur	Shahuwadi	1448.20		1448.20	411.21		411.21
Kolhapur	Panhala	317.55		317.55	243.82		243.82
Sangli	Miraj	417.08	21.48	438.56	390.96	22.26	413.23
Sangli	Walwa	947.02	190.79	1137.81	620.67	67.75	688.42
Sangli	Shirala	1427.79	3.62	1431.41	492.21	2.78	494.99
Total		5590.1	215.9	5806.0	3051.7	92.8	3144.5

(*Source*: Central Groundwater Board and Groundwater Survey and Development Agency, Maharashtra.)

Table 6.5 Difference in (irrigated) cropping pattern among different categories of farmers.

	Total cropped area of farmers using			% of area under irrigated crops of farmers using		
	Canal	Well	River	Canal water	Well water	River lifting
Paddy	7.66		31.56	26.83	0.00	26.16
Wheat	0.175		5	0.61	0.00	4.14
Jowar	0.625	0.175		2.19	1.01	0.00
Barley	0.0075			0.03	0.00	0.00
Maize	1.881		16.25	6.59	0.00	13.47
Groundnut	0.5	0.0043		1.75	0.02	0.00
Green gram	0.125			0.44	0.00	0.00
Bengal gram			6	0.00	0.00	4.97
Chick pea			2	0.00	0.00	1.66
Maize			6	0.00	0.00	4.97
Fodder			1.5	0.00	0.00	1.24
Sugarcane	17.575	17.15	52.32	61.56	98.97	43.37

(*Source*: Authors' own estimates based on primary survey in four locations in Warna river basin.)

The well irrigators in the sub-basin are mostly concentrated in the lower zone (zone 3), owing to the alluvial strata and the influence canal irrigation on groundwater replenishment. Well irrigators are likely to adopt crops, which are water sensitive, owing to the fact that there is greater access to and control over irrigation water as compared to lift irrigation and gravity irrigation. The irrigation wells are shallow open wells. The continuous recharge from canals ensures high water table and sustainable well irrigation. Well irrigation on the other hand prevents waterlogging conditions. Table 6.5 shows that farmers using well as a source of irrigation allocate most of their land for sugarcane, which yields higher returns as compared to paddy and other cereals, as compared to those who access canal water and water from lift irrigation schemes.

6.7.2 Water consumption in domestic sector

The water demand for domestic uses in rural areas is estimated considering a minimum water requirement of 70 L per person per day. The basin being water abundant, such levels of water use are quite probable. Table 6.6 provides the taluka-wise domestic water demand in rural areas. Maximum water

Table 6.6 Annual domestic water use (rural).

Taluka	Rural population A	Domestic water demand for rural area in lpcd $D = A \times 70$ lpcd	Domestic water demand per year (rural)	
			$D \times 365$ days (in 000′ liters)	In billion liters
Shirala	41,577	2,910,390	1,062,292.4	1.1
Walwa	39,799	2,785,930	1,016,864.5	1.0
Miraj	12,649	885,430	323,182.0	0.3
Shahuwadi	64,130	4,489,100	1,638,521.5	1.6
Panhala	15,073	1,055,110	385,115.2	0.4
Hatkanangale	37,920	2,654,400	968,856.0	1.0
Shirol	6304	441,280	161,067.2	0.2
Total	217,453	15,221,710	5,555,924.2	5.6

demand is from Shahuwadi taluka and minimum is from Shirol taluka, both in Kolhapur district. The total domestic water demand in rural areas of the basin is estimated to be 5.6 billion liters per annum or 5.6 MCM per annum. Further, in view of the fact that there are no centralized wastewater collection systems in rural areas, it is assumed that the return flows do get back to the natural water systems.

6.7.3 Water consumption by livestock

The estimates of total water consumption by livestock are based on the animal body weight and annual average of the daily minimum temperature of the region. Table 6.7 presents the estimates of total number of livestock and water consumption by different categories of livestock in Warna sub-basin. Maximum water demand for livestock is from Shahuwadi followed by Shirala, Walwa, Hatkanangale, Miraj, Panhala and least from Shirol taluka. The total livestock water demand in the basin is estimated to be about 1065 million liters (1.065 MCM) per annum.

6.7.4 Water consumption in irrigation (only irrigation water)

Table 6.8 presents the estimates of water consumption by irrigated crops in Warna river basin. The estimation used FAO CROPWAT mode for estimating the PET of crops and effective rainfall, in order to arrive at the irrigation water requirement. The estimation is based on the assumption that the total water depleted in crop production is same as the PET, or in other words, the farmers apply sufficient water to meet the entire ET demand of

Table 6.7 Water consumption by livestock in the basin.

Taluka	Crossbreed cow		Indigenous cow		Buffaloes		Sheep		Goats		Horse		Others		Total water demand (million liters/year)
	No.	Water requirement (000′ liters per annum)	No.	Water requirement (000′ liters per annum)	No.	Water requirement (000′ liters per annum)	No.	Water requirement (000′ liters per annum)	No.	Water requirement (000′ liters per annum)	No.	Water requirement (000′ liters per annum)	No.	Water requirement (000′ liters per annum)	
Shirala	3242	39,982.2	2826	34,736.2	9491	137,773.1	86	98.0	1881	2155.4	424	4853.4	48	55.3	219.65
Walwa	1441	20,816.2	1552	19,297.8	10,725	155,333.6	2491	2855.2	3527	4042.5	7	79.4	291	333.4	202.76
Miraj	1953	22,956.9	2335	27,438.5	2130	29,112.0	378	433.6	1429	1637.5	116	1324.4	84	96.3	83.00
Shahuwadi	2129	31,298.7	6609	82,775.3	14,570	207,141.9	1300	1489.3	8264	9470.2	23	260.7	16	18.5	332.45
Panhala	894	13,131.1	751	9327.9	3420	51,404.7	633	725.4	867	993.3	5	53.6	2	2.4	75.64
Hatkanangale	949	13,712.8	863	10,838.6	6777	97,726.8	3785	4338.0	1928	2209.9	50	574.2	184	211.3	129.61
Shirol	134	1912.4	157	2001.5	1220	17,670.6	267	306.1	280	320.7	7	76.6	6	7.3	22.30
Total	10,742	143,810.3	15,093	186,415.8	48,333	696,162.7	8940	10,245.6	18,176	20,829.5	632	7222.3	631	724.5	1065.41

(*Source*: Authors’ own estimates based on secondary data on livestock population and estimation of water requirement of different categories of livestock.)

Table 6.8 Irrigation water consumption over a decade.

	Irrigated area of the crop (ha)		Water demand (MCM)	
Name of the crop	**2000–01**	**2010–11**	**2000–01**	**2010–11**
Rice	6821.2	9005.7	64.835	84.465
Wheat	1264.5	1195.4	4.986	5.116
Kharif sorghum	2971.1	1611.2	1.396	0.749
Rabi sorghum	5418.1	1528.3	20.857	6.015
Pearl millet	775.6	0.0	0.723	0.000
Maize	1168.1	1494.8	1.197	1.821
Common millet	202.3	0.0	0.082	0.000
Finger millet	1184.8	1191.6	0.529	0.458
Foxtail millet	68.6	0.0	0.064	0.000
Little millet	45.3	0.0	0.020	0.000
Other cereals	223.3	610.3	0.161	0.278
Whole Bengal gram	1361.5	3151.4	1.026	2.793
Red gram	155.0	142.6	0.120	0.108
Green gram	369.9	278.8	0.287	0.237
Black gram	514.5	222.0	0.361	0.178
Horse gram	172.6	0.0	0.139	0.000
Kidney beans	432.0	0.0	0.383	0.000
Cowpea	155.6	0.0	0.103	0.000
Chickpea	8.6	0.0	0.005	0.000
Other pulses	142.2	520.5	0.108	0.375
Sugarcane	9202.2	8295.6	116.396	110.389
Cotton	25.5	1.2	0.079	0.004
Roselle	11.6	0.0	0.034	0.000
Other fiber crops	2.8	0.0	0.010	0.000
Groundnut	5699.7	3251.1	5.171	3.469
Soybean	2149.8	4941.9	3.635	7.253
Coconut	1.6	33.0	0.006	0.117
Sunflower	150.5	94.1	0.683	0.394
Niger seed	214.9	28.0	0.768	0.128
Other oil seeds	4599.7	144.5	16.431	0.548
Total	45,512.8	37,741.9	240.592	224.894

(*Source*: Authors' own analysis based on data provided by district agricultural department.)

the crops and in situations where water application exceeds the shortfall of soil moisture availability, the excess water percolates downs to join the shallow aquifer. The field investigations carried out in four locations, spread across the basin, involving interview of farmers also did not bring out the problem of water availability really affecting irrigated crop production.

Overall, between 2000–01 and 2010–11, irrigated area has decreased from 45,500 to 37,700 ha. This is mainly due to decrease in irrigated area under rainy season crops, which include sorghum, pearl millet, other small millets, and pulses. Further, area under irrigated oil seeds has decreased by 4400 ha. However, irrigated area under paddy, finger millet, Bengal Gram, and soybean has increased. As per the latest cropping pattern (2010–11), irrigation water use in the basin is estimated to be around 225 MCM. Therefore, there is some reduction in irrigation water. The reduction is to the tune of 6.5%.

6.7.5 Summary of water accounts

The total surface inflow in the basin for the year 2009–10 (the year for which the most recent data are available) is the sum of annual outflow and the total effective water diversion from Warna reservoir (1949 MCM) for the hydrological year. It was thus estimated to be 1949.0 + 692.03 MCM, i.e., 2641.03 MCM. As per the mean annual rainfall in the catchment (as shown in Fig. 6.2), the total rainfall falling in Warna sub-basin would work out to be around 3550 MCM per annum. As the rainfall in 2009–10 was a little above the normal, an increase of 10% can be reasonably assumed. Hence, the total quantum of precipitation works out to be 3904 MCM for that year. This makes the total runoff 67% of the annual rainfall. These figures need to be carefully examined as they mainly determine the availability of water for various uses in the sub-basin. But, there are no studies available on Warna, which either estimated or can be used to estimate the runoff (stream flows) possible in the basin. However, there are two important studies available for similar basins in the Western Ghats, which we would like to discuss here.

A recent study in microcatchments inside an east-flowing western ghat river of Kodigibail in North Karnataka for 3 consecutive years (i.e., 2003, 2004, and 2005) showed (surface) runoff coefficients ranging from 21.9% to 40.3% in 2003 (for a rainfall of 2252 mm); 21% to 43.2% in 2004 (for rainfall of 2848 mm); and 28.8% to 67.6% in 2005 (rainfall 3662 mm), for natural forests to degraded forests, respectively. The catchments with (Acacia) plantations had runoff values, higher than that of natural forests but less than that of degraded forests in 2004 and 2005 (*Source*: based on fig. 6.4, a, b, and c in Krishnaswamy et al., 2012: p. 226). However, these hydrological observations were limited to catchments of first order streams—with size varying from 4 to 10 ha. Hence, the stream discharge did not include the base flow component. But, for rivers originating from the western ghat, the base flow component of stream flow (runoff) could be quite substantial. As per the study carried out by National Institute of

Hydrology (1998–1999),[b] the base flow component was estimated to be ranging from 44% for 1000 mm of annual rainfall to 24.5% for 4000 mm of annual rainfall in Netravati river basin, and 11% for 500 mm of annual rainfall to 14.5% for 4000 mm of annual rainfall in Dandavathi river basin (NIH, 1998–1999). Hence, the figures of runoff coefficient estimated by Krishnaswamy et al. (2012) cannot be reckoned with for checking the veracity of our estimates.

The NIH study, which had modeled the rainfall-runoff relationships for five basins, based on daily discharge measurements and the estimated base-flow component of the runoff, further showed that in all the five basins, the runoff coefficient increased linearly with increase in magnitude of rainfall. For the catchment of Dandavathi, which is an east-flowing river of Krishna, the runoff coefficient, as per the rainfall-runoff model, ranged from 0.16 (for 500 mm) to 0.755 (for 3000 mm) to 0.81 (for 4000 mm). Given the rainfall regime of the basin under investigation (600–4000 mm), the climate and the predominantly sloppy terrain, the estimates of basin yield seem to be quite reasonable. However, detailed long-term studies of water inflows and outflows will have to be carried out for the three physiographic-cum-agroclimatic units of the basin by separately considering the rainfall-runoff relationships for these units to obtain more accurate results.

The total annual groundwater replenishment in the basin was estimated to be 58.06 MCM, which includes the additional recharge from irrigation return flows in the command area. We have estimated the additional recharge in the canal command area to be 0.922 MCM. This was based on the assumption that the incremental recharge rate observed in the canal command area (of around 4500 ha) is due to recharge from return flow. The incremental recharge was estimated to be 0.02 m (0.047–0.027) and considering an area of 45 km^2 of recharge area, the incremental recharge from return flow was estimated by multiplying the area by recharge rate (45×0.0206). Hence, the net renewable recharge from precipitation was worked out to be 57.14 MCM.

The estimates of net change in groundwater balance are available from groundwater assessment reports, as $58.06 - 31.44 = 26.62$ MCM. The change in reservoir storage during the year 2009–10 (for only Warna) is 116.03 MCM. The total water consumption in agriculture within the basin area

[b] The NIH study covered five large catchments, viz., Nethravathi, Sithanathi, Malaprabha, Dandavathi, and Barchi. The lowest catchment area was for Dandavathi (118.88 km^2), and the highest was for Nethravathi (3657 km^2) (NIH, 1998–1999).

was estimated to be 224.89 MCM for the year 2010–11 for which land use data were available. This can be assumed as the water consumption in the previous year, i.e., 2009–10 as well as one does not expect any major change in land use and water use in 1 year. The total water use in agriculture, rural domestic, and livestock sectors put together therefore is 231.06 MCM. The water accounting study did not consider urban water use as the town in the downstream of Warna secures its water supplies from Krishna river.

The evaporation loss from Warna reservoir is 22.96 MCM as per the water audit report (GoM, 2010). The nonirrigation use of water from the reservoir is 8.57 MCM (GoM, 2010), and the unutilized flow in Warna at the large drainage point is 1949 MCM (CWC, 2012). The annual storage change (+ve) in groundwater is 26.44 MCM. By applying the continuity equation, the difference between the left- and right-hand side of the equation is 343.93 MCM. This could be roughly the amount of water taken out of the basin for agriculture. The final result of water accounting is shown in Fig. 6.7 and Table 6.9. From Fig. 6.7, it is evident that agricultural water consumption within the basin is only 8% of the total renewable water in the basin.

6.8 Findings

Warna, which is a sub-basin of Krishna river basin of Peninsular India, is considered to be a water abundant sub-basin, as per the classification made by the water resource department of Maharashtra. The annual renewable water resource of the basin, including groundwater, for the year 2009–10 was estimated to be 2698 MCM.

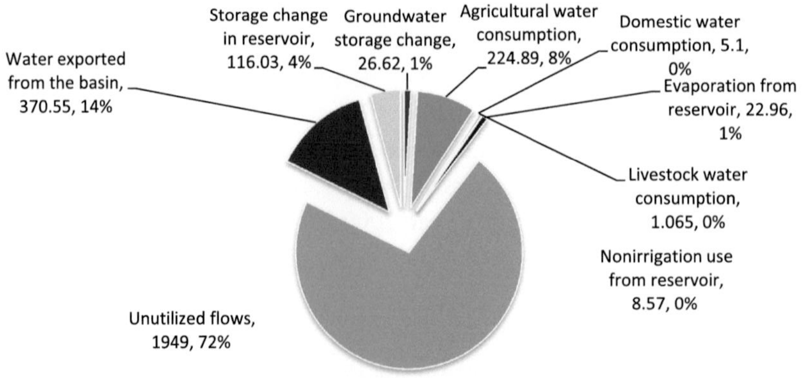

Fig. 6.7 Water accounts for Warna sub-basin: 2009–10.

Table 6.9 Water accounts of Warna sub-basin (2009–10).

Components of water accounts	Amount in MCM
Total inflows	2698.165
Outflows	
Agricultural water consumption	224.89
Domestic water consumption	5.10
Livestock water consumption	1.065
Evaporation from reservoir	22.96
Water diversion for nonirrigation use from reservoir	8.57
Storage change	
Unutilized stream flows	1949.00
Groundwater storage change	26.44
Water exported from the basin	343.93

(*Source*: Authors' own estimates based on secondary data.)

Surface water accounts for a major component of the total renewable water resources of the basin (97.9%). The upper catchment of the basin, though small in size, is the main source of inflows into Warna reservoir, generates nearly 900 MCM of water during the monsoon season every year, by virtue of the excessively high rainfall, humid climate, and rocky terrain in the basin's upper catchments. Groundwater accounts for a small fraction (2.1%) of the total renewable water resources in the basin.

In spite of heavy rainfall, which large parts of the basin receive, gross groundwater recharge is very small, i.e., 58.06 MCM, due to the hard rock geology and sloppy terrain in the upper catchment. This quantity includes recharge due to return flows (0.922 MCM) from surface irrigation in the command areas of lift irrigation schemes and gravity irrigation.

The rapid water accounting study carried out for Warna sub-basin shows that the total amount of water consumed in four major competitive use sectors, i.e., agriculture, domestic use, livestock use, and industrial water use, within the basin is much less than the difference between annual renewable water resources of the sub-basin, the outflows, and the storage change (carry over storage) in Warna reservoir. The difference is to the tune of around 343.93 MCM for 2009–10. This could be the quantum of water exported from the basin for irrigation through river lift schemes and gravity irrigation scheme. However, since the data on the actual amount of area served by the scheme inside and outside the sub-basin are not available, this estimate could not be crosschecked.

Agriculture is the largest user of water in the sub-basin. Paddy and sugarcane account for a major portion of the irrigation water use in the basin, and the total agricultural water use was 224.9 MCM in 2009–10, which came down marginally from 240.6 MCM in 2000–01. There are a wide range of cereals, vegetables, fruits, and cash crops grown in the basin area, by virtue of the vast difference in climatic conditions from the upper to the lower catchment. Nevertheless, sugarcane consumes the largest amount of water.

A quick analysis of the water audit reports of Warna reservoir scheme for 5 hydrological years, i.e., from 2006–07 to 2010–11 shows that the reservoir live storage varies only marginally over the years in spite of significant variations in rainfall in the catchment, as the storage system harnesses only a portion of the dependable flows. Though the audit reports indicate the presence of inflows during lean season in some of the years, analysis suggests that the reservoir has sufficient quantum of storage even at the end of the hot summer, i.e., June 30, most of which is due to the unutilized storage from monsoon flows. The evaporation from Warna reservoir is a small fraction of the total inflows impounded by it—in the range of 2.5%–3%.

The analysis of discharge at the last drainage point of Warna river shows that there is significant amount of outflow contributing to the stream flow of Krishna, 1949 MCM in 2009–10. This is 72% of the total annual flows in the sub-basin. The amount varies from a lowest of 1361 MCM in the driest year (2002–03) to 6739 MCM during the wettest year of the observation period (2005–06). However, there is no stream flow during the lean season. This suggests that all the water released from Warna reservoir for direct lifting from the river for irrigation gets consumed before it reaches the last drainage point. This has serious implications for the in-stream uses of water of the river, particularly fisheries. There are reports of land use changes (particularly agricultural expansion, excessive withdrawal of water from the river, and pollution of river) resulting in large-scale dying of fish in the river, and its undesirable consequences for the livelihoods of the traditional fishing community (Mohite and Samant, 2013).

Since Warna reservoir appears to have sufficient carry over storage every year, the scope for releasing water during the lean season for meeting the environmental flow needs exist.

6.9 Implications of water accounts for water tenure in the sub-basin

Though the total drainage area is only 2097 km^2, the wide spatial heterogeneity in hydrological regime, topography, geology, climatic conditions, and socioeconomic characteristics make water resource availability and conditions with respect to water access and use in Warna sub-basin very complex. The elevation difference between upper and lower catchments is 660 m.

The upper catchment of the sub-basin has hilly terrain and is covered with good forests. The climate in the upper catchment is humid to subhumid. By virtue of the excessively high rainfall, cold and humid climate, and rocky and sloppy terrain, this part of the sub-basin yields large amount of runoff during the monsoon season. It also produces stream flows during lean season on account of base flows. For an annual rainfall of 3600 mm, the runoff from the catchment of 301 km^2 (source: Table 6.1) could be as high as 921 MCM if we consider a runoff coefficient of 0.85.

The upper catchment faces least human interference, and majority of the drainage area of the upper catchment is under a national park. The land use is mainly natural forests, and population is very low. Warna reservoir impounds water from this portion of the catchment, with a catchment area of 301 km^2, which falls entirely in the forests and Chandoli National Park area. The reservoir receives inflows much in excess of the amount needed for realizing the full storage capacity, as is evident from the observed flows in the last drainage point in the basin. This area being part of the western ghat and therefore being protected, land use there is unlikely to change over the years, posing no threat to the hydrological system. This means the rights of those who depend on the water from the reservoir would be quite secured.

Toward the lower catchment, the climatic conditions change rapidly. The major variation is in the rainfall and humidity. While the rainfall decreases sharply, the humidity also reduces. This phenomenon increases the aridity and water requirements for crop production. Since the rainfall is lower, the irrigation water demand for crops also increases gradually toward the middle and lower parts. More importantly, toward the lower parts, major change in the land use is observed, from forests to wastelands, pasture land, and agricultural land. The lower plains have more arable land and higher population density.

The major source of water toward the middle portion of the sub-basin however is flows in Warna river. The flow in Warna river is however regulated by the presence of Warna reservoir in the upper catchment area. Due to the underlying hard rock formations, groundwater availability in the middle portions of the basin is not significant. Therefore, the population in this part of the basin is heavily dependent on the river flows. A large number of KT weirs are constructed by the irrigation department across the river for lifting water for irrigation (under the LISs). The unique spatial pattern in distribution of water endowment also influences the distribution of population in the middle parts. In these parts, the human settlements are all concentrated along Warna river and its tributaries.

The land use in the lowest part of the basin is distinctly different from the upper parts. The lowest part is intensively cultivated. Unlike the upper and middle (transitional) parts, the lower parts consist of alluvium. Because of this, the water availability in the lower parts is more or less uniform. The lower part of the sub-basin is also part of the command area of Warna irrigation scheme. The water released through gravity to this part of the basin, besides adding to the water supplies for irrigation, also adds to groundwater recharge. Recharge occurs from both canal seepage and deep percolation of water from irrigated fields.

The land use and cropping systems in the lower catchment is also different from that of middle catchments, with water-intensive crops such as sugarcane and paddy occupying a significant proportion of the total cropped area. Another important factor, which influences water use in this part of the sub-basin, is the presence of a town and the higher density of population. This increases water demand in the competitive use sectors, viz., agriculture, urban and rural domestic use, and livestock use. Owing to the presence of water in the alluvial formations, water from canals and the river, and the presence of highly productive arable land, agriculture in this part of the basin is intensive and also commercially oriented.

In spite of intensive land use in the lower parts, the sub-basin has significant amount of excess flows contributing to the inflows in the lower parts of the basin. The Warna reservoir, which has a gross storage capacity of around 974.19 MCM, is still not able to utilize all the water in the reservoir for gravity irrigation, as canal network for the scheme is yet not ready. This is evident from the fact that the potential created (4500 ha) is much less than the ultimate irrigation potential of 127,700 ha. However, there is large-scale irrigation occurring in and outside the basin through river lifting and canal lifting. The total irrigation from Wakurde lift irrigation project, which is separated from Warna irrigation project, alone has a potential of 6400 ha. Hence, the gross irrigation from the water stored in the reservoir is around 111,820 ha, with a net irrigated area of 87,000 ha.

Even in situations when the entire canal system within the sub-basin is ready for utilizing the irrigation potential created, the flows in the basin may not be much altered. This is because increased use of water from canals could result in reduced river lifting, and reduced abstraction of groundwater in the lower alluvial parts. This in turn can increase the recharge to groundwater through return flows, which is now quite insignificant. The latter can result in rising of water table and groundwater outflows into the streams. However, the challenge would be in mitigating the water quality impacts. Excessive irrigation from the canals, while increasing groundwater recharge through percolation, can also lead to fertilizer and pesticide contamination of the shallow aquifer. In addition, there could be surface runoff from irrigated fields, which would directly enter the streams and pollute river water. This can seriously threaten the drinking water supply schemes based on river lifting in the downstream areas.

It is important to note that there is a significantly large population in the middle and lower reaches of the sub-basin, depending on the river, whose traditional occupation is fishing. Water quality deterioration resulting from fertilizer residues and untreated effluent from the towns can cause eutrophication and cause death of native fish population, affecting the livelihoods of these communities (Mohite and Samant, 2013). Possibilities might exist for release of unutilized water in Warna reservoir for meeting the in-stream water needs such as fisheries.

Another important potential driver of change in the sub-basin hydrology is the change in land use in the middle portion. Currently, this part of the sub-basin has less area under cropping and more area under natural vegetation such as grasses. With increasing population, this is likely to change. More of such common land would be brought under cultivation, with the result that the runoff generated in this part of the drainage area would be captured in situ in the farms. Farmers normally construct bunds to harvest the runoff over and above the direct storage of a portion of the precipitation, which they get in the soil profile (effective rainfall). This is particularly significant in view of the fact that farmers in both the districts have started allocating increasing amount of land under water-intensive crops such as sugarcane and paddy.

The sub-basin has a net sown area of 45,512 ha from the two districts of Sangli (23,237 ha) and Kolhapur (22,275 ha). If we assume that the entire cultivated land within the sub-basin area gets allocated to sugarcane cultivation, the total annual water consumption from irrigation would be 589.4 MCM. This is 364.48 MCM higher than the current irrigation water consumption in the sub-basin. The volume of water to be diverted to meet this consumptive water requirement would be even higher. Under such a scenario, there could be sharp reduction in the stream discharge from the (sub) basin, especially in low rainfall years.

Increasing urban population is likely to put enormous pressure on the reservoirs located in the western ghat region. This is because of two reasons. First: apart from being plenty, flow of the rivers of originating from the western ghat region has very high dependability given the very low variability of rainfall in the region, and as a result, the reservoirs are likely to receive full storage in every year. This means that the water supply schemes based on such sources are likely to perform much better than the groundwater-based schemes. Second: the quality of surface water from the virgin catchments like that of Warna is comparatively much better than that of water from the aquifers.

Acknowledgments

The authors are extremely grateful to Dr. Charles Batchelor (batchelorch@gmail.com) for his valuable comments on the methodology and earlier version of the report, email discussions, and for sharing some of the published literature and data relevant for the study. The authors are also highly thankful to Mr. James Batchelor for offering some important GIS-based maps on Warna.

References

Allen, R.G., Willardson, L.S., Frederiksen, H., 1998. Water use definitions and their use for assessing the impacts of water conservation. In: de Jager, J.M., Vermes, L.P., Rageb, R. (Eds.), Proceedings ICID Workshop on Sustainable Irrigation in Areas of Water Scarcity and Drought, September 11–12. Oxford, England, pp. 72–82.

Athavale, R.N., Chand, R., Rangarajan, R., 1983. Groundwater recharge estimates for two basins in the Deccan trap basalt formation. Hydrol. Sci. J. 28 (4), 525–538.

Central Water Commission (CWC), 2006. Integrated Hydrological Data Book (Non-Classified River Basins). CWC, Ministry of Water Resources, Government of India, New Delhi.

Central Water Commission (CWC), 2007. Integrated Hydrological Data Book (Non-Classified River Basins). CWC, Ministry of Water Resources, Government of India, New Delhi.

Central Water Commission (CWC), 2009. Integrated Hydrological Data Book (Non-Classified River Basins). CWC, Ministry of Water Resources, Government of India, New Delhi.

Central Water Commission (CWC), 2012. Integrated Hydrological Data Book (Non-Classified River Basins). CWC, Ministry of Water Resources, Government of India, New Delhi.

Government of Maharashtra, 2007. Report on Water Accounting of Irrigation Projects in Maharashtra State 2005–2006. Water Resources Department, Maharashtra.

Government of Maharashtra, 2008. Report on Water Auditing of Irrigation Projects in Maharashtra State 2006–2007. Water Resources Department, Maharashtra.

Government of Maharashtra, 2009. Report on Water Auditing of Irrigation Projects in Maharashtra State 2007–2008. Water Resources Department, Government of Maharashtra.

Government of Maharashtra, 2010. Report on Water Auditing of Irrigation Projects in Maharashtra State 2008–2009. Water Resources Department, Government of Maharashtra.

Government of Maharashtra, 2011. Report on Water Auditing of Irrigation Projects in Maharashtra State 2009–2010. Water Resources Department, Government of Maharashtra.

Groundwater Surveys and Development Agency (GSDA) & Central Ground Water Board (CGWB), 2011. Report on the Dynamic Ground Water Resources of Maharashtra (2008–2009). GSDA and CGWB, Pune and Nagpur.

Joy, K.J., Paranjape, S., 2011. Water resources planning and people's livelihoods in the Northern part of the Western Ghats: the case of Warna basin. In: Paper Presented at the Consultation Meeting on Water Resources of the Western Ghats by WGEEP. KFRI, Trichur.

Krishnaswamy, J., Bonell, M., Venkatesh, B., Purandara, B.K., Lele, S., Kiran, M.C., Reddy, V., Badiger, S., Rakesh, K.N., 2012. The rain–runoff response of tropical humid forest ecosystems to use and reforestation in the Western Ghats of India. J. Hydrol. 472–473 (12), 216–237.

Kumar, M.D., van Dam, J., 2013. Drivers of change in agricultural water productivity and its improvement at basin scale in developing economies. Water Int. 38 (3), 312–325. https://doi.org/10.1080/02508060.2013.793572.

Kumar, M.D., Sivamohan, M.V.K., Narayanamoorthy, A., 2012. The food security challenge of the food-land-water nexus in India. Food Secur. J. 4 (4), 539–556.

Lopez-Gunn, E., 2013. The Governance of Water Tenure: Pilot Case Study in Spain, Ppt Presentation, Prepared for the Study Commissioned by FAO. ICATALIST.

Mohite, S.A., Samant, J.S., 2013. Impact of environmental change on fish and fisheries in Warna river basin, Western Ghats, India. Int. Res. J. Environ. Sci. 2 (6), 61–70.

National Institute of Hydrology, 1999. Rainfall-Runoff modelling of Western Ghat Region of Karnataka. CS (AR)-31/98–99. National Institute of Hydrology, Jal Vigyan Bhawan, Roorkee.

Phuong, D.M., Xuan, N.T., 2011. Sustainable Livelihoods and Biodiversity in Developing Countries, Deliverables 5.1 Report, LiveDiverse GIS Database. National Institute of Agricultural Planning and Projection, Hanoi, Vietnam.

Raju, K.C.C., Kareemuddin, M., Prabhakar Rao, P., 1979. Operation Anantapur, p. 39, Miscellaneous Publ. No. 47, Geological Survey of India. published by Government of India.

Further reading

Government of Maharashtra, 2012. Report on Water Auditing of Irrigation Projects in Maharashtra State 2010–2011. Water Resources Department, Government of Maharashtra.

CHAPTER 7

Water accounting for Luni river basin, Western Rajasthan

M. Dinesh Kumar[a], **Nitin Bassi**[b], **A.J. James**[c]
[a]Institute for Resource Analysis and Policy, Hyderabad, Telangana, India
[b]Institute for Resource Analysis and Policy, Delhi Liaison Office, New Delhi, India
[c]Environmental & Natural Resource Economist, New Delhi, India

7.1 Introduction

Rajasthan is one of the most water-scarce regions in India, with very limited renewable freshwater resources and high aridity. Renewable water availability is least in the western parts of the state, which receives extremely low rainfall and experiences hyperarid climate. Part of the region is Thar desert. Yet, this part of the state has significant agricultural activities, supported by large-scale import of water through IGNP (Indira Gandhi Nehar project) canals irrigating large areas in six districts in the desert, viz., Hanumangarh, Bikaner, Churu, Ganganagar, Jaisalmer, and Barmer. The demand for water for agriculture has been growing with expansion of irrigated crop production, putting pressure on the limited freshwater resources. As surface water resources are very scarce in the river basins of this region and are already overappropriated, groundwater is intensively used in for irrigated crop production and livestock farming. Aquifer mining is a major environmental threat in western Rajasthan.

A project implemented by the government of Rajasthan with the support of the European Union during 2006–13 sought integrated water management solutions for this region, to address the growing demand-supply gap in water resources and protect the integrity of the hydrological system. The project area encompassed 10 districts of the region. Luni river basin is one of the river basins of the region, which covers seven districts viz., Ajmer, Barmer, Jalore, Jodhpur, Nagaur, Pali, and Rajsamand, some partly and some fully. A study was undertaken in Luni river basin to identify the range of IWRM solutions for the region on the supply and demand side and to evaluate the extent to which each one of these solutions would help address the water management challenges.

Current Directions in Water Scarcity Research, Volume 1
https://doi.org/10.1016/B978-0-12-814851-8.00007-0
© 2019 Elsevier Inc.
All rights reserved.

Basin-level water accounting was an important component of the study aimed at understanding the nature of water management challenges and opportunities for the basin. The chapter presents the findings of the water accounting study. It assesses the amount of renewable water resources generated in the basin annually, and then quantifies how much of this water gets used up consumptively for beneficial and nonbeneficial uses in various competitive uses and estimates how much of it goes uncaptured.

7.2 Luni river basin: A bird's eye view

Luni is one of the largest river basins in Rajasthan, which falls fully within the geographical boundaries of the state. The drainage map of the basin is given in Plate 7.1. The basin has a total geographical area of 69,302 km^2.[a] The basin covers the districts of Ajmer, Jodhpur, Nagaur, Barmer, Jalore, and Sirohi in part and Pali in full. The river originates from the western slopes of the Aravalli ranges at an elevation of 550 m above MSL, and after traversing a distance of 495 km in the south westerly direction, it disappears into the marshy land of Rann of Kachchh (Bhuiyan and Kogan, 2009). The rainfall in the basin ranges from as high as 1048 mm in the eastern slopes to a lowest of 221.50 mm in the South western side. The rainfall is highly erratic, and the mean annual rainy days varies from a highest of 38 days in the high rainfall areas to a lowest of 12 days in the low rainfall areas (Tahal Consultants, 2014).

The mean value of maximum daily temperature in the basin ranges from 26.8° to 35° and the highest value of maximum daily temperature ranges from 37.2° to 46.7°. The mean value of minimum daily temperature in the basin ranges from 12.81° to 20.9° centigrade. The lowest value of minimum daily temperature ranges from −2.05° centigrade to 6° centigrade. The annual mean value of daily wind speed ranges from 1.9 to 7.16 km/h. The annual mean of daily relative humidity is 49.2%, with the values ranging from 43.5% to 60.4%. With very low rainfall, high temperature, and low relative humidity, most parts of the basin have arid to hyperarid climatic conditions. The watershed elevation ranges from 1619 to 0 m above mean sea level (Tahal Consultants, 2014).

[a] Past studies and old official records of the government of Rajasthan report a basin area of 37,000 km^2. The report study by Tahal consultants, however, has considered a modified area of 69,302 km^2 based on terrestrial modeling. In our study, we have considered the modified basin area of 69,302 km^2, which includes large areas in the northern side.

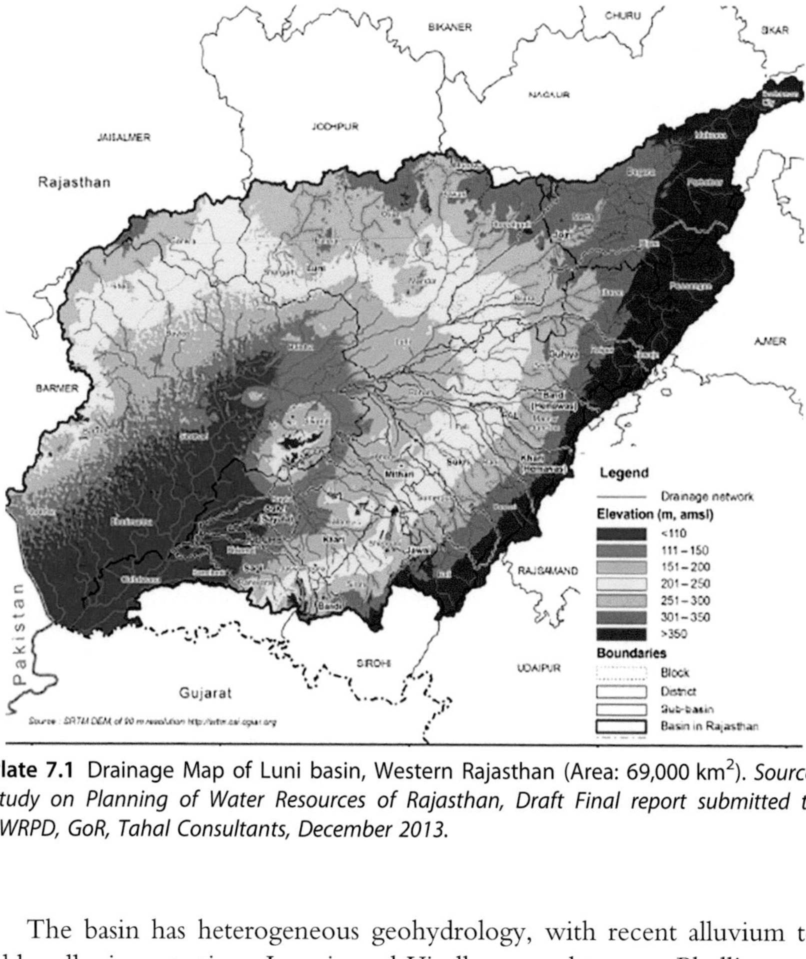

Plate 7.1 Drainage Map of Luni basin, Western Rajasthan (Area: 69,000 km^2). *Source: Study on Planning of Water Resources of Rajasthan, Draft Final report submitted to SWRPD, GoR, Tahal Consultants, December 2013.*

The basin has heterogeneous geohydrology, with recent alluvium to older alluvium, tertiary, Jurassic and Vindhyan sandstone to Phyllites and Schist. The premonsoon depth to water table in the basin varies drastically. The average premonsoon depth to water table is in the range of 20–40 m below ground level in large parts, while many parts have water table in the range of 10–20 m. In certain pockets of Jalore, Nagaur, Barmer, and Jodhpur, the water table depth is in the range of 60–80 m and 80–100 m (Tahal Consultants, 2014).

The soil in the basin is predominantly loamy sand occupying 48% of the geographical area of the basin, and the other soil types are sandy loam, silt loam, sandy clay loam, sand, loam, and clay loam. The annual potential evaporation in the basin ranges from a lowest of 1500 mm near Jawai dam to a maximum of 2600 mm in the northern and north western parts. Luni river basin as 13 sub-basins, the largest one being Luni sub-basin and the smallest is Pali sub-basin. The basin has 2 major and 11 medium reservoir schemes built for irrigation and drinking water supplies (Tahal Consultants, 2014).

7.3 The basin hydrology and groundwater resources

7.3.1 Rainfall in the basin

Topography is a very important factor influencing the occurrence of monsoon rains in Luni river basin. The Aravalli mountain range is higher than the surrounding land and so the moisture-enriched air goes up the slope, showering mostly in the eastern part and the land west of the Aravalli Range receives less amount of rainfall. More than 90% of the annual rainfall occurs during the monsoon season (June–September) alone and, in certain years, monsoon-rainfall accounts for the total annual rainfall (Das, 1996).

Analysis of point rainfall data for 13 locations in the basin for the period 1957–2012 was carried out and the outputs are presented in Table 7.1 for mean and coefficient of variation. The lowest mean annual rainfall was in Barmer (267 mm) and highest in Desuri (638.5 mm) in the South East of Pali. The coefficient of variation in rainfall, which reflects the interannual variability, ranges from a lowest of 39.1% in Ajmer to a highest of 60.4% in Barmer.

The arrival and/or retreat of the monsoon also get delayed in some years, whereas in other years one or both of the events occur early. As a result,

Table 7.1 Min, max, mean, SD, and CV of annual rainfall in Luni river basin (from 1957 to 2012).

Rain gauge station	Location in Luni river basin	Mean (mm)	CV
Ajmer	Northeast	534.9	39.1
Nagaur	Northeast	381.5	52.8
Pali	East	409.1	53.3
Jaitaran	Northeast of Pali	413.1	44.1
Desuri	Southeast of Pali	638.5	45.0
Bali	South of Pali	566.7	49.5
Sojat	Northeast of Pali	405.7	45.3
Raipur	Northeast of Pali	483.4	44.1
Marwar Jn.	East of Pali	494.6	44.7
Jodhpur	North	362.1	46.9
Sirohi	South	616.9	51.0
Jalore	Southwest	407.6	52.3
Barmer	West	267.0	60.4

Source: Authors' own analysis based on data from IMD, 1957–2012.

there is high interannual variability in monsoon in terms of amount and intensity of precipitation, distribution and pattern of precipitation, wind speed, and onset and withdrawal of the monsoon (Singh, 1994). Similarly, the number of rainy days varies from year to year and place to place in the region. On average, there are 21 rainy days in the region; in most years, the bulk of the rainfall occurs in the month of July but in certain years it shifts even to September.

The monsoon rainfall in the Luni Basin[b] follows some patterns with variations in spatial distribution. During the years of low rainfall and drought, the amount of rainfall is found to decrease gradually from east to west or from northeast to southwest (Fig. 7.1A and B). During the years in which the entire country experienced droughts (1985, 1986, 1987, 1999, 2000, and 2002), Luni basin received very low rainfall, ranging from 0.0 to 200 mm). In 1987 and 2002, the entire basin was in the grip of extreme drought (Bhuiyan and Kogan, 2009).

7.3.2 Hydrology and geohydrology

Due to low rainfall and high aridity, the basin produces very low runoff. The Luni river is an ephemeral stream and flows last for 2–3 months in a year, even in a good rainfall year. Due to large-scale water resources development in the relatively better catchments available in the south eastern parts, through small, medium, and large reservoirs, in most years, the basin does

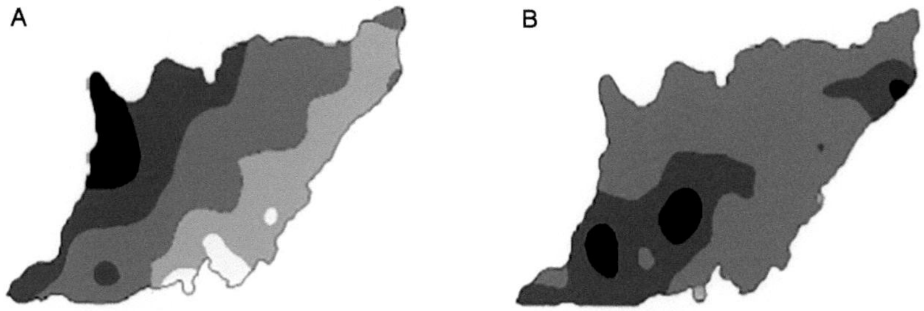

Fig. 7.1 Spatial variation in monsoonal (A) and nonmonsoon (B) rainfall in Luni river basin. *Source: Bhuiyan, C., Kogan, F.N., 2009. Monsoon variation and vegetative drought patterns in the Luni basin in the rain-shadow zone,* Int. J. Remote Sens. *31 (12), 3223–3242.*

[b] The analysis carried out by Bhuiyan and Kogan (2009) of droughts in the basin considered the drainage area, as per the old official records of Rajasthan government, i.e., 37,000 km^2.

not have outflows. The gross storage capacity of all the 13 (11 medium and 2 major) reservoirs built in the basin is estimated to be 560.37 MCM, with a total live storage capacity of 539.17 MCM. The flow of the basin (virgin flow) with 75% dependability estimated for the basin (Draft Final Report of Tahal Consultants, Vol. 3.2), is 196 MCM per annum.

But, the observed flows at Gandhav gauging station for the period from 1970–71 to 2009–10 are provided in a semilogarithmic graph in Fig. 7.2. The highest recorded flow was during 1999–91, with a total annual flow of 2200 MCM. In 14 out of 39 years for which data are available, the stream flow was zero. One important hydrological feature of Luni river basin is that the stream channels are very shallow and infiltration of the soils in the stream beds is very high. Due to this reason, the flow rates (discharge) for the streams reduce toward downstream in the river course.

Plate 7.2 shows the extent of different types of soils in the basin (source: based on Tahal Consultants report, Vol. 3.2, 2c). Loamy sand, sand and sandy loam soils account for 70% of the basin area. These soils have very high infiltration rates. Such a soil cover ensures moderately good recharge to groundwater in the foothills of Aravalli ranges, which receive high rainfall of more than 1000 mm. The average annual renewable groundwater resource in the basin is estimated to be 3.65 cm. This is quite considerable when compared to the fact that the basin is arid to hyperarid.

Groundwater abstraction in the basin has seen significant increase over the past 15 years since 1995. The estimated groundwater draft in the basin in 1995 was 2460 MCM, which went up to 2824.11 MCM in 2007 and then reduced marginally to 2717.5 MCM in 2009 (source: report of Tahal Consultants, 2013, Vol. 3.2, 2c). This is against a renewable groundwater

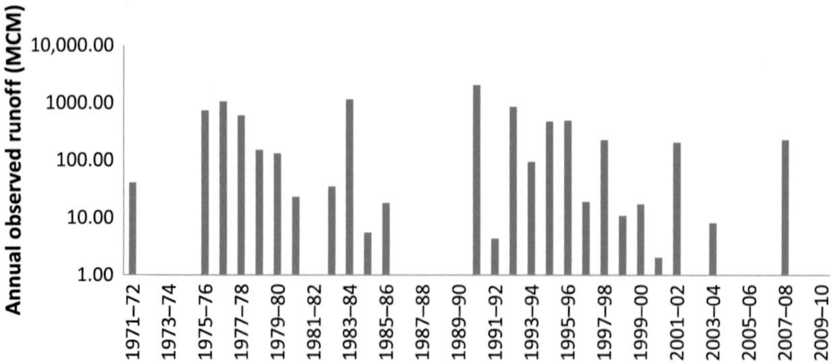

Fig. 7.2 Observed streamflows at Gandhav, Luni river basin (1970–71 to 2009–10).

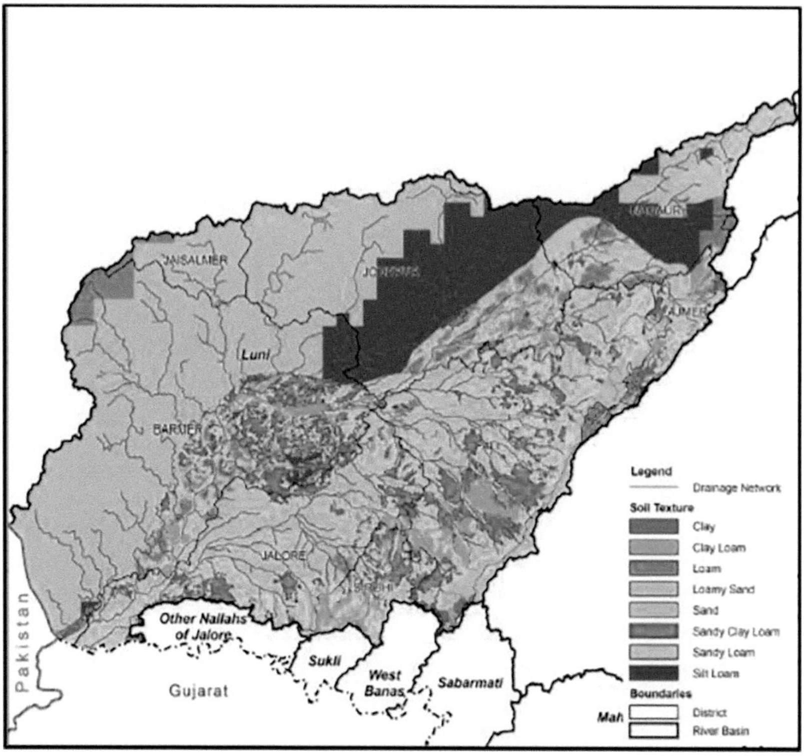

Plate 7.2 Soil cover in Luni river basin. *Source: Study on Planning of Water Resources of Rajasthan, Draft Final report submitted to SWRPD, GoR, Tahal Consultants, December, 2013.*

resource of 2203 MCM per annum. The excessively high draft in 2007 could be because the 5 consecutive years preceding 2007, i.e., 2002–03 to 2006–07 were bad years, with no surface runoff, increasing the pressure on groundwater to meet growing water demands in the basin. The basin is able to sustain such a high level of abstraction by virtue of the groundwater stock available, which is now getting mined very fast.

From the data on stream flows, and groundwater availability, it can be inferred that groundwater is the major source of water in the basin.

But, several of the groundwater bearing formations in the basin also suffer from water quality problems, such as high levels of salinity, nitrate, and fluoride. Large parts of the aquifers in the basin have EC levels ranging from 2250 to 5000 micro mhos/cm. Groundwater in only around 28% of the basin area have EC levels below, 2250 micro mhos/cm, and 72% of the area have EC levels >2250 micro mhos/cm. Only around 22% of the basin area has groundwater with EC levels <45 ppm. Further, only 30% of the groundwater underlying the basin has fluoride levels below 1.5 ppm. This essentially means that only very small areas in the basin have groundwater, which is potable.

As regards suitability of groundwater for irrigation, around 28% of the groundwater underlying the basin area has salinity below the permissible levels for irrigation, i.e., <2250 micro mhos (i.e., <1500 ppm). In around 46% of the area, groundwater has salinity in the range of 2250–5000 micro mhos (i.e., a TDS in the range of 1500–3350 ppm). Draft final report, Vol. 3.2, Tahal Consultants: 103–112). This also means that the return flows from irrigation cannot be recovered in many areas for reuse.

7.4 Socioeconomic drivers of water use in the basin

Luni river basin has a total population of 74.86 lac people, with a population density of 108 persons per km^2 (source: author's own estimates based on data on rural and urban populations in the blocks/districts falling in the basin and the proportion of the geographical area of these administrative units falling inside the basin). Urban population constitutes 36.7% of the total basin population. The economy of the districts falling in the basin is largely agrarian, with crops and livestock farming. Livestock farming is one of the most important economic activities of the region. The total livestock population of the basin is 67.66 lac animal units, nearly 68.6% of which are small ruminants (sheep and goat) (Table 7.3). Some of the districts in the basin also have industries, with dyeing, chemicals, and cement manufacturing.

Crop production in the region is mostly rain-fed, with large area under crop production during the monsoon season. However, with access to wells, irrigated crops are also grown in the region. The total irrigated area in the basin including those which are irrigated during the rainy season is only 24.7% of the total cropped area (Table 7.2). But, in lieu of the fact that rainfall variability is very high in the region, in years of monsoon failure, the crops that are grown in the monsoon season, also have to be provided supplementary irrigation. This situation also upsets the region's water balance, as the recharge and stream flows during such years would also be drastically low. The water accounts for the basin would therefore depend heavily on which hydrological year is considered for the analysis.

As regards industrial production, Pali and Jodhpur are the two most industrialized of all districts located in the basin. The other districts are Jalore, Barmer, and Nagaur. Jodhpur has a total of 24,374 industrial units, of which

Table 7.2 Groundwater resources in Luni river basin.

Name of the sub-basin	Renewable groundwater resources (MCM per annum)	Groundwater draft (MCM per annum)	Stage of groundwater development (%)	Total groundwater stock (MCM)
Bandi	39.46	52.72	133.6	137.2
Bandi (Hemawas)	34.51	51.74	149.9	16.27
Guhiya	84.18	120.06	142.6	81.97
Jawai	129.17	157.73	122.1	272.1
Jojri	116.44	137.61	118.2	522.51
Khari	142.73	140.43	98.4	273.24
Khari (Hemawas)	36.64	38.98	106.4	2.49
Luni	571.7	633.26	110.8	3464.76
Luni WRIS	797.82	726.54	91.1	5084.04
Mithari	64.62	58.98	91.3	80.7
Sagi	80.83	148.74	184.0	342.13
Sukri	55.5	60.77	109.5	29.26
Sukri (Sayala)	50.26	83.85	166.8	399.2
Total for Luni river basin	2203.86	2411.41	109.4	10,705.87

Source: Study on Planning of Water Resources of Rajasthan, Draft Final report submitted to SWRPD, GoR, Tahal Consultants, Vol. 3.2, 2c, December 2013.

Table 7.3 Socioeconomic features of Luni river basin: A quick glance.

Particulars	Rural	Urban	Total
Human population in the basin (2011)	4,739,624	2,747,058	7,486,682
Population density (persons/km^2)			108
Livestock population in the basin (2007)			67,66,726
Total cattle (indigenous, cross bred, bullocks)			11,01,670
Buffaloes			9,52,571
Sheep/goat			46,43,850
Donkeys			12,783
Camels			37,370
Pigs			14,965
Horses/pony			3517
Total rain-fed area (2011–12): lac ha			18.52
Total irrigated area (2011–12): lac ha			6.09

Source: Authors' own analysis based on Census 2011 and livestock census 2007.

15 are large industrial units. There are also 15 medium-scale industries, all located close to Jodhpur. Pali has a total of 13,834 registered industries, of which only 5 are large industries. They consist of two cement industries, one fabric yarn industry, one agro implements industry, and one industrial unit for manufacturing bitumen. Jalore has a total of 4510 industrial units, of which only 2 falls under large and medium category. The industries in the district fall under the following categories: mineral-based industries; agro-based industries; engineering and metal industries; forest-based industries; leather industries; and handloom industries. Barmer has three large and medium industries. In total, it has 2925 industrial units, which include small-scale industrial units. Most of them are textile industries. The other types of industries are agro-based industries, paper industries, and rubber/plastic industries (Table 7.4).

Table 7.4 Details of large, medium, and small industries in five districts of Rajasthan, falling in Luni river basin.

Name of district	Total no. of industrial units	Total no. of large and medium industries	Micro and small enterprises (types)
Jodhpur	24,374	30 (15 + 15)	Engg; agro industries; chemical; Livestock based; building material
Pali	13,834	5	Agro industries; beverages and tobacco; paper; plastics/rubber; metals and engg.; leather industries; textile; mineral and metal based; wood industries
Jalore	4510	5	Mineral-based industries; agro industries; engg. and metal; forest based; leather industries; handloom
Barmer	2925	3	Textile; agro industries; paper; rubber/plastic
Nagaur	8165	3	Agro industries; forest-based industries; mineral-based industries; textile; chemical; engineering; animal product

Source: Brief industrial profile of Jodhpur, Pali, Barmer, Jalore and Nagaur, Micro, Small and Medium Enterprises, Ministry of MSME, Government of India.

7.5 Methodology and analytical procedure

In water accounting for blue water, we can look at: (a) the renewable water resources (annual surface water flows and groundwater replenishment) as the "total inflow" into the basin during a hydrological year; (b) the amount of water that is being used up in various consumptive uses during the same year (various outflow); and (c) the "balance," which is in the form of unutilized water at the drainage outlet of the sub-basin and the changes in groundwater and surface storage occurring during the hydrological year.

The amount of water that is being used up in various consumptive uses during the year consists of evaporation from open water bodies, swamps and ET from crop land, nonrecoverable deep percolation, and the "net" of water used by cities and rural areas for domestic and industrial uses minus the "return flows" to the natural system in the form of wastewater. Here, the outflows from cropland would NOT consider the water directly used by the cropland from rainfall (effective precipitation or the green water use by the crop) and would only consider the consumptive use from irrigation of the crops grown during the three seasons.

The runoff as part of the total inflow (virgin flows) can be estimated by adding up the "observed flows" and the "effective diversion" by the major reservoirs, other storages and diversion points in the basin. The effective diversion would be the total water diverted from the rivers and tributaries for various purposes minus the estimated return flows to the stream. The return flows can be from irrigation commands and urban centers. The "total water diverted from rivers and tributaries" can be estimated using data on reservoir releases, river lifting, reservoir evaporation, and the net storage change during the hydrological year. There are 13 major and medium reservoirs in Luni river basin.

The water accounts for the basin can be estimated as:

$$\begin{aligned} INFLOW_{TOTAL} = {} & CU_{IRRIGATION} + CU_{RURAL-DOMESTIC} \\ & + CU_{URBAN} + CU_{LIVESTOCK} \\ & + CU_{INDUSTRY} + EVAP_{RESERVOIR} \\ & + OUTFLOW_{STREAM} + GWS_{CHANGE} \\ & + SC_{RESERVOIR} \end{aligned} \tag{7.1}$$

$$INFLOW_{TOTAL} = VFLOW_{STREAM} + GWR_{RENEW} + WATER_{IMPORT} \tag{7.2}$$

But, virgin flow ($VFLOW_{STREAM}$) can be estimated as:

$$VFLOW_{STREAM} = OUTFLOW_{STREAM} + EWD \tag{7.3}$$

If we assume that the urban wastewater is reused in agriculture and the return flows from irrigated fields only contribute to groundwater recharge in the command area, return flows from irrigation schemes and urban areas into the streams can be treated as zero. This is normally the case in arid and semiarid regions.

In such situations, the sum of total water released from reservoirs, water lifted from diversion points along the stream/river, evaporation from these water bodies and their annual (+ive) storage change can be treated as EWD. But, in this case, the estimation of renewable recharge should not consider the recharge from command area, as this would lead to double counting.

The consumptive use of water in urban area (CU_{URBAN}) can be treated as 80% of the total water supplied to meet the municipal water needs, whereas all the water supplied to meet the rural domestic water needs can be considered as the $CU_{RURAL-DOMESTIC}$.

Irrigation includes four components, viz., beneficial evapotranspiration by crops (ET); nonbeneficial evaporation from the soil (both from the soil not covered by canopy and the barren soil in the field after crop harvest); nonrecoverable deep percolation (also the water flows into saline formations); and return flows to streams or groundwater system, which can be recovered for reuse. How much of the water applied in the field would be available for these components would be determined by the technical efficiency with which water is applied (Allen et al., 1998; Kumar and van Dam, 2013). Therefore, consumptive water use from irrigation includes three major components, ET, nonbeneficial evaporation, and nonrecoverable deep percolation.

$$\text{Irrigation water consumed in crop production } (CU_{IRRIGATION}) \\ = A \times [\Delta_{IRRIGATION} - \{(\Delta_{APPLIED} - ET) \times F\}] \tag{7.4}$$

Here, $\Delta_{APPLIED}$ is the sum of irrigation dosage ($\Delta_{IRRIGATION}$) and total soil moisture available from rainfall (also known as effective rainfall). For purely irrigated crops of winter and summer, the effective rainfall (P_{EFF}) can be assumed to be zero. A is the cropped area. The factor "F" is introduced to take account for the fraction of the total water applied in excess of crop water requirement, which is available for reuse.

In regions with hyperarid, arid, and semiarid climatic conditions, nonbeneficial soil evaporation, as explained earlier, can be significant.

Again, if the area has deep water table conditions (depth to water table exceeding 100 ft), the deep percolation from the irrigated field might not be recoverable, meaning no return flows to groundwater.

Hence, in the case of deep groundwater table conditions, in semiarid and arid climates, the value of F can be assumed as zero, and in case of very shallow groundwater, the value can be assumed as 1 (one). In the latter case, real water saving from the use of microirrigation systems will be negligible. Given the fact that a large part of the basin has arid climatic conditions, and groundwater table is deep, the value of F can be considered as zero, which means the excess water applied in the field would not contribute to groundwater recharge. In other words, the irrigation in excess of the irrigation requirements would be lost in soil evaporation and nonrecoverable deep percolation. For using this procedure, the value of depth of irrigation should be known from primary survey.

Alternatively, the irrigation application $\Delta_{\mathrm{IRRIGATION}}$ can be estimated as the difference between ET and effective rainfall of the crop (using FAO CROPWAT model), plus the extra water required to take care of field application efficiency.

$$\Delta_{\mathrm{IRRIGATION}} = (\mathrm{ET} - P_{\mathrm{EFF}})/\mathrm{IE}_{\mathrm{APPLICATION}} \tag{7.5}$$

7.6 Presentation of results

7.6.1 Estimation of virgin flows

The time series data on outflows at the last gauging point in the basin, which has a catchment area of 62,228 km^2, are available from the integrated hydrological data book of Central Water Commission (2012). First, the virgin flows were estimated for the basin by adding up the outflows from the basin (at Gandhav gauging station), and the total amount of water stored in the eight major and medium reservoirs spread over the basin, during the monsoon period, and not on the basis of the volumetric water releases, the storage changes over the year. This is because the data on the water release from these reservoirs were not available. Hence, in this case, the reservoir evaporation rates are not required to be considered for inflow estimates on the left-hand side of the water accounting formula. But at the same time, such an approach warrants that the estimates of outflow separately consider the evaporation from the reservoir. Also, as per this approach, the change in storage should consider the water remaining in the reservoirs.

The estimated virgin flows vary from 26.33 MCM (during 1972–73, 1987–88, and 2002–03) to 2144.68 MCM (1990–91). The virgin flow with a dependability of 75% is 133.48 MCM and that with 50% dependability is 179.64 MCM. Fig. 7.3 shows the dependability curve for estimated virgin flows of the basin for the period.

7.6.2 Estimating rainfall-runoff relationships

Data on annual rainfall were available for 13 gauging stations in Luni river basin for the period from 1954 to 2012. The weighted average of the rainfall was estimated for these stations for all the years of observation (Fig. 7.4). The mean annual rainfall was estimated to be 414.5 mm. In volumetric terms, this is 25,762 MCM of water. Since the rainfall varies across space in the Luni basin, ideally Theisen polygon method has to be used. But, this is a time-consuming process. Due to time constraints, we have made some approximations. For gauging stations, which are located close to the basin boundary (three of them), a weightage of 0.25 was given whereas for all the remaining 12 stations, which were interior, a weightage of 1.0 was given.

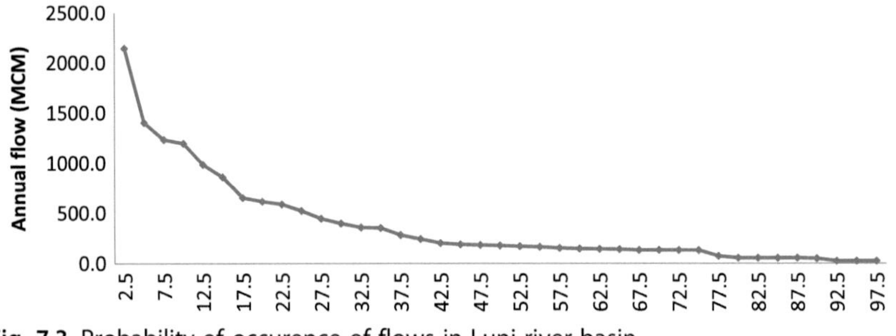

Fig. 7.3 Probability of occurence of flows in Luni river basin.

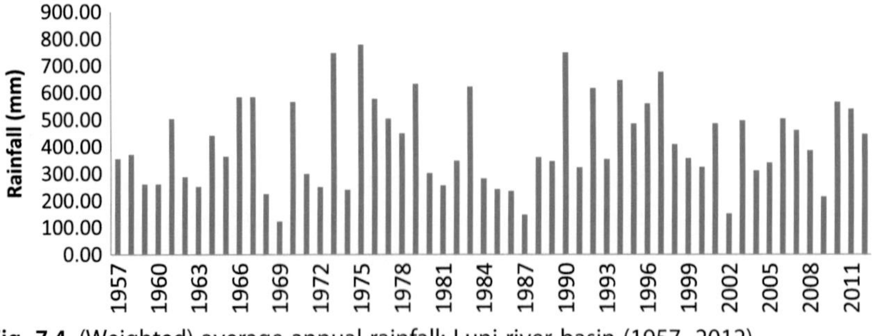

Fig. 7.4 (Weighted) average annual rainfall: Luni river basin (1957–2012).

The virgin flows were estimated for the period from 1971–72 to 2009–10 (as discussed in Section 7.6.1). The mean annual virgin flow is estimated to be 383.7 MCM. This works out to be 13.5 mm, or 3.26% of the total rainfall in the basin. In order to estimate the rainfall-runoff relationship, which can in turn be used to estimate the runoff or future years and also for smaller catchments in the basin, it is essential to estimate the average annual rainfall for the entire basin for the corresponding period.

The rainfall-runoff relationship was estimated to be a power function, indicating higher runoff coefficient for higher rainfall values, or disproportionately higher runoff values for higher values of average annual rainfall. The estimated R square value was 0.54, meaning a reasonably good fit. The rainfall-runoff relationship for the basin is graphically presented in Fig. 7.5. It shows that "X" mm of rainfall generates a runoff equal to $0.0016 \times X^{1.977}$ MCM of runoff.

These runoff rates are extremely low. There are many reasons for this. *First*: Nearly 48.5% of the basin is covered by loamy sand and sand. The topography is flat, with <1% catchment slope.

Second: the climate is hyperarid to arid and the daily temperature is very high in the basin even during the rainy season and humidity low, which keeps the soil moisture depletion rate high, due to which the infiltration rates remains high even after the first few showers. The infiltrating water eventually gets evaporated due to high temperature, rather than percolating down into the deep strata, as 49,485 km^2 of the drainage area (i.e., 71.4%) is barren.

Third: the stream channels (fluvial deposits) are important sources of recharge of shallow aquifers (Sinha and Navada, 2008), with the result that there is a huge transmission loss reducing the runoff volume (Sharma and Murthy, 1998).

Now, the total average annual rainfall of the entire Luni river basin for the year 2011–12 was estimated to be 539.6 mm on the basis of rainfall data

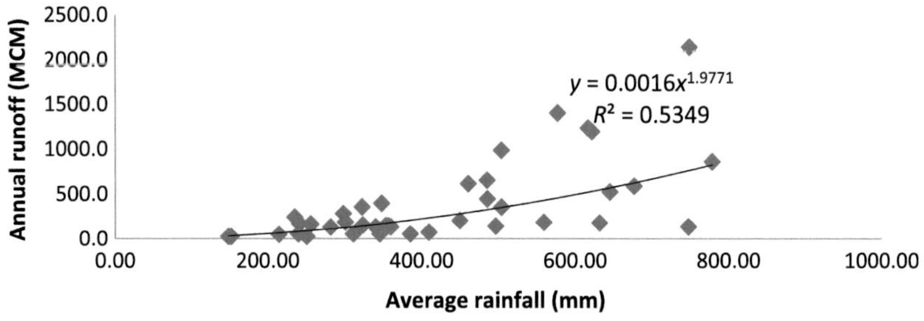

Fig. 7.5 Rainfall–runoff relationship for Luni river basin (1971–2010).

for 13 gauging stations spread over the basin. The corresponding value for runoff was estimated to be 403 MCM using the above rainfall-runoff model. This is the total amount of renewable surface water resources in Luni basin. While a lot of this water would get captured in the reservoirs built in the upper catchment, some might go uncaptured as outflows. However, neither the data on storage nor the data on outflows are available for crossverification.

What is unique about the basin's surface flows is that the entire surface hydrology must have been affected by the large number of water impounding structures. In the absence of these structures, the outflows would not have been higher than what is observed today. This is because of the high transmission loss, which can occur in the stream channels. But, this also means that under such situations (of no surface water-impounding structures), the groundwater recharge happening in the basin would have been much higher than what the estimates show.

7.6.3 Estimating replenishable groundwater

The estimates of annual replenishable groundwater in Luni basin are directly obtained from the report of Tahal Consultants, Vol. 3.2. The assessment was done sub-basin wise. The figures are presented in Table 7.5. The last column of the table also provides estimates of static groundwater resources in the basin. The estimates of groundwater replenishment and draft are average figures, and the actual groundwater replenishment of the basin can change drastically from year to year depending on the rainfall. The total replenishable groundwater in the basin is 2203 MCM per annum. This works out to be around 23% of the static groundwater resources available in the basin.

7.6.4 Imported water

Luni river basin receives water from the adjoining regions. The amounts of imported water with different degrees of dependability are given in Table 7.6. The major sources of water import are Indira Gandhi Nehar Project (IGNP), which supplies water to four districts falling in Luni basin, viz., Jaisalmer, Ganganagar, Churu and Barmer, and Sardar Sarovar Project of Gujarat, which supplies water to Jhalore.

7.6.5 Evaporation from reservoirs

In arid areas, significant amount of water can be lost in evaporation from reservoirs due to excessive evaporation. There are two major and seven medium reservoirs located in the basin. The evaporation from these

Table 7.5 Dynamic and static groundwater resources in Luni river basin.

Name of district	Renewable groundwater resources (MCM per annum)	Existing groundwater draft (MCM per annum)	Stage of groundwater development (%)	Static ground water resources in different sub-basins (MCM)
Bandi	39.46	52.72	133.6	137.2
Bandi (Hemawas)	34.51	51.74	149.9	16.27
Guhiya	84.18	120.06	142.6	81.97
Jawai	129.17	157.73	122.1	272.1
Jojri	116.44	37.61	118.2	522.51
Khari	142.73	140.43	98.4	273.24
Khari (Hemawas)	36.64	38.98	106.4	2.49
Luni	571.7	633.26	110.8	3464.76
Luni WRIS	797.82	726.54	91.1	5084.04
Mithari	64.62	58.98	91.3	80.7
Sagi	80.83	148.74	184.0	342.13
Sukri	55.5	60.77	109.5	29.26
Sukri (Sayala)	50.26	83.85	166.8	399.2
Total for Luni River Basin	2203.86	2411.41	109.4	10,705.87

Source: Study on Planning of Water Resources of Rajasthan, Draft Final report submitted to SWRPD, GoR, Tahal Consultants, Vol. 3.2, 2c, December 2013.

Table 7.6 Imported water in Luni river basin.

Dependability (%)	Volume of water imported (MCM)
25	569.36
50	404.31
75	331.33
90	240.61
Mean	476.61

Source: Study on Planning of Water Resources of Rajasthan, Draft Final report submitted to SWRPD, GoR, Tahal Consultants, Vol. 3.2, 2c, December 2013.

reservoirs depends on the number of days for which water remains in the reservoir, and the potential evaporation in the locality concerned during those days and the reservoir water spread area.

It can be estimated as:

$$\mathrm{EVAP}_{\mathrm{RESERVOIR}} = \sum_{i=1}^{m} \mathrm{RA}_i \times \mathrm{CU} - \mathrm{EVAP}_i \tag{7.6}$$

Where CUE_i is the cumulative evaporation rate for reservoir, *i* for the time period for which water remains in it; RA_i is the average water spread area of the reservoir, "*i*." It is to be kept in mind that the water spread area of the reservoir would keep declining as time progresses and more water is drawn from it, and hence some approximation would be required if this equation is used.

The mean potential evaporation (PE) values for Pali varies gradually from 1700 mm in the southern parts of the district to a highest of 2300 mm in the northern parts (source: Water Resource Atlas of Rajasthan), with a mean value of 2000 mm. The mean potential evaporation for Jaisalmer is around 2600 mm. The reference evapotranspiration (Penman) (ET_0) values for different months for two locations in the basin are given in Fig. 7.6. It shows that ET_0 can be as high as 10 mm in the hottest month of May. The total annual ET_0 for Pali is 2066 mm, while the corresponding value for Jaisalmer is 1967 mm. Hence, the estimated values of the ratio of PE/ET_0 for the two locations are 0.96 and 1.32, respectively.

Using these fractions, the monthly PE values for the two locations can be estimated. The estimated monthly PE values for the two locations are given in Table 7.7. It is evident from the table that evaporation rate would be highest during the month of June in Jaisalmer and May in Pali.

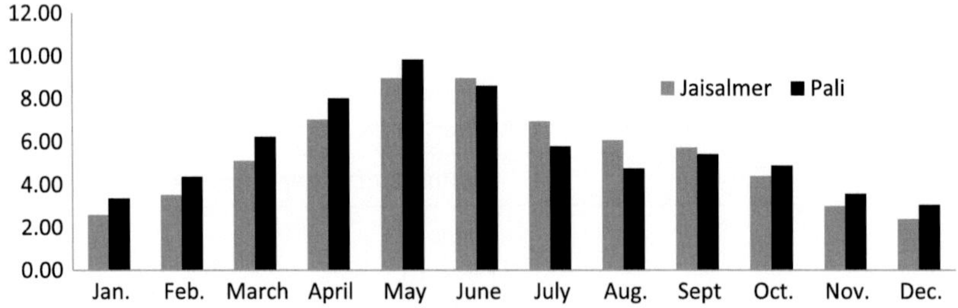

Fig. 7.6 Reference evapotranspiration for two locations in Luni river basin.

Table 7.7 Estimated monthly potential evaporation values for two locations in Luni river basin.

Location/ month	Month											
	Jan	Feb	March	April	May	June	July	Aug	Sep	Oct	Nov	Dec
Jaisalmer	3.41	4.64	6.73	9.26	11.7	11.80	9.17	8.00	7.54	5.80	3.95	3.13
Pali	3.25	4.22	6.01	7.73	9.48	8.3	5.52	4.59	5.23	4.72	3.45	2.95

Source: Rajasthan Water Resources Atlas and IMD data on ET_0.

Based on the inflow data available for the period from 1995 to 2010, the study by Tahal consultants had estimated reservoir evaporation from 13 reservoir projects in Luni river basin. The estimates are given in Table 7.8.

The total evaporation, as estimated by Tahal Consultants from major, medium, and minor irrigation reservoirs, is 35.61 MCM. This is against a total reservoir area of 5930 ha, i.e., 59.3 km^2. This constitutes 0.086% of the total basin area. The average evaporation works out to be 0.60 m. The potential evaporation rate during August to January, the time period during which water is generally available in these reservoirs, for Pali and neighboring districts situated on the eastern part of the basin, where all these reservoirs are located, is around 600 mm. Hence, these values seem to be reliable.

7.6.6 Consumptive water use through irrigation

Since depth of irrigation water applied to the crops is not available for any of the crops grown in the region, we have used the FAO CROPWAT to estimate irrigation water requirement and assumed a factor of 1.25 to arrive at irrigation water application, which is to provide allowance for consumptive uses, which are nonbeneficial—nonbeneficial evaporation and nonrecoverable deep percolation (Allen et al., 1998). This should not be confused with field application efficiency in irrigation, which can be much

Table 7.8 Estimates of evaporation from 13 major/medium reservoirs in Luni river basin.

Name of the major/medium/ minor project	Reservoir water spread area (ha)	Total volume of evaporation (MCM)
Angore dam project	130	1.41
Bandi Sendra project	200	2.36
Bankli bund	500.0	3.34
Bisalpura bund project	0.0	0.12
Giroliya tank	360.0	0.033
Hemawas bund project	30.0	7.14
Jaswant Sagar	410.0	0.62
Jawai bund	1250.0	11.46
Kharda bund	950.0	1.67
Ora bund	10.0	1.68
Phool Sagar Jaliya	10.0	0.037
Raipur Luni project	80.0	0.167
Sardar Samand project	2000.0	5.58
Total	5930.0	35.61

Source: Study on Planning of Water Resources of Rajasthan, Draft Final report submitted to SWRPD, GoR, Tahal Consultants, Vol. 3.2, 2c, December 2013.

lower for surface irrigation in the basin, but can be accepted as a working methodology, as we are ultimately concerned with estimating consumptive water use, and we treat the irrigation water application ($\Delta_{IRRIGATION}$) as the consumptive use (meaning zero return flows from the irrigated field). For this reason, while doing the estimation, the return flow fraction (F) is assumed as zero, irrespective of the method of irrigation. The total consumptive water use in irrigation is estimated by multiplying the depth of irrigation with the area under the crop concerned, for the year 2011–12.

The areas under irrigated kharif and winter crops in the basin are provided in Tables 7.9 and 7.10, respectively.

The estimates of crop wise consumptive water use for irrigation are presented in Table 7.11 for kharif and winter crops.

The total water consumption in agriculture is estimated to be 2404.4 MCM during the year 2011–12. Fig. 7.7 presents the graphical representation of volumetric water use by different irrigated crops in the basin. Castor is the largest consumer of irrigation water in the basin (733 MCM), followed by wheat, which consumes around 497 MCM. Castor, while sown during the kharif season, lasts for most parts of the winter season.

7.6.7 Domestic water use in Luni basin

Luni is an absolutely water-scarce basin. There is very little imported water in the basin, except in Jodhpur city, which receives water from IGNP for its municipal water supply. The rural areas depend on groundwater sources, mostly open wells. During drought years, these wells go dry and only the tube wells function. The rural communities face water shortage for domestic uses. We have therefore assumed a per-capita daily water use of 50 lpcd for domestic uses in the region. After Gleick (1996), this is also the basic minimum required for survival. For nonmetros with planned sewage, we have assumed 135 lpcd (like Jodhpur). For small towns without planned sewage (such as Barmer, Nagaur, Pali), we have assumed a per-capita water consumption of 70 lpcd. This is as per the recommendation in the 12th Plan document for small towns without a centralized sewerage system. From water accounting point of view, though a large share (80%–90%) of this water would be available as domestic effluent (wastewater return flow), it may get depleted in the local sinks (ponds, natural depressions, etc.), and hence is not accounted for separately. However, in the case of Jodhpur, a wastewater return flow of 80% is assumed. This means the actual consumptive water use would be only 20% of the total of 135 lpcd. The remaining

Table 7.9 Irrigated area during kharif season.

	Irrigated area during kharif (2011–12)							
Crops	**Ajmer**	**Barmer**	**Jalore**	**Jodhpur**	**Nagaur**	**Pali**	**Rajsamand**	**Sirohi**
Rice	0.7	0.0	0.0	0.0	0.0	0.0	0.0	0.0
Jowar	0.0	6.3	25.7	25.6	32.0	0.0	0.0	0.0
Bajra	0.4	2501.0	1302.4	5782.5	2397.1	2.0	0.0	8.5
Maize	45.0	0.2	24.0	0.9	4.5	462.0	2.3	49.7
Green gram	0.0	1.2	46.4	240.8	94.4	0.0	0.0	0.0
Moth	0.0	37.5	7.5	115.8	2.2	0.0	0.0	0.0
Chaula	0.0	0.0	3.3	1.8	1.3	0.0	0.1	0.0
Sesame	0.0	1.4	7.5	141.7	13.2	0.0	0.0	0.0
Groundnut	274.8	239.5	4088.6	9447.1	1343.5	0.0	0.0	1976.0
Soybean	0.4	0.0	0.0	0.0	0.0	0.0	0.0	0.0
Castor	61.2	8654.1	62,853.1	6089.2	84.6	1307.0	0.0	13,850.7
Cotton	2135.8	1.2	566.2	2922.5	3928.9	6261.0	484.3	710.2
Sanhemp	5.8	0.0	0.0	0.0	0.0	0.0	0.0	0.0
Sugarcane	3.8	0.0	0.0	0.0	0.1	0.0	20.2	0.0
Cluster bean	2.5	5.6	5.8	518.2	30.5	5.0	0.0	2.0
Chillies	117.4	6.8	321.7	296.4	9.7	464.0	10.6	80.0
Overall	2647.9	11,454.7	69,252.2	25,582.5	7942.1	8501.0	517.4	16,677.1

Table 7.10 Irrigated area under winter (2011–12).

	Irrigated area during winter (2011–12)							
Crops	**Ajmer**	**Barmer**	**Jalore**	**Jodhpur**	**Nagaur**	**Pali**	**Rajsamand**	**Sirohi**
Wheat	9931.3	2584.7	24,135.5	7587.0	6045.4	47,922.0	2840.7	14,849.0
Barley	3419.1	4.7	354.8	32.4	1201.4	4163.0	663.6	484.8
S. Millets	0.0	0.0	422.8	0.3	0.0	0.0	0.0	261.4
Gram	248.0	0.9	0.0	138.9	1151.4	6419.0	23.8	603.2
Green peas	0.0	0.0	0.0	0.0	129.1	0.0	0.0	0.0
Masur	3.1	0.0	1.7	0.0	0.0	0.0	0.0	0.0
Other rabi pulses	36.1	0.0	0.0	0.0	0.0	0.0	0.1	15.4
Rapeseed and mustard	4440.4	1751.2	36,511.6	9456.2	3117.2	34,880.0	214.7	6800.1
Taramira	17.0	25.2	29.8	21.3	19.6	418.0	1.8	12.5
Linseed	6.0	0.5	0.0	0.6	17.5	0.0	0.2	0.0
Coriander	20.6	0.0	0.8	26.6	2.7	171.0	2.6	0.4
Cumin	2830.7	30,875.8	92,594.3	13,467.4	5952.9	14,665.0	3.6	2182.0
Fenugreek	51.7	37.5	145.1	508.0	476.8	1640.0	34.9	7.7
Potato	0.0	0.0	59.7	0.0	0.0	0.0	0.3	128.1
Sweet potato	17.5	4.2	17.4	1.4	2.0	13.0	0.0	0.8
Onion	229.6	25.4	169.1	2091.2	958.1	348.0	2.4	19.8
Tobacco	0.0	0.0	244.6	0.0	0.0	0.0	0.0	0.0
Fennel	229.6	5.1	3286.2	727.2	2365.4	1003.0	0.0	1437.4
Garlic	2.2	0.0	5.8	553.9	16.1	15.0	21.4	6.1
Ajwain	36.3	2.3	350.7	5.0	181.7	20.0	1.9	19.0
Isabgol	4.5	13,570.2	29,405.5	4760.4	3667.2	1029.0	0.0	88.9
Overall	21,523.7	48,887.6	187,735.3	39,377.8	25,304.6	112,706.0	3812.1	26,916.5

Table 7.11 Estimated consumptive water use in irrigation in Luni river basin.

Crops	Consumptive water use in irrigation (in 000′ cu. m.)							
	Ajmer	Barmer	Jalore	Jodhpur	Nagaur	Pali	Sirohi	Overall
Winter season								
Wheat	37,330.4	12,161.8	113,566.6	35,814.3	25,631.0	203,177.3	70,094.8	497,776.1
Barley	9840.7	20.0	1521.1	138.8	4298.6	14,895.2	2074.0	32,788.3
S. Millets	0.0	0.0	1908.3	1.2	0.0	0.0	1059.6	2969.1
Gram	668.8	3.8	0.0	563.2	3886.6	21,667.3	2445.3	29,235.1
Green peas	0.0	0.0	0.0	0.0	435.9	0.0	0.0	435.9
Masur	8.5	0.0	6.8	0.0	0.0	0.0	0.0	15.2
Other rabi pulses	97.3	0.0	0.0	0.0	0.0	0.0	62.2	159.5
Rapeseed and mustard	15,616.7	8850.7	184,529.9	47,887.2	13,374.5	149,652.6	34,436.7	454,348.3
Fennel	563.7	12.6	8068.3	1785.5	5807.8	2462.6	3529.3	22,229.8
Potato	0.0	0.0	833.5	0.0	0.0	0.0	1788.5	2622.0
Cumin	4246.0	46,313.6	138,891.5	20,201.0	8929.4	21,997.5	3273.0	243,852.1
Isabgol	13.4	40,710.5	88,216.4	14,281.3	11,001.6	3087.0	266.7	157,576.9
Monsoon season								
Sorghum	0.0	23.7	97.0	84.7	71.7	0.0	0.0	277.0
Pearl millet	0.6	8457.5	4404.1	20,318.9	5798.9	4.8	29.8	39,014.7
Maize	84.3	1.1	114.4	3.9	13.8	1415.6	211.5	1844.6
Green gram	0.0	4.8	192.8	903.1	255.1	0.0	0.0	1355.8
Moth	0.0	155.8	31.0	434.3	6.1	0.0	0.0	627.2
Chaula	0.0	0.0	13.8	6.9	3.5	0.0	0.0	24.1
Groundnut	642.5	1276.1	21,782.7	45,602.5	4812.9	0.0	9538.2	83,654.9
Soybean	1.2	0.0	0.0	0.0	0.0	0.0	0.0	1.2
Castor	482.5	68,281.6	495,917.4	48,044.8	667.8	10,312.4	109,283.7	732,990.1
Sesame	0.0	2.1	11.2	212.5	19.7	0.0	0.0	245.5
Cotton	6946.6	8.3	4009.8	18,922.5	19,108.9	30,451.2	4598.7	84,045.8
Chillies	1349.8	77.7	3699.0	3408.2	112.0	5336.0	919.9	14,902.6
Cluster bean	6.375	14.28	14.79	1321.41	77.775	12.75	5.1	1452.48
Total in MCM	77,899.375	186,375.98	1,067,830.39	259,936.21	104,313.575	464,472.25	243,617	2,404,444.78

Source: Authors' own estimates based on FAO CROPWAT and secondary data on agricultural land use.

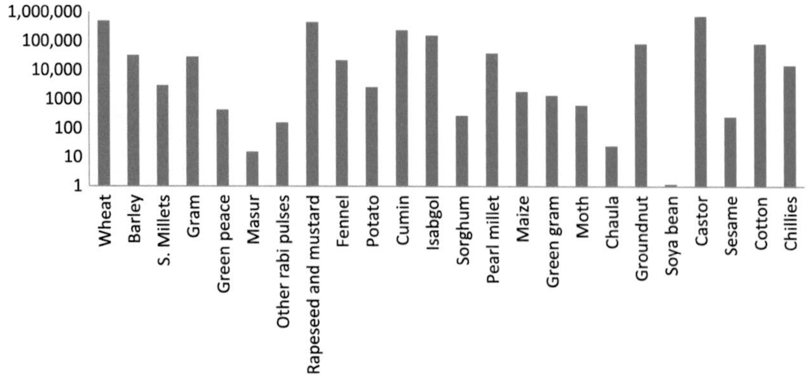

Fig. 7.7 Crop-wise consumptive water use in irrigation: 2011–12.

Table 7.12 Estimated domestic water use (rural) in Luni river basin.

Districts	Estimated population (in 000′)	Domestic water demand Daily (000′ cu. m.)	Domestic water demand Annually (MCM)
Ajmer	345	17.2	6.3
Barmer	564	28.2	10.3
Jalore	1388	69.4	25.3
Jodhpur	373	18.6	6.8
Nagaur	33	1.7	0.6
Pali	1578	78.9	28.8
Rajsamand	89	4.5	1.6
Sirohi	335	16.8	6.1
Udaipur	35	1.7	0.6
Overall	4740	237.0	86.5

Source: Authors' own analysis.

water would get accounted for in irrigated agriculture, as most of the city's wastewater would get diverted for irrigation in peripheral areas.

Rural domestic water demand in the basin was estimated considering a minimum water requirement of 50 L per person per day (lpcd), which is considered to be the basic water requirement for meeting human needs (Gleick, 1996). For estimation of rural population, total population in each district was adjusted as per the proportion of its geographical area falling in the basin. As per the estimates, the overall domestic water demand (rural) in Luni river basin is about 86.5 MCM (Table 7.12).

The norms suggested for water supply in the 12th Five Year Plan document were (1) 135 lpcd for urban areas where piped water supply and

Table 7.13 Estimated domestic water use (urban) in Luni river basin.

Districts	Estimated population (in 000′)	Domestic water demand	
		Daily (000′ cu. m.)	Annually (MCM)
Ajmer	542	65.1	23.8
Barmer	171	20.5	7.5
Jalore	102	12.2	4.5
Jodhpur	1093	131.2	47.9
Nagaur	312	37.4	13.7
Pali	460	55.2	20.1
Sirohi	67	8.1	2.9
Overall	2747	329.6	120.3

Source: Authors' own analysis.

underground sewerage systems are available and (2) 70 lpcd for urban areas provided with piped water supply but without underground sewerage system. In view of the fact that Luni is a scarce river basin, we have considered roughly the mean of the two figures (which comes out to be around 100 lpcd) as the domestic water use in urban areas. Urban population was estimated based on the number of cities and towns falling in the basin. As per the estimates, the overall domestic water demand (urban) in Luni river basin is about 120.3 MCM (Table 7.13). Twenty percent of this (i.e., 24.06 MCM) can be treated as consumptive water use, and the remaining 80% will be available for irrigation in the periurban and rural areas.

7.6.8 Livestock water use in the basin

Livestock water use in the basin was estimated by the following the indicative figures of voluntary water consumption per tropical livestock unit (TLU) for different categories of livestock (as suggested by Pallas, 1986), and the average body weight of different categories of livestock found in the region. Here, we have considered an average body weight of 400 kg for buffaloes and crossbred cows, 250 kg for indigenous cows, 25 kg for goat/sheep, and 450 for camels. The estimates of water use for different types of livestock are presented in Table 7.14.

From Table 7.14, it is evident that small ruminants, which constitute 66% of the total livestock population, account for only 18.3% of the total water use by livestock. Cattle account for nearly 33% of the total livestock water use. Buffaloes account for 46.3% of the total livestock water use in the basin.

Table 7.14 Estimated livestock water use in Luni river basin (2007).

Type	Livestock water demand (000′ cu. m.)								
	Ajmer	Barmer	Jalore	Jodhpur	Nagaur	Pali	Rajsamand	Sirohi	Udaipur	Overall
CB cow	94	2	6	119	37	110	49	8	6	432
Indi. Cow	1173	2091	3427	1345	607	4934	340	1151	183	15,251
Buffalo	1690	822	7683	913	1110	7110	464	1516	165	21,473
Sheep	204	583	956	275	155	1687	21	185	4	4071
Goat	314	944	827	394	277	1281	86	252	28	440 4
Donkey	2	30	18	5	2	17	1	3	0	78
Camel	9	257	127	73	28	155	4	45	1	700
Overall	3486	4730	13,044	3124	2217	15,294	966	3161	388	46,410

Source: Authors' own analysis based on secondary data on weather parameters (IMD), FAO online catalog on water for animals, based on Pallas (1986) and livestock population as per 2007 livestock census.

Industrial Water Use in Luni River Basin

Since no estimates are available on the manufacturing output for various industrial subsectors for the districts in the basin, industrial water use in Luni basin was estimated on the basis of the consideration that on a per-capita basis, the basin would require a minimum of 20 m^3 of water per annum for manufacturing.[c] Hence, water use during 2011–12 (m^3) was estimated on the basis of population of the basin as per Census 2011 as:

$$CU_{INDUSTRY,LUNI} = POP_{LUNI} \times 20 = 149.73\,MCM$$

7.7 The basin water accounts

The final water accounts for the blue water in Luni river basin for the year 2011–12 are presented in Table 7.15. Here, we have compared the total inflows and the total outflows. The total outflow is exclusive of reservoir evaporation. The difference between the two should be equal to the sum of reservoir storage change, evaporation from the reservoir, and the total stream channel outflow. The total of change in storage and (stream) outflow was estimated to be 309 MCM in 2011–12. The year considered for the study was a wet year and it is quite likely that the basin had some outflows in that year.

Table 7.15 Water accounts of Luni river basin (2011–12).

Water resources and use	Volumetric water use (MCM)
Total inflows	2843.60
Annual surface water resources (2011–12)	403.00
Renewable groundwater resources	2203.86
Water imported into Luni river basin	240.60
Total outflows	2741.31
Consumptive water use for irrigation	2404.00
Consumptive water use in the domestic sector (urban and rural)	110.56
Livestock water use	41.41
Industrial water use	149.73
Evaporation from major, medium, and minor reservoirs	35.61
Groundwater storage change	−207.55
Surface water outflows + water remaining in the reservoirs	309.84

Source: Authors' own estimates based on secondary data.

[c] This is based on the estimates for industrial water demand in India for the year 2010 (Kumar, 2010), which indicates an average per-capita water demand of 20 m^3 for industrial uses.

The total blue water inflow in the basin during the year was 2843.6 MCM. Of these, the internal renewable water resource is 2606.86 MCM, from a total rainfall of 33,577 MCM for the drainage area of 62,228 km^2.[d] The remaining water is lost in soil evaporation from the large tracts of barren land (a total 48,750 km^2), ET from the purely rain-fed crops of kharif season, which covered an area of 18,520 km^2 (18.52 lac ha) and the trees and other natural vegetation in the basin.

While the figures of "change in storage" appear to be excessively high, this could be because the figures of imported water are quite tentative. It is quite likely that the during a good rainfall year like 2011–12, less amount of water would have been imported into the basin for irrigation purpose from SSP and IGNP. Since the discharge data for Luni river were not available for the year for which the water accounting exercise was carried out (i.e., 2011–12), the inflow and outflows figures could not be tallied to cross-check the figures of outflows, i.e., surface water outflow + water remaining in the reservoirs.

7.8 Conclusions

The water accounting study for Luni basin suggests that the amount of water utilized in the basin annually (2704 MCM for 2011–12) is higher than the renewable water generated within it (2606 MCM in 2011–12) and a large share of this water is used for irrigation (2404 MCM in 2011–12). Though some amount of surface water flows out of the basin in normal and high rainfall years, all the topographically viable catchments in the basin are fully tapped with 13 major and medium reservoirs capable of storing a total of 560.37 MCM of water. The flat topography of the basin in the lower parts and the very shallow embankments do not permit harnessing this water for beneficial uses using conventional technologies. The high level of water use sustained through groundwater mining and water imported from outside the basin, which help maintain the precarious water balance. The only way to improve water management in this water-scarce river basin is to reduce the consumptive use of water in irrigated agriculture through control of evaporation.

Technologies such as drip irrigation and mulching can be adopted to reduce the nonbeneficial evaporation of water used for irrigating the row

[d] This was estimated using the weighted average rainfall of 539.6 mm rainfall during 2011 and the drainage area of the upper catchment, i.e., 62,200 km^2.

crops. While drip irrigation will be effective in reducing the losses due to nonrecoverable deep percolation, mulching can reduce nonbeneficial evaporation of water applied to the soil. In addition to the groundwater and water from surface reservoirs, there is large amount of rainwater directly used by monsoon crops in the basin area. Mulching can conserve this water retained in the soil profile by preventing evaporation and converting it into beneficial transpiration. Though this intervention can alter the blue water availability in the basin only marginally, it can help improve the efficiency of green water use for crop production. To what extent this intervention can help reduce soil water evaporation would be examined in Chapter 8 of this book using a case study of one of the districts falling fully in the basin.

References

Allen, R.G., Willardson, L.S., Frederiksen, H., 1998. Water use definitions and their use for assessing the impacts of water conservation. In: de Jager, J.M., Vermes, L.P., Rageb, R. (Eds.), Proceedings ICID Workshop on Sustainable Irrigation in Areas of Water Scarcity and Drought, September 11–12. Oxford, England, pp. 72–82.

Bhuiyan, C., Kogan, F.N., 2009. Monsoon variation and vegetative drought patterns in the Luni Basin in the rain-shadow zone. Int. J. Remote Sens. 31 (12), 3223–3242.

Central Water Commission, 2012. Integrated Hydrological Data Book (Non-Classified Basins). In: Hydrological Data Directorate, Information Systems Organization, Water Planning and Projects Wing. Central Water Commission, New Delhi, March 2012.

Das, P.K., 1996. The Monsoons, third ed. National Book Trust, India.

Gleick, P.H., 1996. Basic water requirements for human activities: meeting basic needs. Water Int. 29 (2), 83–92.

Kumar, M.D., 2010. Managing Water in River Basins: Hydrology, Economics, and Institutions. Oxford University Press, New Delhi.

Kumar, M.D., van Dam, J., 2013. Drivers of change in agricultural water productivity and its improvement at basin scale in developing economies. Water Int. 38 (3), 312–325. https://doi.org/10.1080/02508060.2013.793572.

Pallas, P., 1986. Water for Animals. Food and Agriculture Organization, Rome.

Sharma, K.D., Murthy, J.S., 1998. A practical approach to rainfall-runoff modelling in arid zone drainage basins. Hydrol. Sci. J. 43 (3), 331–348. June 1998.

Singh, N., 1994. Optimizing a network of rain-gauges over India to monitor summer monsoon rainfall variations. Int. J. Climatol. 14, 61–70.

Sinha, U.K., Navada, S.V., 2008. Application of isotope techniques in groundwater recharge studies in arid western Rajasthan, India: some case studies. Geol. Soc. Lond. Spec. Publ. 288 (1), 121–135.

Tahal Consultants, 2013. Study on Planning of Water Resources of Rajasthan. Draft Final Report Submitted to the State Water Resources Planning Department, vol. 3.2. Government of Rajasthan. December 2013. 2c.

Tahal Consultants, 2014. Study on Planning of Water Resources of Rajasthan. Main Report-IN-24740-R13-073, Final Report Submitted to the State Water Resources Planning Department, Government of Rajasthan, July 2014.

Further reading

Batchelor, C., 1999. Improving water use efficiency as part of integrated catchment management. Agric. Water Manag. 40 (1999), 249–263.

Government of India, 2011. Dynamic Ground Water Resources of India. as on March 2009, Central Ground Water Board, Ministry of Water Resources, Government of India, Faridabad. November 2011.

Howell, T., 2001. Enhancing water use efficiency in irrigated agriculture. Agron. J. 93, 281–289.

Kumar, M.D., Singh, O.P., Bassi, N., Niranjan, V., Sharma, M.K., 2010. Hydrological and Farming System Impacts of Agricultural Water Management Interventions for Sustainable Groundwater Use in North Gujarat & Strategies for Improving the Poor Farmers' Access to Groundwater. Final Report Submitted to SRTT, Mumbai, Institute for Resource Analysis and Policy, Hyderabad.

Xie, Z.-k., Wang, Y.-j., Li, F.-m., 2005. Effect of plastic mulching on soil water use and spring wheat yield in arid region of Northwest China. Agric. Water Manag. 75 (2005), 71–83.

CHAPTER 8

Downscaling IWRM plans from basin to districts: IWRM plan for Pali district in Western Rajasthan

A.J. James[a], M. Dinesh Kumar[b]
[a]Environmental & Natural Resource Economist, New Delhi, India
[b]Institute for Resource Analysis and Policy, Hyderabad, Telangana, India

8.1 Introduction

The European Union Special Partnership Programme (EU-SPP) sought to assist the Government of Rajasthan from 2007 to 2014 to implement the State Water Policy that was based on Integrated Water Resource Management (IWRM). One of the several activities undertaken by the Technical Assistance (TA) team of the EU-SPP was to downscale basin-level IWRM plans to the district level, i.e., from a hydrologically defined area to an administratively defined boundary. The study had three objectives: (1) to prepare rapid water accounts for Luni river basin and to study the water infrastructure and water access and use in the basin; (2) analyze the droughts in Luni basin and suggest mitigation strategies; and, (3) discuss the methodology for preparation of district IWRM plans and apply the same for Pali district. While the findings of first part of the study have been detailed in Chapter 7 of this volume, this chapter presents the findings on the last part of the study.

The chapter accordingly describes the analytical procedures for district IWRM planning, presents the estimates of water balance for Pali district under the current and future water management scenarios (done for the purpose of illustrating IWRM planning process), and discusses key interventions to close the demand-supply gap in 2025.

8.2 District IWRM planning

Water resources development and management planning under the IWRM approach is basically about meeting the future demands of a region's

Current Directions in Water Scarcity Research, Volume 1
https://doi.org/10.1016/B978-0-12-814851-8.00008-2
© 2019 Elsevier Inc.
All rights reserved.

socioeconomic and environmental systems using the available supplies from the natural water system (here the river basin), without threatening its integrity.

In the context of an administrative boundary like a district, which falls entirely within a river basin, planning for water resources development and management activities using the IWRM concept requires taking cognizance of the opportunities and constraints for augmenting water supplies and reducing water use, giving due consideration to other parts of the basin (in which the district falls) due to their socioeconomic conditions and hydrological dependence. This interdependence affects water demand and water availability calculations, as well as the impacts of a highly variable rainfall and climate, as detailed here.

8.2.1 Water demands

The impacts of intensive water harvesting in an upper catchment district (like Pali, in the Luni river basin) will have negative impact on stream flows in the stretches of Luni river passing through the downstream district of Barmer. Hence, if the entire basin is considered in planning, it would help understand the limits to water harvesting structures in the upper catchment districts.

Conversely, what is being planned in the lower parts of the basin should take cognizance of the situation in the "district" in question in terms of water availability, their future water demands and new development plans, including water use efficiency improvement plans.

8.2.2 Water supplies

While water demand estimation can be done independently for a district, the water to meet the district demand can be located outside the district's administrative boundaries. Further, since there could be losses while water is being transferred from supply sources to the demand sites, the water supply requirement will be higher than the actual demand at site. An example is the seepage losses in canal systems, due to which more water must be supplied from the reservoir than what is required to meet crop water demand. However, this excess water is not permanently lost from the system and would be available for reuse from the shallow groundwater underlying the area through which the canal network passes. Further, not all the water supplied to the demand sites will eventually get consumed, and some of the water could return to the system as "return flows." Such return flows could be significant in surface irrigation commands and could contribute to the groundwater system or drainage channels—which in turn could contribute to the available water resources in the lower parts of the basin as "new water." Such return flows are also significant for domestic demand, particularly in urban areas, and nearly 80%–85% of the water consumed domestically would be available as wastewater flowing into the natural sinks.

On the other hand, demand reduction measures (particularly in agriculture) implemented in an area (district), through the use of efficient irrigation technologies, can have negative implications for water resource availability in the same area or its downstream parts in the long-term; and, this is particularly true if downstream users are heavily dependent on groundwater in the shallow aquifers due to significant reductions in return flows to groundwater in the lower parts of the basin (Batchelor, 1999; Howell, 2001). When considering water use efficiency or productivity at the catchment or larger scale, interactions between different units in the landscape as well as different water users must be considered (Batchelor, 1999; Seckler, 1996).

For this reason, how water gets augmented or used up or saved in one parts of the basin (for example, in a district) can also influence the availability of water resources in other parts of the basin. This means that water demand and supply estimation—which is the basis for water resource planning—should be done at a spatial scale that captures the dynamics of interaction between physical systems affecting water availability, such as surface water-groundwater interactions, land use changes, and stream-flow interactions.

8.2.3 Rainfall and climate variability

Water supplies in a basin to meet water demands are also a function of the climate, which experience very high variability. With the variation in rainfall from year to year, stream flows, groundwater recharge, and soil moisture availability are likely to experience significant variations. This can also induce significant changes in the demand for irrigation water between a dry year and a wet year, as during drought years the demand per unit area of crop could increase and reduce drastically in a wet year. Therefore, it is important to do a temporal analysis for water assessment and planning, involving several years of data with dry, wet, and normal years.

8.2.4 Challenges of district IWRM planning

Five key issues to be explored in district level IWRM planning are the following:

- *Intrabasin water transfers*, i.e., the movement of water from the district under question to other parts of the basin, which are relatively "water scarce"
- *Water imports* from parts of the basin that are "relatively water rich" to the district
- *Multiannual storage of water* in the basin through reservoirs, to store water in excess-rainfall years for supply in low-rainfall years
- *Limits to water resource augmentation* instead of continuing to build such schemes; and
- *Transfer of water savings*, i.e., "saving water" from areas where opportunities for demand reduction are higher, for release to parts of the basin facing severe water deficits.

The next sections analyze the different components of water demand, i.e., for irrigation, domestic and institutional uses, livestock and industry, and then the available sources of water supply, before analyzing demand-supply gaps, and how they can be addressed.

8.3 Analyzing water demand

While it is rather easy to estimate the existing supplies from the available sources, the estimation of potential water supplies to meet the future demand is one of the most complex issues involved in district IWRM planning. This is because of the complex dynamics of interaction between upper and lower catchments, and surface and groundwater. Due to these interactions, water development in the upper catchment can influence water availability in lower catchments and change in water use efficiency improvements in one part of the basin can affect surface or groundwater availability downstream, though with time lapse. Therefore, in water balance studies required for district IWRM planning, it is important to carry out water assessments using basin as the unit of analysis, as it integrates the complex interactions discussed above.

8.3.1 Irrigation water demand

Irrigation water demand estimation should be done for typical rainfall years, i.e., wet years, dry years, and normal rainfall years, to account for the variations in water demand rates and the areas under irrigated crops between years. Given irrigation water demand data for each irrigated crop, including the rain-fed crops, which receive supplementary irrigation during drought years, the irrigation water requirement for crop k at time t is calculated as follows:

$$\Delta_{\text{IRRIGATION},k,t} = \left(\text{ET}_k - P_{\text{EFF},\ k,t}\right) \Big/ \text{IE}_{\text{APPLICATION}} \tag{8.1}$$

where $\text{IE}_{\text{APPLICATION}}$ is water application efficiency for the irrigated field, expressed as a fraction, for which a reasonable assumption can be made, depending on the irrigation method.

Eq. (8.1) suggests that the irrigation demand for the same crop could be different for different years and for different seasons within the same year. Also, irrigation water requirements for rain-fed crops would become significant in drought years, while it might be nil during wet and normal years, depending on the crop duration.

The total water demand (TWD) for irrigation can now be estimated as:

$$\text{TWD}_{\text{IRRIGATION},t} = \sum_{k=1}^{n} \binom{n}{k} \text{A}_{k,\ t}\left[\Delta_{\text{IRRIGATION},k,t}\right] \tag{8.2}$$

where $\Delta_{\text{IRRIGATION},\ k,\ t}$ is the amount of irrigation water dosage to the crop k in year t; n is the total number of crops in the region; $\text{A}_{k,\ t}$ is the area under crop k after a time of t years.[a]

Given that a growing population would cause the demand for agricultural output to increase in the district (and the basin as a whole), there would be greater pressure to increase the area under irrigation. One way to incorporate this in the agricultural water demand estimation is to keep the total cropped area (of individual crops) per unit of population for future years at the same level as that of the present. This means that population growth would be the main driver of agricultural water demand growth, and total future irrigation water demand $\text{A} = \Delta_{\text{APPLICATION}}$.

Total water demand for irrigation can be projected n years into the future as follows:

$$\text{TWD}_{\text{IRRIGATION},t} \times (1 + \varnothing)\hat{n} \tag{8.3}$$

[a] Note that, in a strict theoretical sense, the "total consumptive water use" in crop production is different from "total water applied" and "total water applied to the crop" is also different from "irrigation dosage." This is explained in Annexure 1.

where $t=0$ (year), $\varnothing=$ annual compound growth rate of the population (fraction), n is the numbers of years. If we are considering 2025 as the year for demand projections, n will be 14.

While estimating irrigation water demand, losses in nonrecoverable deep percolation and soil evaporation have been considered. It has been assumed, however, that there is no percolation that contributes to return flows to groundwater systems or stream channels, given the arid climate and geohydrology of most of Pali district. The values of consumptive water use in irrigation and current irrigation water demand (amount to be applied in the field) work out to be same as a result, although they are conceptually different.

Available data from 2011 to 2012 show that wheat and mustard used 86% of the 417 MCM of water used by *rabi* crops, while, among *kharif* crops, irrigated cotton uses most (64%) of the 48 MCM used during the season (Table 8.1).

While estimating the irrigation demand at site, $\Delta_{\text{IRRIGATION},\,t}$ after n years, efficiency improvements can be assumed, which will affect TWD for irrigation as follows:

$$\text{TWD}_{\text{IRRIGATION},t} \times (1+\varnothing)^n \times \frac{\text{IE}_{\text{APPLICATION,CURRENT}}}{\text{IE}_{\text{APPLICATION,FUTURE}}} \tag{8.4}$$

Table 8.1 Crop water demand, Pali district, 2011–12.

Season	Crop	Area under crop (ha)	Crop water requirement (MCM)	Crop water use[a] (MCM)	% to total agricultural water use in season
Rabi	Wheat	47,922	162.54	203.18	49
	Rapeseed and mustard	34,880	119.72	149.65	36
	Cumin	14,665	17.6	22	5
	Gram	6419	17.34	21.67	5
	Barley	4163	11.92	14.9	4
	Total	—	333.55	416.94	100
Kharif	Cotton	6261	24.36	30.45	64
	Castor	1307	8.25	10.31	22
	Chillies	464	4.27	5.34	11
	Maize	462	1.14	1.42	3
	Total	—	38.03	47.54	100
Annual total		—	371.58	464.48	—

[a]Including nonbeneficial evaporation and nonrecoverable deep percolation.

In the current estimation, water application efficiency is assumed to improve from 80% to 90%, over the 13-year period till 2025, which reduces the annual per-capita agricultural water demand from 228 m^3 in 2012 in to 202.7 m^3 in 2025. The total agricultural water demand for 2025 was estimated using the projected population of the district, i.e., 2.384 million persons (based on an estimated annual compounded growth rate of 1.13%) to be 483.3 MCM.

8.3.2 Domestic water demand

Per-capita domestic water demand is not likely to vary much between years, with variation in climate, and hence it is reasonable to assume that the per-capita water demand rate remains the same within shorter time horizons. A change in per-capita income could, however, drive up domestic water demand and this assumption should therefore be modified for longer time durations as the impacts could be significant. Here, domestic water demand considers only water demand for drinking and cooking, washing, and sanitation (bathing and toilet use).

$$\text{Per}-\text{capita domestic water demand} = (1+\varnothing)^n \times \partial \times 365/1000 \qquad (8.5)$$

where ∂ is the average daily per-capita domestic water use (in liters) and $\varnothing$ is the income elasticity of domestic water demand (expressed as a ratio of the increase in domestic water demand per annum against a 1% increase in per-capita annual income).

Assuming a per-capita water demand of 70 lpcd in rural areas and 140 lpcd for urban areas, and an estimated rural population of 1.813 million persons and urban population of 0.578 million persons, the 2025 demand for water for domestic uses in rural and urban areas was estimated to be 46.33 and 29.52 MCM, respectively. This represents a total domestic water demand of 75.85 MCM for Pali district in 2025.

8.3.3 Livestock water demand

Livestock water demand for direct consumption is a function of the animal body weight, feed,[b] ingested salt, lactation, ambient temperature, and an animal's genetic adaptation to its environment. The amount of water

[b] This is a function of animal body weight.

required by a unit of livestock for direct consumption increases with an increase in daily dry matter intake (kg) and, in the case of milch animals, the average daily milk yield (liters), as well. Accordingly, small ruminants such as goat and sheep require less water as compared to large livestock such as camels, buffaloes, and cows. Lactating animals would, however, require more water for consumption than dry animals. For example, indicative water intake by dairy cows could be estimated by the following equation (after Pallas, 1986):

$$Y = 16.0 + 0.71\,i + 0.41\,m + 0.05\,S + 1.2\,T \tag{8.6}$$

where Y is the daily water intake (in liters, assuming 1 L of water weighs 1 kg), i is the daily dry matter feed intake (kg/day), m is the daily milk production (kg/day), S is the sodium intake (g/day), and T is the mean weekly minimum temperature.

Indicative water requirements for different types of livestock (per tropical livestock unit or TLU) are in Table 8.2, which also gives normal figures of animal body weight and the TLU they constitute. Actual water requirements of these livestock need to be worked out based on the actual body weight of the animals, by using the voluntary daily water intake for that category of animal per TLU. For instance, for a buffalo with a body weight of 400 kg, the water requirement would be 38.6 (per unit of cattle) × 400/250, i.e., 77.2 L per day.

Future livestock water demand can be estimated using the observed past annual compounded growth rate of livestock population for each category of livestock, and the population.

Table 8.2 Indicative water requirements for different types of livestock.

Livestock type	Mean live weight (kg)	Tropical livestock units per head	Daily dry matter intake		Voluntary daily water intake by season and average temperature (L/TLU)		
			kg	kg/TLU	Wet	Dry	Dry, hot
Camel	410	1.60	9	5.60	9.4	21.9	31.3
Cattle	180	0.70	5	7.00	14.3	27.1	38.6
Sheep	25	0.10	1	10.00	20.0	40.0	50.0
Goat	25	0.10	1	10.00	20.0	40.0	50.0
Donkey	105	0.40	3	7.50	5.0	27.4	40.0

Source: FAO online catalog on water for animals, based on Pallas, P., 1986. Water for Animals. Food and Agriculture Organization, Rome.

The total livestock water use in Pali district was estimated to be 7.87 m^3 per annum in 2007 (the latest year for which data on livestock population were available).[c] Assuming that in the future that per-capita livestock, the size of holding, the composition of livestock, and water use efficiency in livestock and dairy production remain the same, and hence using only population as a driver of livestock water demand projection, the total water demand for livestock in Pali district for 2025 was estimated to be 18.774 MCM, using the following calculation:

$$TWD_{LIVESTOCK\,2025} = Per - capita\ livestock\ water\ use_{2007} \times Population_{2007} \times (1.0113)\hat{1}8 \tag{8.7}$$

8.3.4 Industrial water demand

The major industries of Pali district are cement plants, textile industries, and small industries. But, in Luni basin, there is a thermal power plant (located near Barmer), having a capacity of 250 MW. The total water consumed in thermal power plants for 1 h of operations (i.e., for 1 MW of power generated) is in the range of 3.5 m^3 per MW of installed capacity. This, however, is only 10% of the total water required for various uses in thermal power plants. The water requirement for different types of industries, excluding thermal power plants, is given in Table 8.3.

The total water demand for industrial production can be estimated based on the total volume of production and the water consumption per unit of production. Total industrial water demand per annum can be estimated as

$$(TWD_{INDUSTRY}) = \sum_{i=1}^{n} V_{PRODUCTION,\ i} \times \varnothing_i \tag{8.8}$$

where $V_{PRODUCTION,\ i}$ is the volume of production for the industry type, i and $\varnothing_i$ is the water consumption per unit of production for industry type i. Given the water consumption per unit of production for any of the 17 industrial types (i.e., from Table 8.3) and, the total volume of output for that industry is known, the annual water demand can be calculated.

Unfortunately, however, data on the quantum of industrial outputs produced (like the total amount of cement, textiles, chemicals, etc.) under each

[c] See details of the livestock population in Annexure 2.

Table 8.3 Industrial production, water use per unit for selected industrial sectors (1996–97).

Category of industry	Water requirement (m^3)/ ton/year (Ø)
Integrated iron and steel	22
Smelters	82.5
Petrochemicals and refinery	17.0
Chemicals—caustic soda	5.5
Textile and jute	20
Cement	4.5
Fertilizer	16.7
Leather products	30
Rubber	6.6
Food processing	6.8
Inorganic chemicals	200.0
Sugar	2.2
Pharmaceuticals	25.0
Distillery (requirement per 1000 L)	22.0
Pesticides	6.5
Paper and pulp	200.0
General engineering	2.2
Total	669.5

Source: GOI (1999) Integrated Water Resources Development: A Plan for Action, Report of the National Commission on Integrated Water Resources Development.

manufacturing type are not available for the districts falling in the basin. An alternative approach had therefore to be used, taking estimates of total industrial water demand for India (available in Kumar (2010) for the period from 2000 to 2025). This gives a total industrial water demand in the country in 2010 of 23.955 BCM, and an average per-capita industrial water demand for the country of 19.5 m^3 per annum. Noting that Seckler et al. (1998) used a per-capita industrial water demand of 40 m^3 per annum for the work on global water supply-demand estimates, a modest figure of 20 m^3 per capita has been assumed for Pali district. Total industrial water demand for 2025 was thus estimated to be 47.7 MCM.

8.3.5 Summary of water demands

The total water demand in Pali district is estimated to be 570.53 MCM in 2011 and projected to rise to 625.62 MCM by 2025 (Table 8.4 and Fig. 8.1).

Table 8.4 Current and projected sectoral water demands (MCM).

Sector	Water use at site (2011)	Demand in 2025
Irrigated agriculture	464.56	483.30
Per-capita irrigation water demand (m^3 per annum)	228.00	202.67
Rural domestic	28.80	46.33
Per-capita (domestic): lpcd	50.0	70.00
Urban domestic (MCM)	20.10	29.52
Per-capita (urban domestic)	100.0	140.00
Industry	41.04	47.70
Per-capita industrial water demand	20 m^3/annum	
Livestock	16.03	18.774
Per-capita livestock water use	7.87 m^3/annum	
Total	570.53	625.62

Authors' own estimates based on Seckler, D., Amarasinghe, U., Molden, D., de Silva, R., Barker, R., 1998. *World Water Demand and Supply, 1990 to 2025: Scenarios and Issues*, Research Report 19, International Water Management Institute, Colombo, Sri Lanka; Kumar, M.D., 2010 *Managing Water in River Basins: Hydrology, Economics, and Institutions*, Oxford University Press, New Delhi; and Census 2001 and Census 2011, and analysis of secondary data, FAO CROPWAT estimates of crop PET.

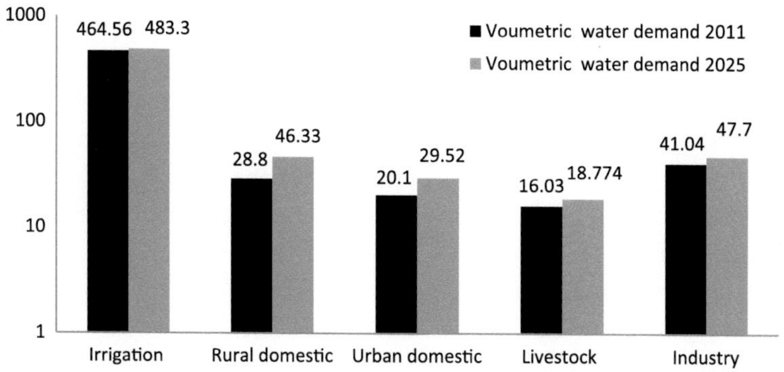

Fig. 8.1 Current and future water demand in different sectors (2025): Pali district (MCM).

8.4 Analyzing water supply

Arriving at a district IWRM plan requires balancing all water demands within the district against the utilizable supplies from various surface and groundwater sources. Not all the water available in a basin or administrative unit can be utilized, whether surface water or groundwater. The first task therefore is to get reliable data on utilizable surface and ground water.

8.4.1 Surface water resources

Estimating stream flows based on simple rainfall-runoff relations can lead to large errors when the interannual variability in rainfall is excessively high and that the runoff coefficient itself changes drastically with quantum of rainfall. In many arid tropics, assessing surface runoff in small river basins is a difficult task, as these basins are not gauged. One reason for this lack of attention to small basins is that most of these rivers are ephemeral and, in most years, do not have any discharge. The Luni river basin has only two gauging points, maintained by the Central Water Commission (CWC), at Balotra and Gandhav, both of which fall outside Pali district.

An alternative approach is needed therefore to estimate surface water availability. One approach involves estimating runoff from rainfall data using the US Soil Conservation Bureau's curve number method, which uses rainfall infiltration rates, based on the soil type, vegetation cover of the soil, and the antecedent moisture. Since the daily rainfall data are available for a large number of stations within the district, runoff for each rainfall event occurring during the monsoon in a year can be estimated using data on land use and land cover available for the area, corresponding to each year. The stream flow series can thus be generated using the historical data of rainfall and land use for each year, the dependability curve for the runoff can be estimated using these data, and "dependable yield" from the catchment can be worked out corresponding to different degrees of dependability. The rainfall-runoff relationship can be estimated, based on the estimated stream flow series and the weighted average of the annual rainfall, for areas with similar rainfall pattern, land use, soil, and climate in the basin or adjoining basins.

The other approach is to estimate the rainfall-runoff relation for the "basin as a whole" using the "estimated annual virgin flows for the basin," and the annual (spatial) average rainfall, and subsequently applying the same for the district of Pali, for the rainfall it receives in each year.

$$\text{Virgin flow VF} = \text{RUNOFF}_{\text{OBSERVED}} + \text{WD}_{\text{EFFECTIVE}} \qquad 8.9$$

where $\text{WD}_{\text{EFFECTIVE}}$ is the effective water diversion, or the total amount of water diverted from the main channel and tributaries in the basin, through various water-impounding structures, which is the sum of the water releases from reservoirs to meet various demands, water evaporated from the reservoirs, and the water remaining in storages at the end of the year from the impounded water, minus the return flows from various demand sites to the stream channels.

For river basins such as Luni, there are several small water harvesting structures in Pali district and other districts of the basin, which are in the upper catchment areas. While estimating virgin flows, it is important to account for the water impounded by these structures. Though these small water bodies generally do not divert any water for any uses, some water percolates from these reservoirs and a significant amount of water evaporates. Therefore, if data are available, it would be advisable to incorporate this in the estimation of virgin flows.

The average annual rainfall of the basin can be estimated by taking the weighted average of the point rainfalls using the Theisson Polygon method as the rainfall gradient across the basin is not very sharp. However, when the gradient of isohyets is sharp, the area between two rainfall isohyets should be estimated and multiplied by the average of the rainfall values represented by the isohyets.

The utilizable yield is to be calculated from the dependable yield. This would depend on the topographical conditions and the technology used for harnessing water, but the actual amount of water that can be tapped depends on the committed flows downstream. For instance, although the district catchment might have a dependable yield of "**X**" MCM per annum, there could be a reservoir downstream of the catchment (designed for a dependable yield of **0.50 X**) but falling outside the district boundary, which depends on inflows received from the upper catchment. This means that the total amount of water that can be utilized from the district catchment would only be just half the total dependable yield of the district catchment. A further possible reduction is on account of conveyance losses (see Annexure 3 for details).

To estimate the surface water potential in the district, the rainfall-runoff relationship estimated for the basin is used in conjunction with the average of the mean annual rainfall values for the rain gauges of Pali district.

First, from point rainfall data from nine rain-gauge stations in Pali, the weighted average of the rainfall was estimated for 41 years for which data were used (1971–2012)—which shows considerable variability (Fig. 8.2).

Next, the rainfall probability curve was plotted (Fig. 8.3) to map rainfall with a probability of occurrence of 3 in 4 years (75% probability of occurrence) and 6 in 10 years (60% probability).

Finally, rainfall with dependability of occurrence of 75% was estimated to be 318 mm and that with dependability of 60% was estimated to be 355 mm, but the average rainfall of 2011–12 (639.77 mm) was used, however, to estimate the utilizable runoff from the district as 109.26 MCM.

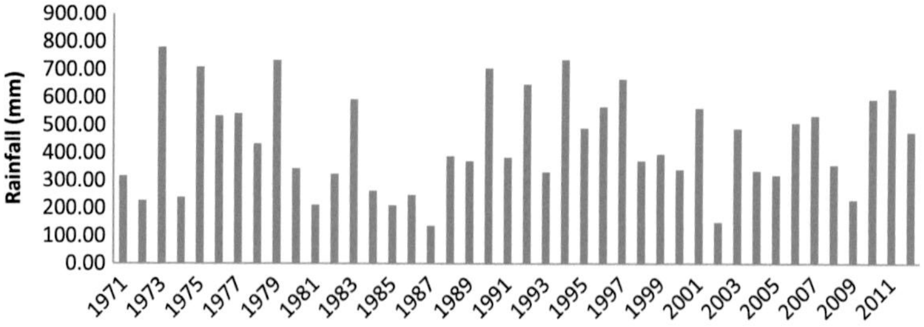

Fig. 8.2 Average rainfall in Pali district: 1971–2012.

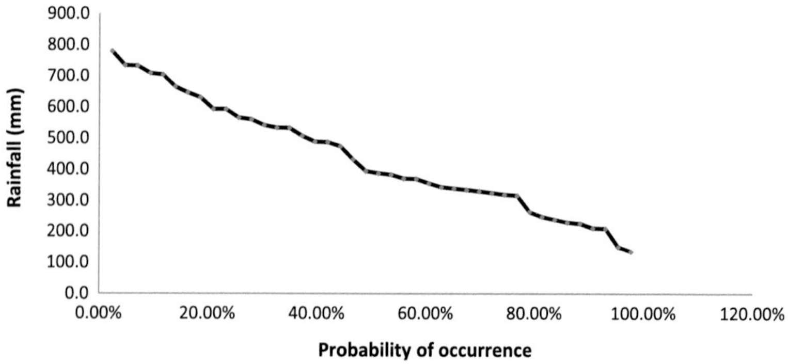

Fig. 8.3 Rainfall probability curve for Pali district.

8.4.2 Groundwater resources

Groundwater resources in western Rajasthan occur mostly in alluvial and sedimentary formations. All this water can be safely abstracted, as the groundwater flow gradients in the region are not steep and the groundwater outflows into natural streams are insignificant. In addition to the renewable groundwater resources, the region also has significant amount of static groundwater resources. Estimates of total amount of renewable groundwater resources in the district are readily available, as are data on the total amount of groundwater resources in the basin.[d]

Groundwater resources in both Luni basin and Pali district are fully developed at present (Rolta report). The wells and water abstraction devices available in the district of Pali can pump out more groundwater than what is permissible under sustainability considerations—as evidenced by current

[d] Tahal Consultants' Report on water resources in Rajasthan, Vol. 3.2.

abstraction rates prevailing in the basin and the district. If further depletion of groundwater is stopped, in deference to the sustainable development criteria, the amount of groundwater that can be tapped from the aquifers of the region would be just equal to or less than the renewable groundwater resources.

The total groundwater stock in the basin is around 10,500 MCM, in addition to an annual renewable groundwater resource of 2000 MCM. Pali district also has some groundwater stock. Therefore, if long-term sustainability of the aquifer is allowed to be compromised and aquifers allowed to deplete at the current rate, the amount of groundwater that can be tapped for various uses in Pali district and Luni basin would be equal to the current rate of withdrawal.

$$\text{Permissible level of abstraction} = GW_{ABSTRACT,\ ANNUAL} = GWR_{RENEWABLE,\ ANNUAL} + \left(\frac{GW_{STOCK}}{n}\right) \tag{8.10}$$

where n is the number of years for which the aquifer can be pumped at the current rate; GW_{STOCK} is the total groundwater stock; and $GWR_{RENEWABLE,\ ANNUAL}$ is the annual groundwater recharge.

Estimating the available renewable groundwater requires an understanding of the groundwater balance, with inflows and outflows. Annual groundwater replenishment (groundwater inflow) in an area can have three components:

1. Recharge directly from precipitation and from local water storage structures
2. Recharge from local water bodies after the monsoon season, when natural recharge from rainfall becomes negligible; and
3. Artificial recharge from recharge schemes.

Groundwater outflows during monsoons include pumping during the season, the net of lateral inflows and outflows (outflow-inflow). The basis for the estimation of groundwater balance is the change in storage, which can be measured through water-level fluctuations. The groundwater balance equation can thus be written as:

$$\Delta_{STORAGE} = RECH_{MONSOON} + (INFLOW_{LATERAL} - OUTFLOW_{LATERAL}) - PUMP_{MONSOON} \tag{8.11}$$

$$\Delta_{STORAGE} = WLF_{POST-PRE} \times S_y \tag{8.12}$$

where $WLF_{POST-PRE}$ is the difference in depth to water level between premonsoon and postmonsoon periods and $\Delta_{STORAGE}$ is the net storage change during monsoon.

Given the topography of the study area, lateral flows can be significant and hence are considered in the estimation of net recharge.

$$RECH_{MONSOON} = WLF_{POST-PRE} \times S_y + PUMP_{MONSOON} \tag{8.13}$$

Note that since whatever recharge occurs from the surface water bodies during June to October is already captured in the water level fluctuation in the wells, there is no need to estimate separately the recharge from water bodies for the monsoon season. Note also that $RECH_{MONSOON}$ as defined in Eq. (8.13) includes recharge from natural water bodies and artificial recharge schemes and thus, to estimate the recharge from rainfall alone (Components 1 and 2 above), the artificial recharge component will have to be segregated.[e]

Available estimates of utilizable groundwater resources in Pali district show that the gross groundwater draft exceeds recharge by 68 MCM annually in the district (Table 8.5).[f]

Though the blockwise estimates of static groundwater resources for Pali show a very modest figure of 47.77 MCM for the whole district, the estimates based on basin and sub-basin wise figures of groundwater resources show a much higher figure of 1242.88 MCM. Given that the rainfall of Pali district is comparatively higher than that of the rest of Luni basin, and that most of Pali district is underlain by deep alluvial formations, the latter figures appear to be more realistic, and therefore used. As per the 2009 estimates of Central Ground Water Board (CGWB), the net utilizable groundwater resource in Pali district is 302.06 (break up not available), considering the recharge during nonmonsoon period and ground water discharge (Government of India, 2011). Since these are more recent estimates (done for 2009), they have been used in the present analysis.

[e] To estimate the recharge from artificial recharge structures, a new method (double difference) must be employed, wherein the water-level fluctuation in two localities (receiving almost the same amount of rainfall)—one WITH recharge structure and the other WITHOUT the recharge structures—are considered and the difference in the average (pre- and postmonsoon) water-level fluctuation taken. This is to nullify the effect of natural recharge on water levels and disaggregate the effect of the structures. The same approach can be used to estimate the recharge from the structures postmonsoon.

[f] Estimates taken from the Report of Tahal Consultants.

Table 8.5 Static and dynamic groundwater resources of Pali district (2005–09).

Name of the block	Net groundwater availability (MCM)	Gross groundwater draft (MCM)	Static groundwater resources (MCM)
Pali	38.63	45.95	5.01
Desuri	22.65	28.67	2.70
Jaitaram	53.91	72.95	11.63
Marwar junction	38.01	55.86	6.71
Pali	8.48	7.63	1.25
Raipur	20.22	27.25	3.78
Rani	28.08	31.56	2.94
Rohat	5.96	5.56	0.68
Sojat	30.97	40.3	5.63
Sumerpur	32.51	31.72	7.44
Total	279.42	347.45	47.77

Source: Study on Planning of Water Resources Planning of Rajasthan, Draft Final Report submitted to the State Water Resources Planning Dept., Government of Rajasthan, December 2013, Vol. 3.2, 2c

Table 8.6 Available water supplies from surface and underground sources.

Water source	Renewable supplies per annum (MCM)	Static fresh groundwater resources in Pali district (MCM)[a]
Surface water (2011–12)	109.26	
Groundwater (average)	302.03	1242.88
Total water resources	411.29	

[a]Own estimate based on available estimates of groundwater in sub-basins falling even partly in the district.

8.4.3 Summary of water supplies

The total utilizable water resources that can be used to meet the demands in the district (based on 2011–12 estimates) are 411.29 MCM, per annum (Table 8.6). While around 75% of the 411 MCM of total annual renewable water supplies are groundwater, Pali also has around 1240 MCM of static fresh groundwater resources.

8.5 Balancing demand and supply

District IWRM planning involves deciding on a variety of water management interventions in water supply (water resource development, water imports, etc.) and water demand (water use efficiency improvement plans) in the district, so that future demands for water in the district—at the aggregate level and for individual sectors—can be met from utilizable future water supplies. While future water supplies can come from either within the district or from outside the district (but within the basin), interventions for water use efficiency improvements must come from within the district. This should, however, be done such that the integrity of the hydrological system (aquifers, stream flows) is not threatened in the long run, while simultaneously ensuring that there are no undesirable consequences for uses and users in other parts of the basin. It is, therefore, necessary that the analysis is done at both spatial and temporal scales, where the different types of interactions of the hydrological system and the interaction between hydrological system and water use systems are well integrated in the analysis of demand and supply balances.[g] Based on the estimated future supply-demand balance for water in the district, the nature and degree of water problems, in terms of both quantity and quality, can be identified.

8.5.1 Water availability-demand gap

Against the total renewable water availability of 411.3 MCM in 2011–12, the district consumed around 465 MCM of water for irrigation alone, nearly 90% of which is during the winter *rabi* (winter) season (Table 8.7).

Table 8.7 Agricultural water demand and water availability, Pali district, 2011–12 (MCM).

Total renewable water availability			411
Groundwater resources	109		
Surface water runoff	302		
Total water demand			570
Total agricultural water demand		465	
Rabi irrigation	417		
Kharif irrigation	48		
Total domestic water demand		48.90	
Total livestock water demand		16.03	
Total industrial water demand		41.04	
Demand-availability GAP			159

[g] The Water Evaluation and Planning (WEAP) system is a user-friendly water balance model that can be used for water resource assessment and basin planning. The water supply-demand balance for any geographical unit can be set up using WEAP model, and the same can be integrated into the basin in which the district falls.

Table 8.7 shows that, even in 2011–12, agricultural water demand is more than 80% of total water demand and total water demands exceeds supplies by 159 MCM: Clearly the district is not able to meet all its water requirements for domestic and productive uses from the available, renewable water resources in the basin. The deficit is currently being met through groundwater mining, but it is quite unlikely that the static groundwater resources in the district (1248 MCM) will be able to sustain the mining—which is currently occurring at the rate of 68MCM per annum—till 2025.

It must, however, be kept in mind that (1) utilizable water supplies can change drastically from a wet year to a dry year and (2) the area under irrigated crops can reduce significantly, thereby affecting major reductions in the actual demand for water. Hence, the analysis will have to be carried out for a long period of time, considering climatic variations and resource dynamics.

Since water demand is projected to increase to around 625 MCM by 2025 (see Table 8.4), and assuming no change in supplies, the current gap of 159.20 MCM will also increase to 214.30 MCM by 2025. Some potential interventions to meet the deficit are discussed now.

8.5.2 Water management interventions

These interventions are suggested based on a consideration of the following: hydrological regime such as rainfall, aquifer conditions, stream flows, and occurrence of hydrological extremes; socioeconomic conditions, viz., cropping pattern, irrigation pattern, and consumptive water use in agriculture and other sectors; climatic conditions; and overall water situation in the region.

- *Rejuvenation of tanks and ponds:* This will impact the supply side of the water balance and therefore will be reflected as an increase in storage capacity of the supply reservoir. Since this essentially will have to come from the natural catchments, it would mean reduction in downstream flows.
- *Construction of (traditional) water bodies:* Traditionally designed water harvesting structures (such as *tanka*s and *beris*) will either capture rainfall runoff or be linked to surface water sources such as canals (in the case of IGNP) or streams. These will have to be incorporated into a water allocation model like WEAP as new supply reservoirs (nodes).
- *Groundwater banking:* The loamy sand, the geological strata dominated by recent and old alluvium, high aridity, and depleted aquifers offer ideal conditions for artificial recharge of groundwater. The vast amount of land

available in the region, which is not cultivated, can be used as infiltration beds. The only constraint is the availability of a dependable source of water for recharge. But, the excess flow available in Luni in very wet years could be a good source of water for recharge. These excess flows could be used for groundwater banking in empty aquifers. Impact of recharge schemes will be on the resource availability and therefore will have to be affected in the estimates in terms of a return flow link to groundwater. Here, the major input will be rate of infiltration of water (which will be shown as a demand site return flow link in a WEAP model). The major sources of water from within the basin or from outside, which can be used for recharging the aquifers, need to be identified.

- *Urban piped water leakage prevention:* A large amount of water is wasted in urban water distribution systems through leakage, and the volume of leakage is a function of age of the infrastructure, length and pressure maintained in water distribution lines. Leakage reduction in urban water systems is an integral part of urban water management, which can not only help defer the investment required to augment the supplies, but also save water. This can be reallocated to other uses within the district or basin. But such measures should be resorted to only when the cost of leakage reduction, i.e., the cost of saving a unit volume of water through leakage reduction measures, is lower than the cost of production of new water source through conventional approaches, which includes the resource cost, the cost of infrastructure to develop and supply, and the environmental degradation cost. Intuitively, in water-scarce regions, the resource cost and the cost of environmental degradation would be high, thereby justifying such investments.
- *Microirrigation and mulching:* Research in north Gujarat has shown that microirrigation and mulching reduced irrigation water used (cubic meters per hectare) by 22%–80% of the original water use in a range of row crops (for more details, see Annexure 4).

8.5.3 Scale effects and economic viability

In order to determine the level at which some of the aforementioned water management interventions (schemes) should be implemented to address the problems that are likely to emerge in future, future scenarios of water balance need to be generated with a variety of potential interventions with different scales. These can be compared against the "base case scenario" of water balance to see the impact of different types of interventions on the overall demand-supply scenario and the resource condition. Scenarios, which produce the highest impact in terms of reducing the demand-supply gap and which have a positive impact on the resource condition (stream flows and groundwater regime) need, can be shortlisted.

Planning of integrated water management interventions cannot also be decided based on only the physical impacts of these potential interventions on future water supplies and demand: The economic viability of these interventions also needs to be examined. In order to identify cost-effective options, their costs per unit must be calculated and compared against potential benefits. Often, however, such information is not readily available. In such situations, these must be generated through heuristic economic analysis. Some key unknowns in such analysis are:

- The cost of harvesting a unit volume of water through traditional and modern techniques (e.g., water harvesting structures or *tanka* systems) against the economic value of the harvested water. In the case of large water-impounding structures, however, it is important to consider only the net hydrological gain, as there could be negative impacts of such interventions downstream.
- Cost of drip irrigation per unit area of crop, against the economic value of water saved from unit area and the incremental benefit from the use of drip irrigation.
- Cost of mulching per unit area of the crop, against the economic value of water saved from unit area and the incremental benefit from the use of mulching systems.
- The net benefit of rejuvenating traditional water bodies, against the those of alternatives.
- Cost of artificial recharge for "groundwater banking" in very wet years against the cost of getting freshwater for domestic water supply in rural and urban areas during water scarcity.

This additional analysis can modify both the types and scale of interventions in individual Gram Panchayats (GP) as well as for the district as a whole: For instance, the area under microirrigation for different crops, area under mulching for different rain-fed crops, the area under cultivation, the number of recharge schemes, the number of *tanka*s, number of tanks and ponds to be rejuvenated, may change—and some may also be dropped, based on their impact on the overall model outcomes vis-à-vis stream-flows, groundwater recharge, and water balance.

Further, based on the overall impact at the district level, the GP-level IWRM plans may have to be adjusted and revised (Fig. 8.4).

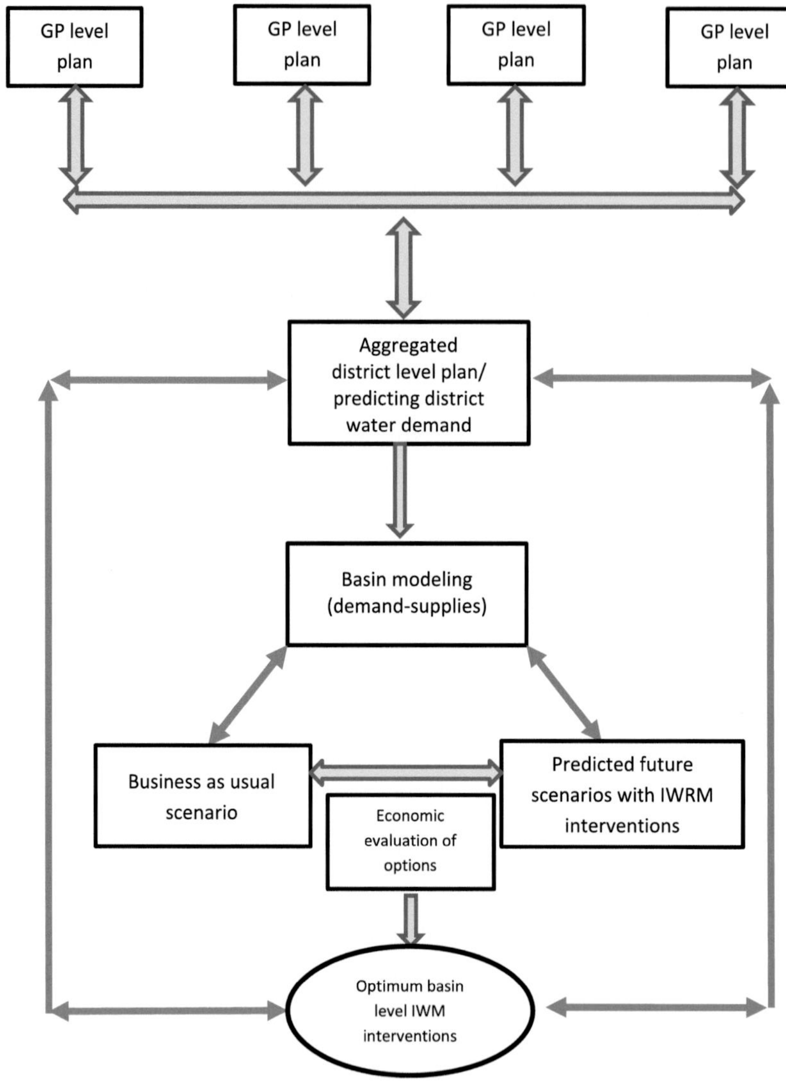

Fig. 8.4 Decision-making framework for district IWRM planning.

8.6 District water management options

There is little scope to eliminate the demand-availability gap in Pali district by reducing either domestic or industrial water consumption, as the overall water consumption in these sectors is much lower than that in agriculture—which is around 80% of total water demand (Table 8.7). Outside of bringing

Table 8.8 Potential impact of drip irrigation on row crops in Pali district (100% coverage): 2011–12.

Crop	Irrigated area (ha)	Total water use under traditional methods of irrigation ('000 m^3)	Extent of reduction in water use (%)	Reduction in irrigation water use (MCM)
Castor	1307	10,312	22	2.25
Cotton	6261	30,451	51	15.50
Chillies	464	5336	69	3.69
Fennel	1003	2462	30	0.73
Total	9035			22.17

Source: Based on Kumar, M.D., 2010. Managing Water in River Basins: Hydrology, Economics, and Institutions, Oxford University Press, New Delhi and INCID, 1994. Drip Irrigation in India. Indian National Committee on Irrigation and Drainage, New Delhi.

water from other river basins, two clear options to close the deficit by 2025 are (1) expand microirrigation and mulching and (2) restrict further increases in irrigated area.

8.6.1 Expand microirrigation and mulching

If all the crops that are amenable to drip irrigation are covered under micro-irrigation systems in the entire district, the impact in terms of reduction in water demand would be around 22 MCM (Table 8.8).

The analysis assumes that (1) all the water used for irrigation (which is only 10% higher than the ET demand) eventually gets depleted and (2) the use of drip irrigation for these row crops reduces nonbeneficial evaporation (of water from the soil not covered by canopy) and nonrecoverable deep percolation, thereby achieving "real or wet water saving."[h]

Plastic mulching for rain-fed *kharif* crops can give a potential "green water" saving of nearly 30 MCM per annum, given the current cropping pattern (Table 8.9).[i]

Even with the combined reduction of around 52 MCM, Pali will still not be able to meet all its domestic and productive water requirements from the available, renewable water resources in the basin (see Table 8.10).

[h] Seckler (1996) details the distinction between real water saving and notional water saving.

[i] Only crops with distant spacing between plants and between rows have been considered, and not crops such as wheat, pearl millet, sorghum, and maize.

Table 8.9 Impact of plastic mulching on rain-fed crops in Pali district (2011–12).

Crop	Irrigated area (ha)	Effective increase in water availability (MCM)
Maize	17,622	8.81
Groundnut	1353	4.80
Castor	9548	0.50
Cotton	990	0.68
Cluster bean	30,320	15.20
Overall	59,833	29.92

Source: Based on Xie, Z.-K., Wang, Y.-J., Li, F.M., 2005. Effect of plastic mulching on soil water use and spring wheat yield in arid region of Northwest China, Agric. Water Manag., 75, 71–83.

Table 8.10 Water management scenarios for the district: 2011–25.

	Demand-supply gap	
Gap between demand and renewable, annual utilizable supplies	2011	2025
Under the "base case" scenario	159	214
With drip irrigation technology (for row crops)		192
With plastic mulching (for rainy season crops) and drip irrigation		162

8.6.2 Reduce irrigated area

This option is especially for crops that are not amenable to pressure irrigation systems and mulching. The availability-demand gap of 159.20 MCM by 2025 can be achieved in either of two ways:

- *Without expanding pressure irrigation and mulching,* through a 45% reduction in the area under both wheat and mustard grown using conventional methods of irrigation.
- *With expansion of pressure irrigation and mulching wherever possible* AND reducing the areas under irrigated wheat and mustard by about 30%.

This assumes of course that the demand for water in the agriculture and other sectors will not grow in the future, either due to population pressure or industrial growth—or that the availability decreases. If either of these happens and the gap widens, the area under major irrigated crops (mustard and wheat) may have to be reduced by 45% from current levels to close the gap.

Both options to bring future water demands in line with available water supplies require strong government-led action on (agricultural) water demand management, without which not only will the mining of static groundwater continue—to the detriment of future generations—but rising competition (and conflict) for increasingly scarce surface water resources will also affect poor and marginalized farmers adversely.

Annexure 1: Water use hydrology of irrigation

Water use hydrology of irrigation is crucial in water resource planning as it affects not only water demand but also utilizable supplies of water (if water demands and supplies are considered on a reasonable time scale) due to the interactions between the irrigated fields and groundwater. Water use hydrology of irrigated field is complex, however, and depends on several factors including the method of irrigation, crop type, soils, climate, and the geohydrology. The interaction between surface water (irrigated fields) and shallow groundwater must first be defined, based on the extant irrigation practice. In the case of efficient irrigation technologies used for crops, the deep percolation to shallow aquifer can be assumed to be negligible. In the case of traditional methods of irrigation, however, the deep percolation rate can be reasonably assumed as the fraction in excess of the total water applied to the crop (which is the sum of soil moisture and the irrigation water applied) and the ET requirement of the crop.

$$\mathrm{DP}_{\mathrm{RECH}} = (\Delta_{\mathrm{APPLIED}} - \mathrm{ET}) \times F$$

where, $\Delta_{\mathrm{APPLIED}}$ is the sum of irrigation dosage and total soil moisture available from rainfall (also known as effective rainfall). For purely irrigated crops of winter and summer, the effective rainfall (P_{EFF}) can be assumed to be zero.

$$\text{Water consumed in crop production} = \Delta_{\mathrm{APPLIED}} - \{(\Delta_{\mathrm{APPLIED}} - \mathrm{ET}) \times F\}$$

In the case of deep groundwater table conditions (i.e., depth to water table exceeding 100 ft) in semiarid and arid climates, the value of F can be assumed to be zero, while in case of very shallow groundwater, the value can be assumed to be 1. In the latter case, real water saving from the use of microirrigation systems will be negligible.

$$\text{Maximum water saving from microirrigation} = \{(\Delta_{\mathrm{APPLIED}} - \mathrm{ET}) \times (1 - F)\}$$

Annexure 2: Livestock population in Luni river basin (2007).

Livestock (2007)	Districts in Luni river basin									
	Ajmer	Barmer	Jalore	Jodhpur	Nagaur	Pali	Rajsamand	Sirohi	Udaipur	Total
Cattle										
Indigenous	83,239	1,48,441	2,43,273	95,454	43,089	3,50,186	24,166	81,685	12,976	
Exotic	4182	99	262	5287	1651	4859	2174	374	273	19,161
Total cattle	87,420	1,48,540	2,43,535	1,00,742	44,740	3,55,045	26,340	82,059	13,248	11,01,670
Goat	1,71,924	5,17,495	4,53,096	2,15,945	1,52,005	7,01,932	47,212	1,38,202	15,308	24,13,118
Sheep	1,11,964	3,19,494	5,24,083	1,50,419	85,129	9,24,553	11,718	1,01,108	2265	22,30,732
Buffalo	74,990	36,459	3,40,806	40,505	49,255	3,15,418	20,571	67,234	7332	9,52,571
Pig	4388	969	2409	159	335	6505	182	0	19	14,965
Camel	465	13,692	6774	3908	1508	8289	234	2424	77	37,370
Donkey	323	4920	2977	765	256	2778	147	570	46	12,783
Horse	125	402	1362	91	128	1154	104	127	9	3502
Pony	1	0	0	0	2	12	0	0	0	15
Total	4,51,600	10,41,971	15,75,042	5,12,534	3,33,358	23,15,686	1,06,508	3,91,724	38,304	67,66,726

Source: Based on Livestock Census of Rajasthan 2007.

Annexure 3: Conveyance losses in irrigation and piped water systems in Pali district

The amount of water to be supplied to meet the demand in a sector or in a locality could be higher than the actual "demand" at the site, because of the losses in conveyance of water. Thus, conveyance losses need to be considered in estimating water supply requirements. The total water supply requirement (WS_{REQ}) can therefore be expressed as:

$$WS_{REQ} = W_{DEMAND}/(E_{CONV})$$

where E_{CONV} is the difference between the total amount of water supplied and the amount of water expected to be lost in conveyance for that level of water supply and supply schedules[j] against the total water supplied from the source, expressed as a fraction.

Such "losses" could be reusable depending on the local specific situation and characteristics of the water supply system.

Conveyance losses are generally low for groundwater-based irrigation systems, due to the short distance between the source and the point of demand (farmer's fields) even when open channels are used to convey water. Losses are generally high in surface water-based canal irrigation systems due to the long distance between the reservoir/diversion structure and the command areas, and the high infiltration from the (generally large-sized) canals.

Rural domestic water supply systems are generally of two kinds in this region: those based on local groundwater sources, where water is supplied to the village where the source is located and pipelines are used for distribution of water to different settlements; and those where water is taken from a distant surface water source (either a large reservoir or a diversion structure or river/canal lift scheme) to many villages through pipelines. Conveyance losses are higher in the latter due to the large distance of water transport, compared to local groundwater-based schemes.

Conveyance losses in piped water supply systems are, however, bound to be less than that of large surface irrigation schemes that use canals. "Technical losses" in long-distance piped water distribution systems used in drinking water supply schemes are due to leakages (from cracks and joints), while nontechnical losses occur due to illegal tapping of water.

[j] Note that the conveyance loss will change with a change in supply volumes and supply rates.

Losses in canal-based water distribution and delivery systems used in surface irrigation schemes, however, could be from seepage (depending on the type of lining, i.e., whether open channel, brick lining, cement lining, or concrete lining) and evaporation.

Technical losses from water conveyance systems end up, in most cases, in natural water courses or groundwater systems or lakes from which the water can be regained. Sometimes, however, this water can also end up in the sea/oceans or saline formations—like the Great Rann of Kachchh (in the case of Luni river) or the Little Rann of Kachchh (in the case of Banas river)—from where it cannot be recovered.

Ideally, in the case of irrigation water supply, the supply requirement needs to be estimated based on the demand at site and conveyance efficiency. In Pali district, however, groundwater is the major source of irrigation and only a miniscule area is under canal irrigation. Also, farmers using well irrigation do not use long open channels for conveyance as their wells are close to their plots. And since they use rubber and plastic (HDPE and PVC) pipes due to water scarcity, the conveyance efficiency is close to 100%.[k]

Annexure 4: Microirrigation and mulching

Microirrigation

The use of microirrigation (MI) systems (e.g., drips and sprinklers) as well as plastic mulching for water demand management will reduce water application rates for irrigated crops. Necessary modifications will, however, have to be made in the return flow fraction since the deep percolation rate under MI will differ from that of traditional irrigation methods.[l]

Row crops for which drip system can be used in the region are castor, groundnut, tomato, onion, maize, and cluster bean, and the total area under irrigated crops that are amenable to MI system needs to be assessed. In order to apply water saving technologies as a new intervention in the water balance model (e.g., WEAP), and to assess their impacts on overall water balance,

[k] Another important issue to be noted and resolved is that even when the conveyance losses are significant (like in gravity irrigation systems), there are dangers of factoring in the "conveyance efficiency" to arrive at the supply requirement. This is because such an approach can overestimate the supply demand gap in a "static analysis," as the return flows to groundwater or to the stream flows contributed by the "seepage losses" are not factored in the estimation of resource availability, which leads to underestimation of the supplies.

[l] This rate was imputed from data available from research in Rajasthan on the water saving impact of microirrigation.

it is important to know the physical impacts of these technologies. Research in north Gujarat has shown that drip irrigation reduced irrigation water used (cubic metres per hectare) by 22%–80% of the original water use in a variety of crops (see Table A4.1).

Mulching

The impact of mulching on water demand in rain-fed crops is in terms of reduced soil evaporation (E) values of these crops and an increase in the total water available to meet the transpirative demands from the soil profile (Xie et al., 2005). With mulching, the water demand of the crop would be equal to transpiration instead of evapotranspiration. In terms of demand calculation, it means that the supplementary irrigation dosage can be reduced. Row crops, which are amenable to mulching, and which are grown during the rainy season in the district, need to be identified. In order to assess the potential impact of mulching on water balance, we look at the extent of

Table A4.1 Potential change in irrigation water use for different crops after adoption of MI.

			Irrigation water use (m^3/ha)		
Season	**Crop**	**WST**	**Before adoption of WSTs**	**After adoption of WSTs**	**% change**
Monsoon	Cluster bean	Sprinkler	2549	1305	49
	Castor[a]	Drip	4160	3250	22
	Groundnut	Sprinkler	5603	5258	6
	Chili	Drip	11,500	3540	69
	Eggplant	Drip	5967	1180	80
	Green gram	Drip	840	540	36
	Cotton	Drip	7151	3510	51
	Fennel	Drip	2455	1728	30
Winter	Potato	Sprinkler	13,965	12,721	9
Summer	Pearl millet	Drip	8368	5031	40
	Millet	Drip	11,339	8776	23

[a]Data on water saving for castor were taken from INCID (1994).
Source: Kumar, M.D., Singh, O.P., Bassi, N., Niranjan, V., Sharma, M.K., 2010. Hydrological and Farming System Impacts of Agricultural Water Management Interventions for Sustainable Groundwater Use in North Gujarat & Strategies for Improving the Poor Farmers' Access to Groundwater, Final Report submitted to SRTT, Mumbai, Institute for Resource Analysis and Policy, Hyderabad.

Table A4.2 Estimated irrigation water requirement of selected (irrigated) Kharif crops.

Crop	Ajmer	Barmer	Jalore	Jodhpur	Nagaur	Pali	Sirohi
Pearl millet	150.0	338.2	338.2	351.4	241.9	240.0	350.6
Groundnut	233.8	532.8	532.8	482.7	358.2		482.7

Source: Based on estimation of irrigation water requirement of selected crops grown during the Kharif season from estimated values of PET and effective rainfall.

irrigation water use in crops grown entirely during the rainy season in the region. Making the fairly realistic assumption that by the application of mulch the irrigation needs for the crops grown entirely during rainy season in a normal year can be brought down to zero, the extent of irrigation water use for rainy season crops during a normal year can be considered as the total amount of rainwater that can be saved through mulching (Table A4.2).

From Table A4.2, it is evident that the irrigation water requirement of these two crops is every high and, in a few districts, higher than the rainfall of the district in question. This can happen only if the duration of the crop is long, with the result that the crop stands in the field much beyond the last rains but are reported as irrigated kharif crops. Under such circumstances, crops will not able to harness all the moisture in the soil profile, which is directly available from rainfall and the full crop water requirement will have to be met through irrigation. Therefore, only those crops should be considered wherein the estimated irrigation requirements are the lowest—or crops that are also grown extensively under rain-fed conditions.[m]

Here, the irrigation water requirement in pearl millet (grown extensively under rain-fed conditions in the basin) can be considered for all districts and for all crop areas that receive irrigation during the rainy season. However, this figure can only be applied to only one irrigated crop grown during the rainy season, i.e., groundnut.

$$\text{SAVING}_{\text{RAIN-WATER}} = 0.15 \times \sum_{i=1}^{m} \text{A}_i$$

where A_i is the area under the rainy season crop i.

The coefficient will, however, have to change as the rainfall conditions (in terms of duration of rainy season and the quantum of rainfall) change. For other situations, the equation for estimating green water saving can be derived as follows:

$$\text{SAVING}_{\text{RAIN-WATER}} = \Delta_{\text{IRRIGATION}-} \{\text{T} - P_{\text{EFF}}\}.$$

[m] Crops that can utilize the maximum amount of rainwater during growth stage.

where T is the crop transpiration and P_{EFF} is the effective rainfall during the crop growing period.

For crops that are purely rain-fed, it is assumed that a total of 50 mm of water could be saved through mulching. This value is quite reasonable. For instance, a study in northwest China on the effect of plastic mulching on crop water use, yield, and soil evaporation for spring wheat (unirrigated) showed that soil evaporation reduced by 0.49 mm per day, while the crop consumptive use (transpiration) increased (by 49 mm) resulting in 82% higher yield and 51% higher water use efficiency (Xie et al., 2005). In water balance models, this can appear as increase in water supplies (green water) during normal years and reduction in pumping for the irrigated rainy season crops.

References

Batchelor, C., 1999. Improving water use efficiency as part of integrated catchment management. Agric. Water Manag. 40 (1999), 249–263.

Government of India, 2011. Dynamic Ground Water Resources of India (as on March 2009). Central Ground Water Board, Ministry of Water Resources, Government of India, Faridabad. November 2011.

Howell, T., 2001. Enhancing water use efficiency in irrigated agriculture. Agron. J. 93, 281–289.

INCID, 1994. Drip Irrigation in India. Indian National Committee on Irrigation and Drainage, New Delhi.

Kumar, M.D., 2010. Managing Water in River Basins: Hydrology, Economics, and Institutions. Oxford University Press, New Delhi.

Pallas, P., 1986. Water for Animals. Food and Agriculture Organization, Rome.

Seckler, D., 1996. Wet and Dry Water Savings: The New Era of Water Resources Management. Research Report # 1. International Water Management Institute, Colombo, Sri Lanka.

Seckler, D., Amarasinghe, U., Molden, D., de Silva, R., Barker, R., 1998. World Water Demand and Supply, 1990 to 2025: Scenarios and Issues. Research Report 19. International Water Management Institute, Colombo, Sri Lanka.

Xie, Z.-k., Wang, Y.-j., Li, F.-m., 2005. Effect of plastic mulching on soil water use and spring wheat yield in arid region of Northwest China. Agric. Water Manag. 75 (2005), 71–83.

Further reading

Bhuiyan, C., Kogan, F.N., 2009. Monsoon variation and vegetative drought patterns in the Luni Basin in the rain-shadow zone. Int. J. Remote Sens. 31 (12), 3223–3242.

Central Water Commission, 2012. Integrated Hydrological Data Book (Non-Classified Basins), Hydrological Data Directorate, Information Systems Organization, Water Planning and Projects Wing. Central Water Commission, New Delhi, p. 2012.

Das, P.K., 1996. The Monsoons, third ed. National Book Trust, India.

Gleick, P.H., 1996. Basic water requirements for human activities: Meeting basic needs. Water Int. 29 (2), 83–92.

Kumar, M.D., van Dam, J., 2013. Drivers of change in agricultural water productivity and its improvement at basin scale in developing economies. Water Int. 38, 312–325. https://doi.org/10.1080/02508060.2013.793572.

Kumar, M.D., Singh, O.P., Bassi, N., Niranjan, V., Sharma, M.K., 2010. Hydrological and Farming System Impacts of Agricultural Water Management Interventions for Sustainable Groundwater Use in North Gujarat & Strategies for Improving the Poor Farmers' Access to Groundwater. Final Report Submitted to SRTT, Mumbai, Institute for Resource Analysis and Policy, Hyderabad.

Rathore, M.S., 2005. State Level Analysis of Drought Policies and Impacts in Rajasthan, India. Working Paper 93, Drought Series Paper No. 6. International Water Management Institute, Colombo, Sri Lanka.

Sharma, K.D., Murthy, J.S., 1998. A practical approach to rainfall-runoff modelling in arid zone drainage basins. Hydrol. Sci. J. 43 (3), June 1998.

Singh, N., 1994. Optimizing a network of rain-gauges over India to monitor summer monsoon rainfall variations. Int. J. Climatol. 14, 61–70.

Tahal Consultants, 2013. Study on Planning of Water Resources Planning of Rajasthan. Draft Final Report submitted to the State Water Resources Planning Dept. vol. 3.2. Government of Rajasthan, p. 2c. December 2013.

CHAPTER 9

Water accounting for Kanpur metropolitan area in Ganges river basin

Nitin Bassi[a], S.V. Suresh Babu[b], M. Dinesh Kumar[c]
[a]Institute for Resource Analysis and Policy, Delhi Liaison Office, New Delhi, India
[b]WWF-India, New Delhi, India
[c]Institute for Resource Analysis and Policy, Hyderabad, Telangana, India

9.1 Introduction

Domestic water supply in most Asian cities is highly unreliable and often inept to take care of needs of the growing population. The major reasons for this condition are poorly maintained water supply and treatment infrastructure; contamination of the supplied water; high "unaccounted for water losses"; and inefficient water pricing, which doesn't reflect the production cost and scarcity value of the resource (Bassi and Kumar, 2012; Kumar, 2014). The degradation of water supply sources due to pollution is exacerbating the problem. Though substantial amount of money has been pumped into improving the water supply infrastructure in cities of the developing world in recent times, these efforts by and large are still inadequate to meet the growing urban water demands.

In India too, ensuring drinking water security is a major challenge for the water utilities in large cities (Bassi and Kumar, 2012; Kumar, 2014). The challenge is compounded by the water demand from industrial and other competing use sectors (Kumar, 2014). Though a high proportion of population has an access to piped water supply, it is often highly unreliable and supplied water is of poor quality. Further in most of the urban centers, distribution losses alone are in the order of 30%–50% of the total water supplied (ADB, 2007), which further reduces the water available for the end user. The condition is even worse in the informal settlements and slums in the urban areas where basic water and sanitation infrastructure is almost nonexistent (Bassi and Kumar, 2012).

Current Directions in Water Scarcity Research, Volume 1
https://doi.org/10.1016/B978-0-12-814851-8.00009-4
© 2019 Elsevier Inc.
All rights reserved.

The other important aspect of challenges in urban water supply is the treatment of the generated wastewater. Less than 31% of the domestic wastewater from cities and towns is treated in cities of developing countries, compared to 80% in the developed world. Situation in India is not quite different where capacity exists to treat only 37% of the 61,754 million liters of sewage generated per day from urban centers (CPCB, 2016). However, the sewage treatment capacity in the metropolitan cities (population of 1 million and above) is substantially higher (51%) than in Class I cities (population above 100,000 but other than metropolitan cities) and Class II towns (population between 50,000 and 100,000) where it is 18% and 9%, respectively (CPCB, 2009). The actual amount of sewage treated is further low due to inadequate collection system and nonfunctional treatment plants. Thus, a large proportion of the sewage is discharged into water bodies without any treatment. This increases the pollution load in rivers and lakes and makes their water unfit for further use (mainly for domestic use) without treatment. As a result, pressure on groundwater resources to meet the city growing water demand also increases, which further leads to its overabstraction and depletion.

The city of Kanpur, which is one of the important urban centers in the State of Uttar Pradesh, has a similar problem concerning water supply. It is the tenth most populous city in India and one of the key urban and industrial conglomerations on the banks of river Ganges. The city is heavily dependent on the river for meeting its water needs and it also acts as a sink for most of the sewage generated in the city. As per the estimates from the Municipal Corporation, Kanpur city has a domestic water requirement of about 600 million liters per day (MLD), but the existing infrastructure can supply only about 385 MLD. The gap in the supply and the entire industrial water requirement is met by tube wells and hand pumps. However due to overabstraction, groundwater is depleting and most of the wells tapping shallow aquifers do not give sustained yield during peak summers. Also, both the groundwater and surface water resources in the city are heavily polluted. The city produces about 340 MLD of sewage of which only 170 MLD is treated (CPCB, 2013b). The unplanned disposal of domestic waste combined with discharge of untreated or partially treated effluents by the industries (leather, textile, jute, and chemical) directly on land or in the river is resulting in pollution of surface and subsurface water. The average total coliform count in the stretch of river Ganges in Kanpur was observed to be as high as 240,000 MPN/100 mL in summer months. Further, traces of heavy metals such as Cd, Cr, Cu, Fe, Pb, As, Zn, and Mg have also been observed in both surface and groundwater.

The water resources of the city of Kanpur are under tremendous threat due to their overappropriation to meet the growing municipal water demands. The absence of required wastewater treatment infrastructure and poor adherence to environmental protection norms by the industries pose a major threat to the river ecology and aquifers. It also means a huge risk to business, economy, and people's livelihoods, which are dependent on it. This risk is essentially shared between different water users and would require collaborative multistakeholder action at the city level for mitigation. In this backdrop, a study was planned to undertake comprehensive water accounting of Kanpur metropolitan area (KMA); identify present and future risks related to quantity and quality of water supply for domestic and industrial consumers; and analyze the existing institutional and regulatory framework related to urban water management in Kanpur city.

Based on a complete understanding of the water accounts—water supply, water consumption, leakage, wastewater outflows, and the amount of storm water generated of the city—alternative scenarios of urban water management (water supply augmentation, demand management, waste reduction, and pollution control) to reduce the water supply and pollution risks can be generated. Further, based on an understanding of the weaknesses of the existing institutions and policies governing water use and pollution control, institutional and policy alternatives can be evolved, keeping in view the nature of interventions for reducing the risks.

9.2 Methodology

First of all, to estimate the water accounts, water requirements of various competing use sectors in the city were estimated. This was done using data on current population, and current unitary demand in the water sector (for domestic, commercial, and industrial uses) in per-capita terms. The total annual and monthly water withdrawal from surface and groundwater resources in the city by the utility and private parties (housing stocks, hotels, industrial units, etc.) to meet various water demands was estimated as the inflows. The total losses in the system (comprising the technical and administrative losses) were arrived at from the estimates of water use from all approved connections (both metered and unmetered), and total water supplied from source. The "wastewater return flows" from the municipal areas and industrial units into the natural sinks, such as surface water bodies and shallow groundwater, were assessed using the data on estimated effluent load generated by various industries and municipal corporation; the existing treatment capacity in the city; and the location of the various natural sinks of untreated wastewater.

In order to assess the water-related risks, the risk assessment framework of World Wildlife Fund (WWF), which considers the physical risk (due to water supply gap and water pollution in future) and economic risk (due to regulatory regime and organization reputation), was used. The economic risk was linked to the physical risk. To assess the physical risk, the water supply requirements was estimated for projected future population and economic activities, based on historical growth in population, economic conditions, commercial enterprises, and volume of industrial outputs in the city. The likely future increase in volume of polluting effluents was also estimated, based on realistic assumptions of the potential increase in treatment capacity.

Using data on stream flows, groundwater recharge, water withdrawals, and effluent disposal, the present condition of the natural water system (hydrological system) was ascertained—in terms of depth to groundwater table, water-level fluctuations, degree of groundwater stress, alterations in stream flow regimes, and surface and groundwater quality. From this, what the natural system can yield on a sustainable basis in future (sustainable extraction levels) under a "business as usual scenario" was worked out. Subsequently, the likely future water withdrawal needs from both the surface and groundwater systems were compared against the projected "sustainable extraction levels," to assess the risk in water supply. Likewise, the future scenarios of wastewater generation, and reuse of untreated wastewater were compared against the future "carrying capacity" of the land and water ecosystems to assess the "pollution risk." Accordingly, the "economic risk," in terms of the size of economic activity, which is at stake, was assessed.

Existing institutional and policy framework governing the use of water in the city was analyzed to examine their impacts on water use, including pollution of freshwater bodies. The analysis would include the water supply and water-pricing policies for various sectors; presence of sewage tax; pollution control norms and the administrative mechanisms for their enforcement; urban land ceiling Act; building bylaws; and special rules on rainwater harvesting in urban areas. The exercise also involved consultations with the representatives of the city water utility, the State Pollution Control Board, and other stakeholders.

9.3 Kanpur metropolitan area

Kanpur metropolitan area (KMA) is an urban region on the southern bank of river Ganges in the State of Uttar Pradesh and is one of the oldest industrial centers of North India. KMA includes area under Kanpur city, Nagar Palika of Unnao and Gangaghat, and Nagar Panchayat of Akbarpur and Bithoor. As per the Census of India 2011, overall spread of the KMA is about 340 km^2 with an estimated population of about 3.2 million.

The climate in KMA is subtropical with average temperature varying from as low as 8°C during winters to 41°C during summers. The average annual rainfall is about 800 mm and exhibit high interannual variability. About 90% of the rainfall is received during the monsoon months (June to September). The relative humidity ranges from less than 30% in the summer season (March to May) to about 70% during monsoon.

9.3.1 Demographic changes

To analyze the demographic changes over time, details on only Kanpur city were considered as reliable information on the historical changes in KMA was not available. Kanpur city comprises of area under Kanpur Municipal Corporation including its outgrowth, Kanpur Cantonment Board, and Census Towns of Chakeri, Northern Railway Colony, and Armapur Estate. Spatially, Kanpur city has grown by only 4 km^2 over the last 50 years, i.e., from 296 km^2 in 1961 to 300 km^2 by 2011. However, population has almost increased by more than three times from mere 0.96 million in 1981 to 2.9 million in 2011. But, population growth during the last decade has registered a decline. Against a decadal growth of 33% recorded during 1991–2001, it was only 8% during 2001–11 (Fig. 9.1). As a consequence of exponential growth, population density has gone up from 3243 persons/km^2 in 1961 to 9666 persons/km^2 in 2011, thus putting more pressure on the existing natural resources in the city.

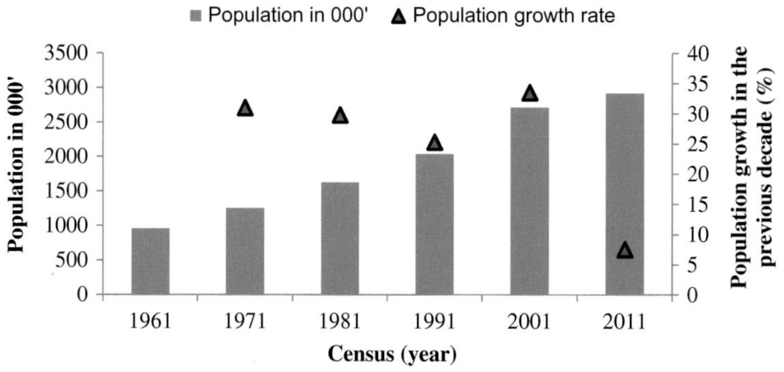

Fig. 9.1 Population growth in Kanpur city. *(Source: Authors' own analysis based on data presented in Census of India, 2011. Decadal Variation in Population Since 1901. Ministry of Home Affairs, Government of India, New Delhi.)*

9.3.2 Industrial growth

Kanpur is a heavily industrialized metropolitan area. By March 2011, there were more than 16,000 registered industrial units set up with a total investment of around US$ 138 million (Table 9.1). Out of these, nearly 99% are small-scale industries. Also, number of leather processing industrial units was found to be highest. However, almost 47% of the industrial units are either sick or closed (JNNURM, 2006).

Kanpur has a diverse industrial set-up, which has evolved over the years. By the turn of the 20th century, Kanpur had become an industrial center for textile and leather goods (Bansal and Bansal, 1984; Singh, 1990). Due to the lack of rail connectivity in the region, these industries flourished on the banks of the river Ganges to utilize the navigable channels of the river (Singh, 1990). The period from early 1900s to the independence of the country in 1947 saw the emergence of new industries such as sugar, jute, iron and steel, chemicals, etc. (Singh, 1990). Post the country's independence saw a period of consolidation for the existing industries in the region (JNNURM, 2006). Since 1984–85, there has been a steady increase in the

Table 9.1 Details of various industrial units in Kanpur metropolitan area (up to March 2011).

Sr. no.	Industries	No. of units	Investment (in million US$)
1	Food products and beverages	1221	6.0
2	Cotton textile	371	6.0
3	Hosiery and garments	2445	38.5
4	Wood products	424	5.4
5	Paper products	613	20.8
6	Leather products	2428	10.3
7	Rubber and plastic	611	14.9
8	Chemical	657	6.0
9	Mineral based	141	1.8
10	Basic metal and metal products	168	1.5
11	Engineering units	1215	12.8
12	Electrical machinery and transport equipment	181	2.5
13	Repairing and servicing	3189	3.2
14	Others	3011	8.3
	Overall	16,675	138

Source: Authors' own analysis based on data presented in GoI (Government of India), 2011. Brief Industrial Profile of District Kanpur Nagar. MSME-Development Institute, Ministry of MSME, Government of India, Kanpur.

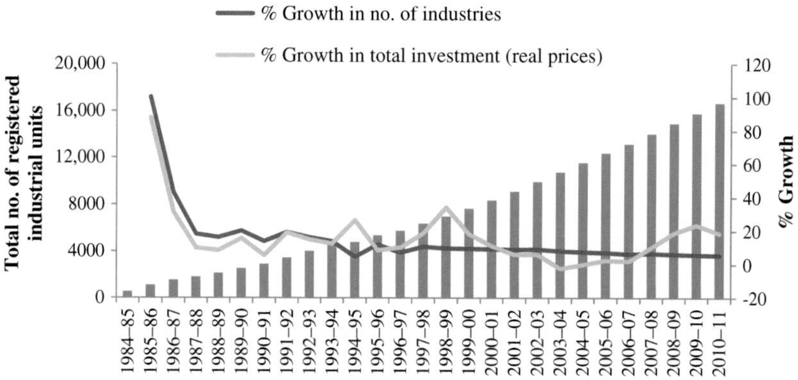

Fig. 9.2 Industrial growth in KMA. *(Source: Authors' own analysis based on data presented in GoI (Government of India), 2011. Brief Industrial Profile of District Kanpur Nagar. MSME-Development Institute, Ministry of MSME, Government of India, Kanpur.)*

number of industries in Kanpur (Fig. 9.2). However, from 1993 to 1994 onwards, industrial growth rate (in terms of new industries opened every year) has slowed down in comparison to the previous years. Against an average annual growth rate of 2.9% between 1984–85 and 1993–94, it has dipped to 0.47% between 1994–95 and 2010–11. Lack of infrastructure development and inadequate power supply is being identified as one of the reasons for closing down of numerous big and small industrial units since the 1990s (Mishra, 2006). Investment patterns have also shown similar trends. However, post 2004–05, it almost doubled by 2009–10. This indicates that investors are now looking for medium and large-scale industries to have better economies of scale. But it will also put pressure on existing water resources in city to cater to the water demands of large-scale manufacturing units.

Nevertheless, significant growth was witnessed in the small-scale industries segment, which increased from mere 7000 units in 1998 to around 16,000 by 2011. During this period, industries based on leather, light engineering, hosiery and garments, chemicals, and food processing have grown. Number of tanneries alone has increased from 175 units in 1994 to 402 units by 2015. More than 50% of them use chrome tanning process. Untreated effluent from these units is the main source of pollution of the Ganges. Though it is mandatory for tanneries to have an effluent treatment plant within their premises, most of them discharge effluent with only primary treatment. Further, many of these primary treatment plants are nonfunctional. In 1994, UPJN installed a 36-MLD common effluent treatment plant to treat the effluent from tanneries along with the domestic wastewater to make it more amenable to biological treatment. But since then effluent discharge from tanneries has increased from 9 to 26 MLD, making the plant operation difficult.

9.4 Status of water sources, water supply, wastewater and its treatment in KMA

9.4.1 Surface water: River ganges

Kanpur metropolis forms a part of Ganges sub-basin in the Central Indo-Gangetic Plain. Average annual renewable surface water availability was worked out using the observed streamflows in the stretch of river Ganges passing through KMA. As per the data available from the hydrological station located at Kanpur, the long-term average annual discharge at Kanpur is about 1679 cubic meter per sec (cumec) (O'Keeffe et al., 2012). This means that average annual renewable surface water availability in Kanpur stretch of the river is about 53 billion cubic meters (BCM) per annum. However, there is significant variation in water availability across seasons with average flow being less than 1000 cumec during the lean period (November to May).

As per the water quality monitoring conducted in 2011 by the CPCB, the quality of water in river Ganges deteriorates significantly after passing through Kanpur city (Table 9.2). While the dissolved oxygen (DO) content reduces substantially in the downstream in comparison to the upstream stretch, the biological oxygen demand (BOD) gets doubled. The EC also increases substantially. This indicates presence of greater amount of organic and inorganic effluent entering the river water. Though fecal coliform at the upstream stretch is above the permissible amount (<2500 MPN/100 mL), it increases to almost five times in the downstream stretch. The worsening of river water quality is mainly due to the untreated wastewater from the city, which is drained into the river. Thus, the river water, which leaves the city, is heavily polluted and unusable for further use without treatment and poses a health hazard for the downstream users.

9.4.2 Groundwater

In KMA, ground water occurs under unconfined to confined conditions. Depth to ground water level varies from 6 to 16 meter (m) during premonsoon. During post monsoon, groundwater level shows a minor rise due to replenishment of aquifer by the rainfall. Ground water flow is in the eastern direction of KMA, confirming the effluent nature (getting inflows from

Table 9.2 Results of water quality monitoring of river Ganges at upstream and downstream of Kanpur city (2011).

		Water quality monitoring station					
		Kanpur U/S (Ranighat)			Kanpur D/S (Jajmau pumping station)		
Sr. no.	Parameter	Min	Max	Mean	Min	Max	Mean
1	Temperature (°C)	15.0	32.0	25.8	15.0	32.0	25.8
2	Dissolved oxygen (mg/L)	6.0	12.8	8.6	4.0	11.5	6.9
3	pH	7.4	9.1	8.2	7.1	8.8	8.0
4	Electrical conductivity (μmhos/cm)	48.7	560.1	315.7	54.9	637.3	380.9
5	BOD (mg/L)	3.0	5.5	4.3	6.6	9.6	8.4
6	Nitrate (mg/L)	0.5	4.0	2.0	1.1	4.7	2.8
7	Fecal coliforms (MPN/100 mL)	900.0	28000.0	6666.7	4000.0	93000.0	38941.7
8	Total coliforms (MPN/100 mL)	1100.0	150000.0	31991.7	23000.0	240000.0	151333.0

Source: CPCB (Central Pollution Control Board), 2013b. Pollution Assessment: River Ganga. Central Pollution Control Board, Ministry of Environment and Forests, Government of India, New Delhi.

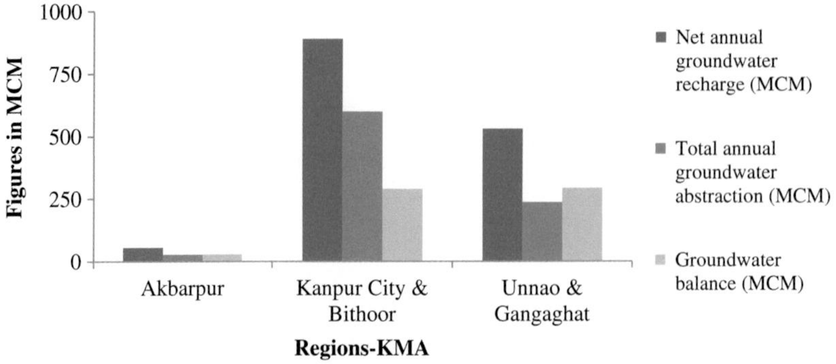

Fig. 9.3 Estimated groundwater availability in KMA. (Source: *Authors' own analysis using data from Central Groundwater Board Data, Government of India.)*

groundwater) of river Ganges in this stretch. Nevertheless, long-term ground water-level data from the seven observation wells in KMA show a general declining trend with depth to groundwater level going down by up to 9.6 m.

Annual groundwater availability in KMA was estimated from the "per unit area replenishable annual groundwater resources" derived from the corresponding data for the three districts of Kanpur (Kanpur Dehat, Kanpur Nagar, and Unnao) where parts of KMA falls. As per the estimates, net annual groundwater availability in KMA is about 1475 million cubic meters (MCM) and the total annual draft (including for irrigation) is about 863 MCM. This indicates that annually, 612 MCM of groundwater is available for further use. It is highest in Unnao and Gangaghat region followed by Kanpur city and Akbarpur (Fig. 9.3).

Table 9.3 presents the groundwater quality in two industrial areas of Kanpur city. EC was found to be high in Jajmau industrial area where most of the tanneries are located and nitrate was found to be high in Panki industrial area. Apart from fluoride, other parameters were found to be within permissible limits. Fluoride content was found to be high in groundwater in Jajmau in comparison to the Panki industrial area.

9.4.3 Water supply system

Piped water supply in Kanpur city was started in 1892. Responsibility of construction of water supply and sewerage system in Kanpur city is with the Uttar Pradesh Jal Nigam (UPJN), whereas its operation and maintenance is under the purview of Jalkal Vibhag (JV), which is under the

Table 9.3 Groundwater quality in Kanpur city.

		Location of observation wells					
		Jajmau industrial area			Panki industrial area		
Sr no.	**Parameter**	**Min**	**Max**	**Mean**	**Min**	**Max**	**Mean**
1	Temperature (°C)	27.0	27.0	27.0	27.0	27.0	27.0
2	pH	7.8	7.8	7.8	6.9	6.9	6.9
3	Electrical conductivity (μmhos/cm)	796.4	796.4	796.0	548.2	548.2	548.0
4	BOD (mg/L)	2.4	2.4	2.4	0.8	0.8	0.8
5	Nitrate (mg/L)	2.5	3.2	2.9	34.5	34.5	34.5
6	Fluoride (mg/L)	3.0	3.0	3.0	1.2	1.2	1.2

Source: CPCB (Central Pollution Control Board), 2013a. Status of Water Quality in India—2011. Central Pollution Control Board, Ministry of Environment and Forests, Government of India, New Delhi.

Table 9.4 Different sources of water, installed capacity, and actual supply in Kanpur city.

Sr. no.	Type	Source of raw water	Number	Installed capacity (MLD)	Actual supply (MLD)
1	Surface water	Ganges Channel, Bhaironghat	1	310	210
		Lower Ganges Canal	1	130	50
2	Ground water	Tube wells	135	110	80
		Hand pumps	9830	50	50

Source: Based on data present in Kanpur city development plan.

Kanpur Municipal Corporation (KMC). Against an installed capacity of 600 MLD to draw raw water, about 390 MLD of water is actually supplied through the formal water supply system in the city. The system mostly uses water from river Ganges to meet the domestic and nondomestic water demand. About 130 MLD of abstracted groundwater is also supplied through the same system (Table 9.4). Additionally, 30 MLD of water (mainly from surface water sources) is supplied in the Unnao Municipal limits.

Out of the total water supply, only 255 MLD is treated at the three water treatment plants (Table 9.5). However, groundwater is supplied without any treatment. JV supply water only in the area under the city municipal limits and also sell bulk water to the Kanpur Cantonment Board. For the purpose,

Table 9.5 Existing raw water treatment capacity in Kanpur city.

Sr. no.	Location of treatment plant	Year of installation	Installed capacity (MLD)	Actual running capacity (MLD)
1	Benajhaber	–	310	210
2	Gujaini	2005	30	20
3	Ganges barrage	2005	200	25
	Overall		540	255

Source: Based on data present in Kanpur city development plan.

area under Kanpur Municipal Corporation is divided into four service districts, namely, City, West, South, and East. There is no formal water supply system in rest of the city. In these areas, households depend on private tube wells to meet their domestic and other water demand. Further, all the water requirement by the industries is met by abstraction through private tube wells.

Presently (2011), water supply at the end user level by JV for domestic users having connections is estimated to be around 97 liters per capita per day. This is after accounting for the 30% administrative and the technical losses (UFW) on account of water theft, and old and leaky pipelines. This is much below the norms (135–150 lpcd) suggested in the CPHHEO manual on water supply and wastewater treatment. Water supply in the inner core area is much less than the 97 lpcd, as the pipes are leaky and often supply contaminated water. Hence, water supply in these areas is being augmented by local community-based tube wells or hand pumps. Though there are a large number of private wells in the residential and industrial areas, they are unaccounted and no such data exist with urban local bodies or parastatal agencies.

Under the Jawaharlal Nehru National Urban Renewal Mission (JNNURM), a pan-Indian city modernization scheme (2005–14), about US$ 100 million was sanctioned to support the existing water supply system of KMA. Among others, it included construction of two new water treatment plants (200 MLD each) and about 1746 km of water distribution system covering Kanpur city inner old area (under Phase-I by 2010) and remaining parts of the city (under Kanpur Water Supply Scheme Phase-II by 2012). However, the execution of work was very slow and missed several completion deadlines. As per the last available information, works under Phase-I were completed in June 2016 whereas those under Phase-II are yet to be completed.

9.4.4 Wastewater and its treatment

Estimated sewage generation in Kanpur city is about 339 MLD and that in Unnao town is about 24 MLD (CPCB, 2009). However, the total wastewater generation (including runoff) in KMA is estimated to be about 767 MLD. Apart from use of a minor amount of treated wastewater in irrigation, rest of it and the untreated wastewater is drained into the river Ganges in north and river Pandu (a tributary of Ganges) in south of the Kanpur city. Table 9.6 presents details on wastewater being generated from different regions in KMA. *Sismau* and *Permiya Nalas* contribute to about 50% of total wastewater, which is discharged into river Ganges and river Pandu.

Presently, sewerage system exists only in Kanpur city inner core area but not in the city periphery. Overall, it covers only 29% of the city area, which is being administered under four districts and five different zones. Total length of the main and trunk sewers is 74 km, whereas branch sewer lines are 875 km. There are total 13 sewage pumping stations and 30,000 manholes. Sewer lines are cleaned by sewer jetting machines, sewer clearing machines, and also manually as per requirement. In other regions of the KMA, sewerage system is nonexistent.

Coverage by sewerage network is highly uneven across the city and is mostly concentrated in the city drainage district (Table 9.7). In South drainage districts too, some parts are covered under the sewerage network. However, in east and west drainage districts, sewerage network is totally nonexistent.

There are four Sewage Treatment Plants (STP) in Kanpur city, and majority of them are in Jajmau where tanneries are located (Table 9.8). Out of these, one 210-MLD STP constructed under JNNURM was commissioned only in 2014 after several delays. Also, there is one treatment plant (19 MLD) in Unnao town. In rest of the KMA, storm water drains carry sewage and

Table 9.6 Amount of wastewater being contributed by different regions in KMA.

Sr. no.	Catchment region	Discharge point	Estimated discharge (MLD)
1	Kanpur city	River Ganges	598.19
2	Unnao	River Ganges	77.76
3	Other areas in KMA	River Pandu	91.27
	Overall	–	767.22

Source: Based on data present in Kanpur city development plan and CPCB (Central Pollution Control Board), 2013b. Pollution Assessment: River Ganga. Central Pollution Control Board, Ministry of Environment and Forests, Government of India, New Delhi.

Table 9.7 Coverage of sewerage network in Kanpur city.

Sr. no.	Waste water generating districts	Area covered by sewerage network	Characteristics of existing sewerage system	Area profile
1	City drainage district	58%	Underground sewerage with its outfall into river Ganges at Jajmau	Tannery cluster is located in this district. Except for the primary treatment plants, which these industries claim to have installed, effluent is discharged directly into the river. Three wastewater treatment plants mainly to treat domestic waste water are located here.
2	South drainage district	30%	Some pockets are covered under the sewerage system	Rest of the area waste water is disposed into open drains. Industrial effluent from Panki area meets the river Pandu separately through industrial drains.
3	West drainage district	20%	Very limited sewerage facilities	Waste water flows into the Pandu river through open drains.
4	East drainage district	0%	No sewerage network	Primarily comprises developing areas

Source: Based on information presented in Kanpur city development plan and ASCI (Administrative Staff College of India), 2013. City Sanitation Plan for Kanpur. Administrative Staff College of India, Hyderabad, India.

industrial wastewater directly into Pandu river or Ganges river. The five plants put together have a wastewater treatment capacity of only 400 MLD, which is only 52% of the total wastewater generated in the KMA. However, actual treatment is further less due to the operational constraints of the plants and only 27% of the wastewater generated is actually treated. Three other STPs, with a combined capacity of 100 MLD, which were to be commissioned during the JNNURM were not completed as of March 2016.

Table 9.8 Details of sewerage treatment plants in Kanpur city.

Location of STP/CETP	Year of commissioning	Technology used	Type of waste water fed	Installed capacity (MLD)	Actual treatment (MLD)	Use of treated water
Jajmau	1989	Up flow Anaerobic Sludge Blanket (UASB)	Domestic	5	4.5	Treated effluent flows into river Ganges
Jajmau	1994	UASB	Domestic and industrial	36	19	Treated effluent irrigates 2200 ha of sewerage farm. Untreated wastewater is discharged into the river Ganges
Jajmau	1999	Activated Sludge Process (ASP)	Domestic	130	100	
Bingawan	2014	UASB	Domestic	210	65.5	Untreated wastewater is discharged into the river Ganges

Source: Based on data presented in CPCB (Central Pollution Control Board), 2013b. Pollution Assessment: River Ganga. Central Pollution Control Board, Ministry of Environment and Forests, Government of India, New Delhi; CPCB (Central Pollution Control Board), 2015. Inventorization of Sewage Treatment Plants. Central Pollution Control Board, Ministry of Environment and Forests, Government of India, New Delhi.

9.5 Institutional framework for urban water management in KMA

In KMA, various agencies are involved in urban water supply works and services. In Kanpur city municipal area, UPJN plans and implements water supply and sewerage related works, whereas JV looks after operation and maintenance of water supply and sewage treatment plants. Kanpur Development Authority and Uttar Pradesh Housing Board are also responsible for planning of water supply works in colonies developed by them (Table 9.9). In other areas of KMA, UPJN is responsible for planning, implementation, and operation and maintenance of water supply works.

However, presence of multiple agencies has not led to efficient management of urban water services as indicated by the shortfall in domestic water supply and existing wastewater treatment capacity in the metropolis. In fact, even after an expenditure of nearly US$ 82 million under "Ganga Action Plan" (GAP-I and GAP-II) and "JNNURM" on drainage and sewage works, Kanpur city is still struggling to intercept wastewater from the open drains and divert it for treatment. Further, large areas are without any sewerage network. The KMC even find it difficult to run the existing plants as there is extensive load shedding and prefer to simple bypass the wastewater and discharge it without any treatment into the river (CSE, 2014).

While the slow pace in capital works can be attributed to lack of funds, operation and maintenance issues are due to capacity constraints within the KMC. Resistance from UPJN to transfer its function to KMC has also led to

Table 9.9 Various agencies for the provision of urban water supply in Kanpur city.

	Institution responsible		
Service	**Planning**	**Implementation**	**O&M**
Water supply	Uttar Pradesh Jal Nigam (UPJN) for the city municipal area Kanpur Development Authority (KDA) & Uttar Pradesh Housing Board (UPHB) for colonies developed by them District Urban Development Agency (DUDA) for slum area	JV & UPJN DUDA for Slum	JV & UPJN
Sewerage	–	JV & UPJN	JV & UPJN
Storm water drainage	Kanpur Municipal Corporation (KMC) & UPJN	KMC & KDA	KMC

Source: Based on information presented in Kanpur city development plan.

its improper functioning (IIT, 2013). For its financial sustainability, the JV used to depend on water and sewerage charges, which it collected from various consumers. However, only those households, which are within the city municipal limits have formal connections and rest of the areas depend on groundwater, which is presently unaccounted. Most of the industries also have their own tube wells and thus do not depend on JV or KMC for meeting their water requirement.

Further, the technical losses (UFW) on account of old and leaky pipelines from Benajhaber water works are estimated to be about 30% of the total water supplied. The most significant losses are due to huge amount of water wastage and negligible revenue collection from the public utilities (e.g., parks, fire-fighting, etc.) and stand posts, which take away about 10% of the supplied water. Though the water and sewerage recovery has increased to 93%, large proportion of it (nearly 71%) is spent on establishment cost, leaving a meager amount for maintenance of existing old water supply and sewerage system.

The current Central Government has set a target to rejuvenate Ganges under the "Namami Ganga" (Clean Ganges) program, which is aimed at integrating efforts to clean and protect river Ganges in a comprehensive manner with the involvement of the states, urban local bodies, and panchayats. Around US$ 3.4 billion has been earmarked for the program duration (2014–19). The immediate priority is on: rehabilitation and upgradation of existing sewage treatment plants; creation of more sewerage infrastructure; and abatement of industrial pollution by enforcement of strict environmental norms. Further, there is also a program on urban renewal and retrofitting for 100 cities in India called as "Smart Cities Mission," which has its focus on adequate water supply and sanitation among others. The total budgetary outlay is about US$ 15 billion and Kanpur is one of the selected cities. KMA is expected to receive significant funds under these two programs. However, for effective implementation of any such program, various parastatal agencies and utilities involved in water supply and sewerage services need to better coordinate with each other and also put emphasis on timely completion of any new projects taken up.

9.6 Water accounts of KMA

9.6.1 Estimated water demand

During 2001–11, population growth in the KMA has slowed down. Industrial growth has also been affected during the period. Considering the population in 2011, domestic water demand in KMA is estimated to be around

Table 9.10 Water demand by different industries in KMA.

Sr. no.	Category	Estimated daily water demand (KLD/unit)	Total no of units considered	Estimated total water demand (MCM)
1	Textile, dyeing, and printing	213	2816	218.83
2	Tannery	56	2428	49.78
3	Chemical based	82	657	19.61
4	Paper based	1338	3	1.47
5	Dairy	210	2	0.15
6	Distillery	2200	2	1.61
7	Fertilizer	14,000	1	5.11
8	Food products	275	2	0.20
9	Machinery parts	193	2	0.14
10	Sugar	6180	5	11.28
	Overall	617	5918	308

Source: Authors' own analysis based on data presented in GoI (Government of India), 2011. Brief Industrial Profile of District Kanpur Nagar. MSME-Development Institute, Ministry of MSME, Government of India, Kanpur.

175 MCM per annum. This is arrived at by considering the urban water supply norm of 150 liters per capita per day (lpcd) provided in the Central Public Health and Environmental Engineering Organization (CPHEEO) manual on water supply and treatment. Annual water requirement from industrial units is estimated to be around 308 MCM. This is estimated based on the average water requirement by each unit under different categories (Table 9.10). However, only major polluting industries, for which data and information were available, were considered for the estimation. Also, as 47% of the industries were either sick or closed, the estimate is quite close to the actual industrial water demand.

9.6.2 Estimated water supply and use

The total water supply and use in the form of daily water withdrawal from surface and groundwater resources in the KMA by the utility and private parties (housing stocks, hotels, industrial units, etc.) is estimated to be around 529 MCM per annum. Sixty-seven per cent of the water supply by JV is through lifting of river Ganges water and remaining is through groundwater abstraction. Unaccounted for Water (UFW) losses are nearly 30% of the formal water supply.

Details of surface water withdrawal by the public utility are already discussed in the previous section. Estimates of groundwater abstraction in

KMA through private wells are based on the assumption that all the unmet domestic and industrial water demand is fulfilled by withdrawing groundwater through private wells and handpumps.

9.6.3 Summary of annual water account

Water account of KMA is presented in Table 9.11. It considers the total water demand for domestic and industrial purposes, water supply and use, and wastewater return flows (carried by storm water drains), which all are estimated in the previous sections. Overall, there is a huge deficit of around 376 MCM of water (1030 million liters per day) between demand and supply through public supply sources. This unmet demand is met by private groundwater sources (Fig. 9.4).

Around 86% of the total water withdrawal from the natural system by the JV and Unnao Municipality emerges as a domestic sewage in various storm water drains and sewerage channels. Overall, about 280 MCM of

Table 9.11 A summary of annual water account for KMA.

Sr. no.	Particulars	Unit (MCM/annum)
A	**Water demand (2011)**	
A1	Domestic sector	175
A2	Industrial units	308
	Total (A)	**483**
B	**Water supply and use (2011)**	
B1	Withdrawal from surface water resource by the JV	95
B2	Withdrawal from groundwater source by the JV	47
B3	Water supply by Unnao Municipality	11
B4	Estimated withdrawal from groundwater resource through private wells for domestic uses	68
B5	Estimated withdrawal from groundwater resource through private wells for industrial uses	308
	Total (B)	**529**
C	**Administrative and technical losses in water supply by JV and Unnao Municipality (2011)**	**46**
D	**Wastewater return flows (2011)**	
D1	Estimated domestic sewage generation	132
D2	Wastewater carried by storm water drains (other than domestic sewage but including runoff)	148
	Total (D)	**280**

Source: Authors' own analysis.

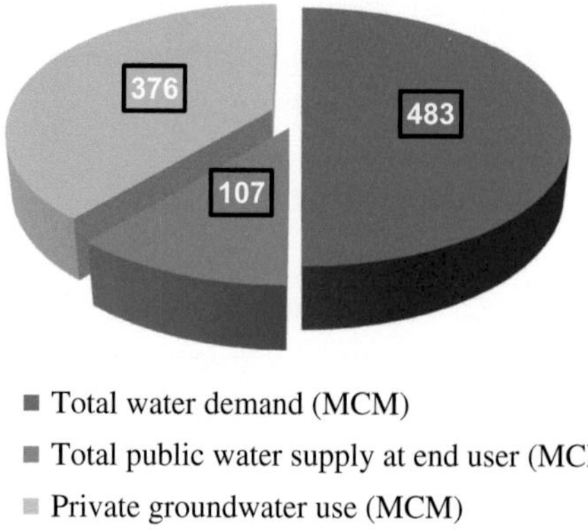

Fig. 9.4 Overall water demand, supply, and use in KMA. *(Source: Authors' own analysis.)*

wastewater (including runoff) is generated in KMA. However, only 18% is treated and close to 230 MCM is discharged into river Ganges and Pandu without any treatment.

9.7 Water supply management challenges

Risk assessment framework of WWF is used to assess the present and future physical and economic risks (regulatory and reputational) related to quantity and quality of water supply for domestic and industrial consumers.

9.7.1 Physical risks

Physical risks are analyzed from the point of view of future water requirement and the capability of existing water resource base (surface and groundwater) to fulfill it. Various aspects of it are discussed in the next subsections.

9.7.1.1 Future water requirement

Future domestic water requirement was worked out on the basis of guidelines given in the CPHEEO manual on water supply and treatment, which suggests a minimum water supply of 150 lpcd for the urban areas that are connected to a sewerage system. Additionally, 15% of the total water supply is also considered for Unaccounted for Water (UFW) losses. This has an inbuilt assumption that there would be retrofitting of the existing water supply pipelines, with replacement of old leaky pipes by new ones. Future population estimates are based on the average decadal growth rate

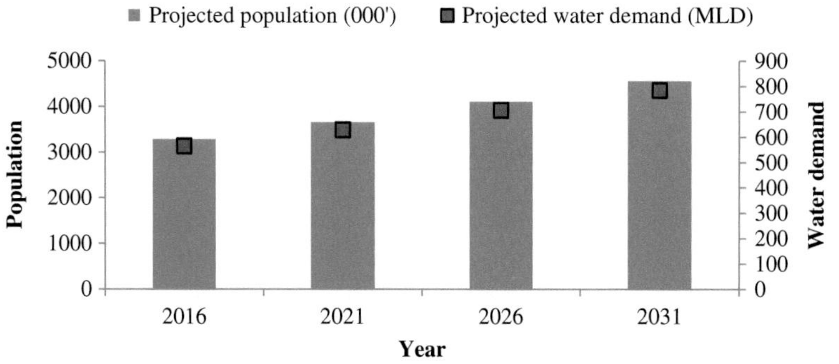

Fig. 9.5 Projected domestic water demand in Kanpur city. *(Source: Authors' own analysis.)*

observed between 1961 and 2011. Thus, considering a norm of 172 lpcd and population growth in the Kanpur city, it is estimated that the water demand for domestic uses will rise to about 785 million liters per day (MLD) by 2031 (Fig. 9.5). However, industrial water demand in KMA is assumed to remain same (844 MLD) as industrial growth (in terms of number of industries) has slowed down significantly and there is increasing trend of sick or closed industrial units.

With the construction of the Ganges Barrage, a permanent and reliable source of water supply is available to provide 1600 MLD of raw water. This is sufficient to cater to the needs of the Kanpur city up to 2031. However, only one pumping and one treatment plant with a capacity of 200 MLD is installed at the site, which is operated and maintained by UPJN. Its capacity utilization is only 12.5% and is expected to increase when more households are connected with the formal water supply system and industries falling within its supply zone will take connections. Further, with a view on availability of raw water at the Barrage site, more raw water pumping and treatment plants can be added at the site. Nevertheless, proper feasibilities studies should be undertaken to explore the possibility of expanding or renovating existing distribution system, which can supply water to majority of users in Kanpur city and for bulk water transfer to Unnao municipality and other areas comprising KMA.

9.7.1.2 Availability of renewable water resources

As discussed in the previous sections, average annual renewable surface water availability in Kanpur stretch of the river is about 53 BCM per annum. However, there is significant variation in water availability across seasons with average flow being less than 1000 cu m/s during the lean period (November to May). Further, estimated net annual groundwater availability in KMA is about 1475 MCM and the total annual draft is about 863 MCM.

Considering the renewable surface water resources and net annual groundwater availability in the metropolitan area, there is not much constraint in meeting existing or future domestic or industrial water demand of KMA in terms of water quantity. However, required physical infrastructure for water conveyance and distribution needs to be in place.

9.7.1.3 Water quality

As highlighted in previous subsections, both surface water and groundwater get heavily polluted by the domestic wastewater and industrial effluents being discharged into the river or on the landmass or into the numerous open drains. While the river water pollution will affect the domestic water supplies, any further adverse changes in groundwater will affect both domestic water supplies and industrial production. This will result in a major repercussion on industrial growth and its contribution to region's economy.

In future, wastewater generation from KMA will further increase with population growth and increased industrial activities. However, it took over 18 years for KMC to install existing sewage treatment facilities (including drainage systems) and since their completion in 2005 no additional STPs have been completed. Further, most of the effluent treatment plants installed by industries provide only primary treatment.

Even if one assumes that in 2016, the proposed plants were completed (addition of about 200 MLD of wastewater treatment capacity) and are operating on their full capacity, a significant amount of wastewater will be discharged into rivers without any treatment from the Kanpur city (Fig. 9.6). As a result, water resources especially surface water will be under risk of getting severely polluted, which will make it unfit for consumption without expensive treatment. This will have a significant cost implication on water supply utilities as the raw water treatment cost will increase. The polluted river water would also have a significant impact on the riverine ecosystem and health of downstream users who depend on it for meeting their domestic water demands.

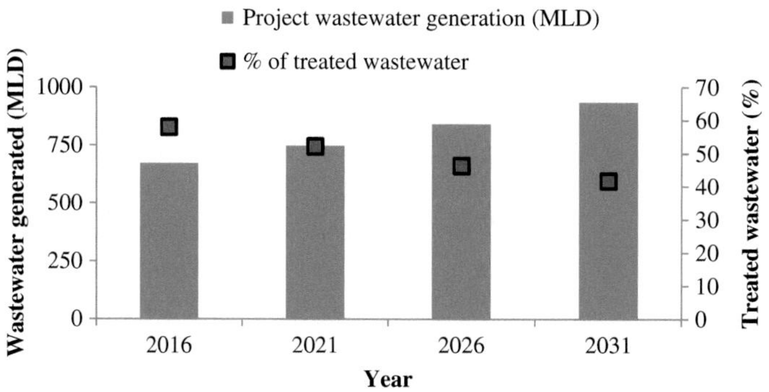

Fig. 9.6 Projected wastewater generation and its treatment. *(Source: Authors' own analysis.)*

9.7.2 Economic risks

Economic risk can affect any new investments in the region and are mainly associated with the changes in macroeconomic conditions like government regulation or any other adverse events. They are discussed in the following subsections.

9.7.2.1 Regulatory risk

Regulatory risks are general due to change in regulations and law that might affect an industry or a business. In the present case, they are analyzed from the point of view of industries, which are flouting standards pertaining to wastewater quality and are either penalized or closed (voluntarily or involuntarily). These risks are more so important in case of industries operating in KMA because some of these units (especially tanneries located on the banks of river Ganges) have been identified as one of the major sources of water pollution.

The risks of closure or penalties for industries flouting wastewater discharge standards has increased especially after the establishment of the National Green Tribunal (NGT) in 2010, which is empowered to deal with cases relating to environmental protection and conservation of natural resources including enforcement of any legal right relating to environment and giving relief and compensation for damages to persons and property.

In the year 2014, as per the directions of NGT, the Uttar Pradesh Pollution Control Board (UPPCB) inspected the status of water consumption and wastewater generation by the industries located in KMA. As per the inspection, 15 industries, which do not have an effluent treatment plant, have been closed. Further, 43 industries, which were not having consent[a] and were not achieving the prescribed wastewater discharge standards, were ordered to be closed. 99 other industries were found to be under self-closure. Additionally, some 134 industries, which were not having consent though having a proper pollution control system, have been issued show cause notices. Though 19 out of them have applied for the consent, which was subsequently granted by the UPPCB, 109 industries have been ordered to be closed. The electricity and water supply to all the noncomplying industries was also disconnected. Further, UPPCB is being instructed by NGT to regularly monitor and report effluent water quality of all the 19 operating industries (*Source*: NGT order 2014). Thus, these units are also at a risk of closure if not found adhering to the water pollution control norms.

The inspection also found that 700 odd tanneries in Kanpur, which are located on the banks of river Ganges, are one of the largest sources of pollution of river Ganges. The pollutants discharged by them are found to be injurious to human health, biodiversity, and ecology of the river. Though one 36 MLD Effluent Treatment Plant (ETP) has been installed at the tannery cluster near Jajmau, since then number of tanneries have increased by three times and thus a significant proportion of effluent from them is discharged without any treatment directly into river Ganges. Further, some units, which have installed ETPs in their respective units, are practically unoperational because of nonavailability of electricity and cost factor. In the light of this, shifting of these tanneries from the bank of river Ganges is also under consideration. This gain presents a huge regulatory risk to tanneries as there is a possibility that all of them may have to start there operation afresh and may also have to adhere to "polluters pay principle," which would have huge financial implications. This may also make them less competitive than the industries operating from other cities.

[a] Under the provisions of the Water (Prevention & Control of Pollution) Act, 1974 and the Air (Prevention & Control of Pollution) Act, 1981, an entrepreneur running or establishing any industry or process, and discharging effluent/emitting pollutants into any water resources or on land/air and polluting thereby the environmental water/air is required to obtain consent, which needs to obtained in two phases: (1) "consent to establish," which is to be obtained prior to establishing any industry or process; and (2) "consent to operate," which is required once the industry or process plant is established along with the necessary pollution control systems. This consent is given for a particular period, which needs to be renewed regularly.

9.7.2.2 Reputational risk

Reputational risk is associated with losses resulting from damages to a firm's reputation, in lost revenue; and increased operating, capital, or regulatory costs due to an adverse event. In KMA, this risk is generally for tanneries, which are exporting large volume of products but are under watch from pollution control bodies due to their significant contribution to water pollution of river Ganges. Often various national newspapers carry report on the risk, which these tanneries pose to the health of river Ganges. Under such circumstances, other companies or individuals may not like to do business with such industrial units, which are continuously under watch of a Government regulator or have been imposed heavy penalties.

Presently, tanneries attract the highest amount of investment per unit among various industries. They also contribute to 13% of the average annual industrial turnover and employ nearly 25% of the workforce (Table 9.12).

Table 9.12 Industrial cluster wise annual turnover and employment in KMA.

Sr. no.	Industry cluster	No of functional units	Average amount of investment (US$ thousand per unit)	Annual turnover (US$ million)	Total no. of people employed
1	Cotton hosiery	467	5.4	39.4	10,680
2	Readymade garments	400	15.4	30.8	10,000
3	Soap & detergent	100	15.4	92.3	4000
4	Paint	50	76.9	15.4	2000
5	Harness & Saddlery	225	30.8	61.5	15,000
6	Footwear	1500	3.1	155.8	45,000
7	Leather and its products	1635	3.8	45	100,000
8	Steel furniture	685	3.1	27.8	50,000
9	Bakery products	378	3.5	7.7	2500
10	Cold storage	100	307.7	7.7	2000
11	Pharmaceutical	75	107.7	33.8	2300
12	Corrugated paper	170	18.5	16.2	2600
13	Textiles	180	1.5	3.1	2500
14	Plastic	100	30.8	123.1	5000
15	Pan Masala	55	32.3	769.2	8000
16	Agro based	174	76.9	1538.5	2500
17	Tannery	420	384.6	461.5	86,000
	Overall	6714	-	3428.8	350,080

Source: Based on data presented in GoI (Government of India), 2011. Brief Industrial Profile of District Kanpur Nagar. MSME-Development Institute, Ministry of MSME, Government of India, Kanpur.

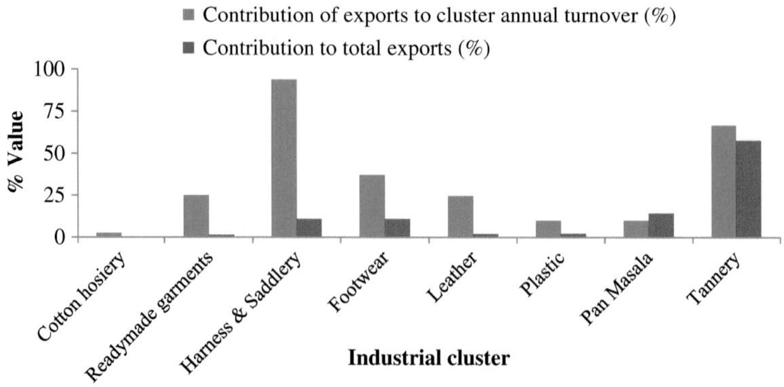

Fig. 9.7 Contribution of exports (value) to annual turnover. (Source: *Authors' own analysis based on data presented in GoI (Government of India), 2011. Brief Industrial Profile of District Kanpur Nagar. MSME-Development Institute, Ministry of MSME, Government of India, Kanpur.)*

With many tanneries facing a risk of closure due to nonadherence to water pollution control norms, there will be a significant impact on region's economy and people's livelihood.

Further, leather exports account for nearly 67% of the annual turnover of tanneries and it also accounts for nearly 58% of the total industrial exports (which is about US$ 538 million) from Kanpur metropolis (Fig. 9.7). Therefore, closure of such units or lack of demand from importers of such products will create a huge setback for the industrial growth in KMA. Thus, there is a huge reputational risk to industrial sector in Kanpur unless they all take proper consent and adhere to wastewater discharge standards.

9.8 Major findings

The major findings from the desktop study are as follows:

1. Kanpur is a growing metropolitan area. Between 1961 and 2011, the population of KMA has increased at an average decadal growth rate of 25%. However, it slowed down to a mere 8% during the last decade (2001–11). Nevertheless, the average population density of the city had nearly increased threefold, i.e., from 3243 persons/km^2 in 1961 to 9666 persons/km^2 in 2011.
2. Kanpur is a heavily industrialized metropolitan area. Total number of industrial units has increased from a mere 537 in 1984–85 to more than 16,000 in 2010–11. Presently, nearly 99% are small-scale industries and the highest numbers of industries are from the leather tanning category.

Number of tanneries has increased from 175 units in 1994 to 402 units by 2015. However, of late, industrial growth has also slowed down. Against an average annual growth rate of 2.9% between 1984–85 and 1993–94, it was 0.47% between 1994–95 and 2010–11. The population growth in the city seems to have been driven by industrial growth.

3. The city has two major sources for water supply, viz., the river water and groundwater. There are also formal and informal water supplies, the latter being managed from privately owned wells.
4. Against an installed capacity of 600 MLD of raw water, only about 390 MLD of water is actually supplied through the formal water supply system in the Kanpur city. Out of this, 33% (130 MLD) is constituted by groundwater. Additionally, 30 MLD is supplied in the Unnao Municipal limits. Almost the entire surface water, which is supplied by the utility, is treated whereas groundwater is supplied without any treatment.
5. The per-capita water supply at the end-user level for households with formal water connections is about 97 lpcd. This is far less than the per-capita water supply norms of 150 lpcd as per the CPHEEO. In other areas of the metropolis, the households depend on private tube wells to meet domestic water needs. Entire industrial water demand is also met by groundwater. It could be reasonably assumed that the domestic water requirement is fully met from tap water and water from private wells. The existing annual water demand for domestic and industrial consumption is estimated to be about 483 MCM. Since the formal water supply is only 153 MCM per annum and 30% is considered to be lost as UFW, the current deficit of around 376 MCM of water (1030 MLD) could be considered to be met by pumping from privately owned tube wells in the city.
6. The total wastewater generation (including storm water runoff) in KMA is estimated to be about 767 MLD. Apart from use of a minor amount of treated wastewater in irrigation, rest of it and the untreated wastewater is directly drained into the river Ganges in north and river Pandu in south of the Kanpur city.
7. Presently, sewerage system covers only 29% of the Kanpur city whereas it is nonexistent in remaining areas. Further, capacity exists to treat only 52% of the total generated wastewater. Actual treatment is further low at 27%.
8. Under the JNNURM project, US$ 100 million was sanctioned to support the existing water supply system. However, the work execution

was very slow and largely remained incomplete as per the last available information (August 2012). Also, construction of three out of the four sanctioned STPs under JNNURM could not be completed.

9. Average annual renewable surface water availability in Kanpur stretch of the Ganges river is about 53 BCM. Further about 612 MCM of surplus groundwater is also available annually. Therefore, there is not much physical risk for meeting existing or future domestic and industrial water demand of KMA in terms of water quantity (estimated to be about 1629 MLD by 2031). However, required infrastructure for water conveyance and distribution needs to be in place.
10. Nevertheless, water quality is an issue. This is mainly due to limited infrastructure to treat raw water and also because a substantial portion of wastewater is discharged into river or on land mass without any treatment. Therefore, the river water, which is the major supply source, is heavily polluted and can have adverse effect on human health. Groundwater is also contaminated and polluted, which can affect the domestic water supplies and industrial production. In the wake of the Utility not expanding the infrastructure for supply of treated surface water to cover more areas, the dependence on groundwater would only increase.
11. Regulatory risk was found to be high for those industries, which do not conform to pollution control norms and are thus under risk of closure. Already, many industries have been shut by the State Pollution Control Board for noncompliance of effluent discharge norms. Nevertheless, this will accentuate financial burden on the small-scale industries that operate on low profit and can make them less competitive. The long-term negative impact of increasing regulatory risk is evident from the fact that industrial growth in the city had come down remarkably during the past 15–20 years.
12. Reputational risk, associated with losses resulting from damages to industries' reputation, is about US$ 538 million, which is equal to the estimated annual value of the exports from various industrial clusters in KMA. Tanneries with an average annual export worth US$ 308 million (58% of the total exports) have the biggest risk as they are under continuous watch of the State regulator for their failure to comply with wastewater discharge standards.
13. Presence of multiple agencies has not produced desired results in terms of efficient management of urban water services as indicated by the shortfall in domestic water supply through public sources and inadequacy of the existing wastewater treatment capacity in the metropolis.

14. The aging water supply infrastructure coupled with high establishment costs has rendered the resource available for maintenance of the existing systems highly inadequate. Capacity constraints within KMC have also been identified as the other major reason for poor O&M of the water supply and sewerage system.

9.9 Conclusion

The water accounting study carried out by us suggests that there is not much physical risk for meeting existing or future water demands vis-à-vis quantity of the supplied water. However, water quality is an issue as the required infrastructure to treat raw water and wastewater is not in place. This scenario poses a major threat to the river ecology and poses a risk to livelihoods of people dependent on the river. The economic risk for the industries alone is estimated to be US$ 538 million per annum. To mitigate these risks, various parastatal agencies involved in water supply and sewerage services need to coordinate better and also ensure enforcement of strict environmental norms for the industries. With Government of India having substantial budgetary allocation for Clean Ganges Programme and Sustainable Cities Mission, some resources need to be allocated for undertaking similar type of water accounting exercise in other major cities along river Ganges to assess their water risk and water security challenges and to prioritize investments for water supply and wastewater treatment infrastructure, and institutional reforms (laws, policies, and institutional coordination) accordingly.

References

ADB (Asian Development Bank), 2007. 2007 Benchmarking and Data Book 0f Water Utilities in India. Asian Development Bank, Manila, Philippines.

Bansal, U.R., Bansal, B.B., 1984. Industries in India during 18th and 19th century. Indian J. Hist. Sci. 19 (3), 215–223.

Bassi, N., Kumar, M.D., 2012. Addressing the civic challenges perspective on institutional change for sustainable urban water Management in India. Environ. Urban. ASIA 3 (1), 165–183.

Census of India, 2011. Decadal Variation in Population Since 1901. Ministry of Home Affairs, Government of India, New Delhi.

CPCB (Central Pollution Control Board), 2009. Status of Water Supply, Wastewater Generation and Treatment in Class-I Cities and Class-II Towns of India. Central Pollution Control Board, Ministry of Environment and Forests, Government of India, New Delhi.

CPCB (Central Pollution Control Board), 2013b. Pollution Assessment: River Ganga. Central Pollution Control Board, Ministry of Environment and Forests, Government of India, New Delhi.

CPCB (Central Pollution Control Board), 2016. CPCB Bulletin: Volume I. Central Pollution Control Board, Ministry of Environment and Forests, Government of India, New Delhi.
CSE (Centre for Science and Environment), 2014. Ganga: The River, Its Pollution and What We Can Do to Clean it-Briefing Paper. Centre for Science and Environment, New Delhi.
IIT (Indian Institute of Technology), 2013. Ganga River Basin Environment Management Plan: Interim Report. IIT Consortium, India.
JNNURM (Jawaharlal Nehru National Urban Renewal Mission), 2006. Kanpur City Development Plan—Final Report. JPS Associates Pvt. Ltd, New Delhi.
Kumar, M.D., 2014. "Thirsty Cities", How Indian Cities Can Meet Their Water Needs. Oxford University Press, New Delhi.
Mishra, S., 2006. Killing Kanpur. India Today December 4, 2006.
O'Keeffe, J., Kaushal, N., Bharati, L., Smakhtin, V., 2012. Assessment of Environmental Flows for the Upper Ganga Basin. World Wide Fund for Nature-India, New Delhi.
Singh, S.N., 1990. Evolution of Kanpur as an industrial centre. In: Singh, S.N. (Ed.), Planning & Development of an Industrial Town: A Study of Kanpur. Mittal Publications, New Delhi.

Further reading

ASCI (Administrative Staff College of India), 2013. City Sanitation Plan for Kanpur. Administrative Staff College of India, Hyderabad, India.
CPCB (Central Pollution Control Board), 2013a. Status of Water Quality in India—2011. Central Pollution Control Board, Ministry of Environment and Forests, Government of India, New Delhi.
CPCB (Central Pollution Control Board), 2015. Inventorization of Sewage Treatment Plants. Central Pollution Control Board, Ministry of Environment and Forests, Government of India, New Delhi.
GoI (Government of India), 2011. Brief Industrial Profile of District Kanpur Nagar. MSME-Development Institute, Ministry of MSME, Government of India, Kanpur.

CHAPTER 10

Addressing the scale issues in watershed development

V. Ratna Reddy[a], Geoff Syme[b]
[a]Livelihoods and Natural Resource Management Institute, Hyderabad, Telangana, India
[b]School of Business and Law, Edith Cowan, Hobart, TAS, Australia

10.1 Background

Watershed development (WSD) has a long history as a soil and water conservation technology. While most of the developed and developing countries have been implementing WSD to protect their river basins and watersheds, Indian policy has given WSD a much bigger role of providing stability to its vast regions characterized with low rainfall and low irrigation (poorly endowed). WSD has become critical in developing agriculture in these poorly endowed regions, which account for 60% of the cropped area in India. Over the years, about a quarter of these areas have been covered under the WSD program. However, the impact of WSD on the productivity or stabilization of agriculture has been marginal (Reddy et al., 2010). While inadequate implementation is often identified as the root cause of the poor performance of the program, a number of other reasons have been flagged. For instance, lack of people's participation and collective action is identified as a reason for poor implementation and impact of the program.

While WSD is a technology meant for soil and water conservation that would strengthen the natural resource base for the farming systems and improving its resilience, the focus has been on improving crop yields and agricultural incomes. More importantly, the interlinkages between different and dependent natural systems such as biophysical and hydrogeological systems have been totally neglected. Biophysical aspects such as rainfall, soils, and land use, determine the nature and intensity of impacts. Similarly, the hydrogeological features of a watershed determine groundwater storage potential and its sustainability in the short and medium terms. However, these aspects are hardly taken into account while designing or assessing

Current Directions in Water Scarcity Research, Volume 1
https://doi.org/10.1016/B978-0-12-814851-8.00010-0
© 2019 Elsevier Inc.
All rights reserved.

the impacts of the WSD programs. In the absence of information on these aspects, WSD interventions (type as well as intensity) have been uniform across locations. This results in: (i) interventions that may not be effective as they are not in line with the hydrogeology and biophysical requirements; (ii) impact assessments that are not comprehensive as they do not take the externalities associated with hydrogeology (groundwater) into account leading to under- or overestimation of the cost-benefit ratios; and (iii) variations in the impacts of WSD at different locations (upstream and downstream) not being captured in the context of micro-WSD (about 500 ha).

The WSD policy has clear goals in terms of improving the natural resource base and increasing sustainable production through the creation of a host of physical structures designed to capture and retain water resources and minimize erosion. It has generally been practiced at a micro (i.e., village) level (about 500 ha), with prioritization of treated areas based upon a range of factors that include the level of land degradation, degree of groundwater stress, coverage of "scheduled caste" and "scheduled tribe" households, location in relation to the ridge and valley, and the availability of drinking water. Consequently, not all areas within a catchment can be covered. Further, there has been some concern that this land area may be somewhat too small and may lead to less than optimal hydrological, economic, and equity outcomes. Therefore, it has been suggested that the area of application of WSD should be increased. As a result, guidelines have been created for mesoscale WSD of above 5000 ha in an endeavor to improve the outcomes for WSD while creating implementation efficiencies. Perhaps the most significant motivation for this change has been increased efficiency in terms of administration (Government of India, 2008).

10.1.1 Scale aspects

The microwatershed approach enables amicable integration of land, water, and infrastructure development, particularly because of the homogeneous nature of soil, water, and overall physical conditions within the microwatershed. Theory and experience have shown that facilitating collective action in small, village-level watersheds has fewer constraints compared to more complex communities. Moreover, organizing collective action at the microwatershed level generally proved to result in lower costs, and in improved use of financial and human resources, particularly for the management of common resources.

The recent generation of watershed management projects has been mostly successful in its integrated and participatory approach to sustainable conservation and development in upstream areas. This has given some impetus to scaling up. However, the microwatershed approach encounters adversity when it comes to scaling up. Operating at the microwatershed scale does not necessarily aggregate up or capture upstream-downstream interactions. A mix of upstream interventions would only have a considerable impact downstream if prioritized and planned within the larger watershed perspective and with understanding of the spatial and hydrological links between the perceived externalities and their underlying factors (e.g., land and water use).

Watershed management projects are generally anticipated not only to provide local on-site benefits at the microwatershed level, but also to offer positive externalities in the form of valuable environmental services downstream. They are also seen as means to provide corrections of downstream negative externalities within the larger watershed. Therefore, investment in upstream cannot be justified by their on-site benefits alone and can only pass economic reasoning when downstream benefits are embodied. In moving from a micro- to mesoscale of WSD, the key issue of evaluation of mesoscale WSD was initially approached primarily as one of "scale." The issue of scale has been of ongoing interest in hydrology for some time (e.g., Bergstrom and Graham, 1998; Klemes, 1983; Merz et al., 2009; Sivapalan et al., 2004). This chapter describes the background to WSD and the conceptual issues associated with scale when endeavoring to evaluate the potential outcomes of mesoscale WSD programs and their significance from a design and implementation perspective.

Before specifically addressing mesoscale WSD, it should be noted that while scale is commonly used and understood to reflect the reach or size of land area that is under consideration, Gibson et al. (2000) have distinguished between the concepts of scale and level. Scale is referred to as describing the nature of what is being measured. In this case, scale would refer to the qualitatively different realms of hydrology, land use, livelihoods, and equity. Level relates to the breadth of the dimension. For hydrology the level units could be, for example, basin, sub-basin, meso and micro. For equity, level could be represented as perceived fairness of WSD among groups within villages, the level of personal investment in time or money in WSD, the degree of deservingness as measured on current income, relative outcomes for land owners versus landless and so on. Thus, there may be multiple dimensions and levels within one scale; therefore, this chapter deals with both scale and level in terms of their integration to provide a sustainable and resilient WSD program.

Conceptually, the issue of dealing with levels has been well discussed. The ultimate unit of hydrological analysis is seen to be the catchment level, while the unit for hydrogeologic analysis is at the basin or aquifer boundary level (for limited cases they may coincide). The components of the surface water or groundwater "catchment" should be managed to avoid negative externalities (or spillage) on their surrounding areas either in terms of water quality or quantity, livelihoods impacts, or, as far as possible, preferred land use. It must be noted, however, that the size of some groundwater systems may make this impossible in practice.

Ideally, institutions should also be fundamentally based on the catchment level with appropriate formal and informal structures at matching levels with the WSD implementation (e.g., Lee, 1993; Scoones, 1999; Folke et al., 2007; Dore and Lebel, 2010). This institutional functioning is not restricted to formal and informal group entities but includes those rules, which govern how the catchment is managed (Olsson et al., 2007). At a whole-of-catchment level, there is the hope that the benefits arising from WSD will diffuse from the programmatic sites to others who through observation will see the benefits of WSD.

There are also other issues, which make the simple hierarchical approach difficult in the view of some. In hydrology, some favor a "bottom-up" approach in which water flows are modeled from an initially small scale outwards, whereas others prefer to start at a catchment level and hierarchically scale their modeling down to the area of interest. Needless to say the micro-approach does not necessarily lend itself to simple upscaling to the macro-level and the statistical approaches at the macrolevel don't necessarily scale down to the microlevel given the variability and nonlinearity that is so evident at that scale. Nevertheless there has been some progress on this problem (e.g., Viney and Sivapalan, 2004; Savenije, 2009).

In addition, level is among a number of things that cannot be viewed in their own right for sustainable water management. Hydrological processes depend on level of rainfall, soil type, slope, land use, and specific groundwater/surface-water interactions, among other things. Thus, the area of land chosen for WSD policy needs to be considered as only one of a number of interacting variables if the application is to make hydrological sense. If these relationships are not considered, there is no automatic reason to consider that a mesoapproach will lead to better results than a microapproach. Further, many hydrological processes are nonlinear as are the impacts of WSD itself, which, are in part, a function of those processes. Not only are they nonlinear in a spatial sense, but the externality effects of interventions such as WSD take time to emerge so that there is a need to also consider the temporal aspect. Similar arguments can be made in relation to economic, livelihood, equity, and resilience issues. How then should one confront the issues and interactions between scale and level when designing and evaluating WSD delivery in an integrated manner?

10.2 WSD scale and its evaluation

The microwatershed approach enables amicable integration of land, water, and infrastructure development, particularly because of the homogeneous nature of soil, water, and overall physical conditions within the microwatershed. Theory and experience have shown that small village-level watersheds, because of their ongoing social cohesion, are conducive for promoting the collective action that is critical for watershed management. Moreover, collective action at the microwatershed level has generally proved to result in lower costs, through efficient use of financial and human resources, particularly for the management of common resources. As a result, watersheds with active participation are also found to be performing better than others.

However, the microwatershed approach encounters problems when it comes to upscaling. Operating at the microwatershed level does not necessarily aggregate or capture upstream-downstream interactions. Watershed management projects are generally anticipated not only to provide local on-site benefits at the microwatershed level, but also to offer positive or negative externalities in the form of valuable environmental services or disservices down stream. They can also provide a means of correcting downstream negative externalities within the larger watershed. Therefore, investment in upstream areas cannot be justified by their on-site benefits alone and can only be justified in an economic sense when downstream benefits/disbenefits are embodied (e.g., Oates, 1972; Ostrom, 1999).

However, watershed management programs have usually paid attention to on-site interventions and their benefits. Whether these actions were of benefit also to the downstream location or were the best possible approach to minimizing negative externalities was often not ascertained. Similarly, stakeholder involvement and participation normally covered on-site requirements of local farmers, and the spatial dimension was tackled through community-based planning of their region. The WSD institutional approach only focused on the microwatershed, with no conscious attempt to create cooperation across the watersheds or between upstream and downstream populations. The success of the WSD project was assessed on-site, and the individual level outcomes (income increase, land area treated, yield increase) were, in general, aggregated across the watershed area.

Despite their apparent objective of improving natural resource conditions in a watershed, WSD programs may improve the conditions in upstream at the cost of resources in the downstream areas. Research has revealed that the microwatershed approach may result in hydrological problems that would be minimized by operating at a macrowatershed level. For example, in India, hydrological research cautions that watershed projects may be aggravating precisely the water scarcity they intend to overcome. The study by Batchelor et al. (2003) reported that capturing water in upper watersheds came at the expense of lower watershed areas. On the basis of the data from macrowatershed level (covering many villages), they document cases where water harvesting in upper watersheds reduced water availability downstream. With the lowering of the watertable downstream, deepening of wells was needed, which the poor could often not afford, leading to inequitable distribution and use of water (Calder, 2005).

The closure of catchments may also occur at the regional level such as for the Krishna River Basin in southern India (Biggs et al., 2007), whereby WSD is one of a number of factors for diminishing coastal outflows. In this case, watershed development checks the movements of surface runoff downstream that may be transferred to the groundwater system where downstream migration may or may not be constrained. It indicates that hydrologically what is good for one microwatershed can be bad for others downstream. Thus, whereas addressing social and economic considerations favors small microwatersheds as the unit of operation, approaching this hydrological problem calls for working in large macrowatersheds.

While successful WSD projects have overcome the inherent constraints to collective action through mechanisms such as obtaining commitment through private contributions to infrastructure development, localizing rainfall and groundwater data collection and interpretation, and enhancing the inclusiveness of local water user groups, they have not conquered two outstanding barriers. First, projects with high investment in social organization may not be replicable beyond a small number of cases. Second, operating on the basis of a feasible social unit (a village microwatershed instead of a macrowatershed that crosses administrative boundaries), in fact, trades one set of problems for another. This would involve working simultaneously to promote watershed governance capacity both within and between microwatersheds. However, there may potentially be difficult tradeoffs between these two approaches. For example, production benefits enhanced by WSD at the village level may need to be curtailed if a wider perspective is adopted. Changes in reliability of water supply may also occur differentially between microwatersheds. Resolving the tradeoffs is necessary for the widespread success of the watershed development program, but there are no obvious solutions.

The difficulty of managing watershed interventions at diverse levels so as to achieve the larger-level objectives of downstream impacts is further complicated because of participatory approaches, which basically give the option of interventions to the communities rather than the planners (Reddy et al., 2004). While there is significant literature examining a wide range of coordinating mechanisms, including property rights formulations, market mechanisms, and social agreements for collaboratively based management, in the Indian context (e.g., Ebrahim, 2004), these have yet to be applied at various scales in the WSD context.

Basically WSD management is critically dependent on ecological/environmental and social aspects—ecological/environmental because the technology per se strengthens the resource base and hence the environmental benefits have the potential to be greater when watersheds are implemented in a holistic manner (i.e., covering the entire watershed, which is often very big in scale). This is beneficial in the long run and also helps environmental sustainability and, hence, resource-dependent livelihoods. Nevertheless, the ability to do this depends on the successful development of integrated institutions often at a regional level and sometimes across political boundaries.

Social aspects arise as watershed management transcends across households, communities, or villages covering both private and public (common) lands and water bodies. Interhousehold/community/village coordination or cooperation is necessary for successful implementation and management of watersheds. As a result, collective action and participation of communities has been treated as mandatory and become part of the guidelines since 1994. The theory of collective action suggests that the smaller the group, the greater the chances for collective action or cooperation—e.g., see De Groot and Tadepally (2008) for an Indian perspective. That is, while microwatersheds have achieved the much needed participation, to a large extent the environmental gains have been limited due to the limited coverage. Nevertheless, such an assertion may be too simplistic for mesolevel WSD as there is still the possibility of shared values through similar socioeconomic and cultural status (e.g., Brown and Purcell, 2005). Huchtemann and Frondel (2010), for example, have suggested a settlement clustering pattern for identifying compatible social units on the same catchment with similar environmental conditions to create efficiency in transboundary water management.

As the shift from micro- to mesoscale is taking place, there are tradeoffs between environmental impacts and social or institutional requirements. Generally it is thought that enhancing the match between the hydrological, social, and institutional scales of operation will enhance the effective governance of the catchment. In framing the appropriate scale for adaptive governance, Cash et al. (2006) suggest that there is a need to deal with the hierarchy of constitutions (institutional scale), the levels of engagement (network scale) and the linkages between general and specific knowledge (knowledge scale). These three considerations will be pre-eminent when considering evaluating the mesoscale.

With the new mesoapproach, there is potential for significant efficiencies to be made by ensuring that WSD funds are invested in ways to achieve better outcomes by considering mesolevel issues. However, there is little economic analysis to underpin this. Moreover, often the wide range of technical interventions utilized in watershed development may not always be the most appropriate or effective measures. In addition, often, serious equity issues arise out of water harvesting benefits being realized in one part of the watershed by one group of farmers, and those who may not necessarily have access to the increased surface water or groundwater resources in another part of the watershed. Equity issues have to date not been satisfactorily incorporated into assessment of WSD.

10.2.1 Internalizing externalities

One of the most important characteristics of watershed management is the ability to improve the management of externalities, which generally emerges because of land and water interactions. There exist a number of approaches to "internalizing externalities"—that is, compensating those who generate positive externalities and taxing those who cause negative ones. These approaches include attaching the adoption of conservation practices to other benefits such as access to credit (Pagiola, 2002), and practices like cost sharing—full subsidy to the cost of adoption, or partial subsidy. Investment subsidies, particularly cost sharing, have been the most frequently applied procedures. However, one study on the Indian experience observed investment subsidies to be the least effective mechanism (Kerr et al., 2007). Experience also suggests that subsidies, if not sustained, do not realize long-term changes in conservation practices. Once the projects end and the subsidies cease, land users have often resumed their previous land uses where they disregard the conservation measures they had adopted, or even actively destroy them (Lutz et al., 1994).

In order to avoid the apparent problems of a "compensation" approach, watershed management programs also resort to a variety of nonfinancial approaches to persuade stakeholders to adopt the recommended conservation practices. While some have recommended to provide alternative income generation activities to compensate for lost income because of conservation practices, some have relied on a hoped-for "demonstration effect"—assuming that conservation practices could eventually demonstrate their usefulness to stakeholders who would approve them of their own accord once their benefits had been established (Pagiola, 2002). Others have employed approaches such as awareness generation, moral obligation, and regulatory limits and fines. Generally these approaches have not proved effective (Enters, 1997; Pagiola, 1999). The alternative income-generating activities approach has had mixed results; the demonstration effect has often failed because the assumption that conservation practices were lucrative to upland stakeholders was often not the case. Regulatory approaches are often very difficult to implement and may entail high costs on poor land users by forcing them to adopt land uses that generate lower returns.

Apart from the aforementioned approaches, market-based contracting approaches—payment for environmental services (PES)—have also been used in some cases, particularly in Latin America in small-scale initiatives involving water services. Presently, several countries are already experimenting with such systems (Landell-Mills and Porras, 2002). The basic principle behind such approaches is that those who supply environmental services should be compensated for their service and that those who receive the services should pay for their provision. This approach has the added advantage of providing supplementary income sources for upstream poor land users and, thus, helping them improve their livelihoods. In most cases, although a PES approach is apparently attractive, putting it into practice is far from simple. Application of these approaches requires the presence of several building blocks (Pagiola and Platais, 2007) and systematic evidence and scientific information for designing appropriate policies.

10.2.2 Institutional issues

Harmonizing upstream activities with management objectives at the broader watershed level is obviously a major challenge, as upstream-downstream linkages are multifaceted and the information essential to understand the interactions has, until recently, proved complicated and costly to accumulate. However, development of dynamic modeling at the basin level coupled with more affordable monitoring tools such as remote sensing, allows for enhanced understanding of watershed properties with better capability to define upstream-downstream relations, functions and management impacts. Moreover, if watershed management is to be justified by its beneficial impact on the downstream environment, institutional arrangements are needed to endorse interaction among microwatershed groups at the mesolevel within large macrowatersheds, and to determine and monitor outcomes and impacts. It could involve specific mechanisms to facilitate the interaction such as new legislation or new arrangements for sharing upstream-downstream costs and benefits. A variety of institutional mechanisms exists, from simply maintaining an information system that identifies externalities, through the formation of platforms for dialog between upstream and downstream communities, to building higher-level watershed planning institutions. Preferably, the institutional framework should be capable of incorporating the microwatershed management plans through mesoplans to the broader scale of the watershed as a whole (Reddy et al., 2010). This would involve developing something like a "nested platforms" approach at a macrowatershed scale.

The important question regarding the tradeoff between operating at a hydrological level, which is sustainable, versus a social level that is perceived to be equitable between key groups and interests is its intractability. In the early days of watershed projects, disregarding the values and functioning of the social unit resulted in failure of the projects as they could not accomplish effective watershed governance. Of late, the pendulum has swung in the opposite direction and now most projects operate at the village level, disregarding hydrological linkages between microwatersheds. The downstream effect in general, with catchment closure as an extreme manifestation, has appeared in part by overlooking these hydrological linkages, and it illustrates the need to deal with them by working at a broader scale.

10.2.3 Integrating hydrology and livelihoods

Scale issues become even more complicated when the focus shifts from bigger (ultimate) scale to the smaller scale. Technically, the river basin or complete watershed level is the ultimate one where interactions of various systems and impacts can be modeled to a large extent. Importantly, the availability of data at that level is generally not a constraint. At the next (second) level down, there is the sub-basin or subwatershed level. Some aquifers could be big enough to fit this level. At the third level is the meso- or megawatershed level or the aquifer level. The experience shows that no two systems are coterminous at this level, i.e., the boundaries of mesowatersheds do not match with aquifer boundaries at this level, and hence it is difficult to assess the surface and subsurface flows and the transboundary problem may need to be considered. Lack of availability of technical data at this level is also often an issue.

While it is a challenge to characterize the surface and subsurface flow processes reasonably well, making generalizations of such outcomes could be more challenging, if not impossible. From a surface-water hydrology perspective, level is not necessarily the primary issue—as processes can be modeled at whatever level is needed. Sometimes there is a need to model at a finer level. This often means that there are no appropriate gauges available for calibrating models, leading to the need to make predictions at ungauged locations. This may be relatively easy if there is a suitable gauge nearby. If not, the uncertainty in the model predictions increases (e.g., Bardossy, 2007), and this needs to be allowed for in any output from integrated modeling required for holistic evaluations of WSD.

There is a problem when the social and economic units do not line up with hydrological boundaries. From a hydrological viewpoint, modeling part of a catchment is possible, though it can lead to complications and increased uncertainty. For example, one of the simplest methods is to add the effective rainfall from each watershed or HRU (hydrological response unit), and then convolve this with the unit hydrograph for the point in the stream network we are trying to model. From a groundwater perspective, it is more common than not that only part of an aquifer is modeled, but it relies on having enough information to set the necessary boundary conditions. If the social/economic level matches with the hydrologic divides, then more meaningful results can be achieved. It is also argued, at the conceptual level by some hydrologists, that, all other things being equal, both micro- and mesoscales may have the same effect on the gross water balance for the watersheds.

10.2.4 Incorporating the wider socioeconomic catchment context

Managing a mesolevel watershed brings up socioeconomic issues of a different nature than those faced at microwatershed levels. While farmers are aware of the role of politically dominant groups upstream in influencing water availability downstream, such iniquitous interventions are mostly assigned to fate and ignored at a microlevel approach. Managing mesolevel watersheds from a top-down perspective presents the opportunity to rectify some of these inequities through external interventions and, more importantly, presents scope for examining alternative institutional arrangements among upstream and downstream water users. Another social benefit of a mesolevel approach is the opportunity to evaluate presence of persistent patterns of inferior livelihoods among the backward groups of the society that cut across smaller village-level boundaries.

Economic implications of management interventions also differ when applied at meso- and microlevels. First, mesolevel interventions are better able to harmonize sustainability of rural livelihood with that of groundwater sustainability. At a microlevel, village-level heterogeneities may disguise wider patterns present across a much larger level. For instance, presence of significant seasonal migration and concomitant remittances in a water-starved village might present a picture of a sustainable livelihood, whereas in reality, depleting groundwater and the vagaries of urban demand for rural labor could actually be increasing the vulnerability of such villages on a longer time frame (or level). Secondly, it is much easier to direct social development efforts such as formation of cooperative societies, through economic incentives in drought-prone villages, once their increased vulnerability has been identified in the larger watershed region. This, however, does not imply that microlevel interventions are any less desirable. In the absence of any distortions, efficiency in allocation of resources at a household level is achieved best when individuals act in their self-interest of maximizing profits. However, the common property nature of groundwater resources requires inducement toward collective actions so that challenge of a lack of well-defined property rights is mitigated. The issue of scale in watershed management, in light of the aforementioned, should not be perceived as one of the contradictions between micro- and mesolevels, but rather as an opportunity to exploit the best outcome through their simultaneous management.

10.3 Socioeconomic implications

In the absence of appropriate or optimum design and implementation of watershed interventions, the expected positive socioeconomic impacts may not be evident. Moreover, they would vary across locations depending on the aquifer geometry, type and nature of aquifer, rainfall-runoff and recharge, surface-groundwater recharge, extent of surface water storage, and land use pattern. The externalities of hydrogeology could be captured in the upstream-downstream context at the mesoscale, when watersheds are placed within a hydrological unit. The common issue at all levels is the lag between the implementation and impact assessment could influence the impacts, as measured from the experience of the households—the greater the lag, the higher the risk of households missing the linkages between the interventions and their impacts.

Technically, watershed interventions are expected to strengthen the natural resource base and improve the resilience of the farming system. Hence, the resilience of the household is taken as an indicator of watershed impact. Resilience can be measured in terms of household capacity to withstand a number of droughts, is positively associated with the rainfall (Hydrological unit-HUN), location (downstream), and watershed (treated area). That is, the HUN with better rainfall is more resilient than the one with lower rainfall; downstream locations are more resilient than upstream and midstream locations; and villages treated with watershed interventions are more resilient than untreated (control) villages. These propositions were examined in a study of two regions (HUNS) in Andhra Pradesh with relatively high and low rainfall in an extensive multidisciplinary study (Reddy and Syme, 2015).

While these hypotheses tested positive in most cases, there are deviations to this logical pattern: the extremely poor performance of the upstream village in the low-rainfall zone (HUN) despite being a model watershed (acclaimed as a best-implemented watershed) and the unexpected poor performance of a watershed in the relatively better rainfall zone (midstream village in HUN) despite the shifts to high value horticultural crops, defy standard explanations (for details see Reddy et al., 2015). The explanation for these deviations lies in the hydrogeology of the locations. The hydrogeology of the upstream village in the low-rainfall zone (HUN) is very shallow (basin) and does not support any on-stream interventions for groundwater recharge. As a result, despite well-constructed and maintained check dams, this village could not benefit from groundwater recharge. Hence, this village continues to depend on shallow wells and the situation worsens during years with less than normal rainfall.

On the other hand, the case of the midstream village in the better rainfall zone (HUN) is that the land use pattern is not in line with the groundwater potential. This village is characterized by a moderately shallow basin with limited groundwater potential. Due to the nature of the aquifer, groundwater swells and depletes faster during good as well as bad rainfall years. In the absence of this hydrological information, horticultural crops were promoted, and when the demand for water surpassed supply the potential wells started failing and the crops started drying up. This was because groundwater was exploited beyond its potential (sustainable yields or rainfall-recharge coefficient). Therefore, as long as there is balance between demand and supply, cultivation of water-intensive crops such as the horticultural crops is sustainable, as observed in parts of the low-rainfall zone.

These two cases clearly demonstrate the role and importance of hydrogeology and land use practices in explaining and understanding the watershed impacts. In the absence of such information, the impacts are often attributed to the quality of watershed implementation or at the most to rainfall variations (if any). This clearly indicates the need for considering the biophysical aspects while designing and implementing the watersheds. With such integration of designing, implementation and assessment becomes convenient and comprehensive when watersheds are placed in the context of a hydrological unit (Reddy and Syme, 2015).

Watershed is a technical intervention that needs to be designed by taking the hydrogeological and biophysical aspects into account. Hitherto the experience is that these aspects are rarely taken into consideration at the design stage. Impacts are also assessed in isolation of hydrogeology and the biophysical aspects. As a result, impacts are attributed purely to watershed interventions. For instance, variations in rainfall between watersheds could influence the impacts substantially. Apart from rainfall, other biophysical attributes such as soils, land use, and irrigation vary between the two locations.

From the hydrogeology perspective, the aquifer geometry in the areas can be categorized into three zones: (i) moderate-to-deep weathering and fracturing zone suitable for artificial recharge measures by water spreading methods; (ii) areas with deep fractures suitable for artificial recharge methods by injection methods; and (iii) areas with very shallow basement not suitable for any intervention. Based on the drainage order, minipercolation and percolation tanks can be proposed on the first- to third-order streams, and when the area gets into plane topography check dams are effective. The interventions required for a watershed tapping only the weathered zone are different from watersheds tapping the fractured zone, while groundwater is tapped from deep aquifers in both the watersheds. In these areas, injection wells are more suitable for recharge. In the upstream locations of low rainfall watershed, the aquifer is shallow in nature. In these areas, groundwater recharge is affected through water-spreading methods such as check dams, percolation tanks, and farm ponds. The trends of water-level fluctuation are the key for assessing the groundwater recharge, which directly depicts the changes in groundwater storage in a given area. Furthermore, the shift in cropping pattern from irrigated crops to horticulture in the upstream areas also impacted the groundwater storage and its sustainability.

It is observed that the existing interventions do not match the required or the most appropriate interventions for these two HUNs (Table 10.1). Stream interventions such as check dams are the main focus of the watershed development in these regions, though farm treatments are also evident. The concentration of check dams is more in the upstream regions, which are not very effective in improving the groundwater recharge.

Our observation in the upstream village of HUN1 clearly indicated that despite the construction of check dams, there is no improvement in the groundwater situation. Attempts to install bore wells have failed and the normal rainfall supports only swallow wells. Though swallow wells (only about 2–3) overflow during excess rainfall years, the hydrogeology is not conducive for deep water storage. Hence, the construction of check dams has not helped beyond marginal improvement swallow wells. In fact, during the below-normal rainfall years, the village even faces drinking water shortage. This is mainly because of the limited groundwater potential in this village due to its very shallow basement. Hence, attempts to improve groundwater situation through on-stream interventions are efficient investment decisions, and on-farm interventions with judicious land use planning are expected to be more beneficial as well as sustainable. This is very much reflected in the declining income from agriculture and the resilience of the households (the proportion of households reporting high resilience is small).

In the case of HUN2, which is characterized by moderate shallow basement in the midstream and downstream locations, the situation is better as the check dam interventions have helped in improving the groundwater, though the aquifer storage potential is limited to moderate shallowness. Unfortunately, in the absence of such information, horticultural (sweet lime) crops were promoted on a large scale. While these crops sustained as long as rainfall was normal and above normal, they could not be protected during the below normal rainfall years; hence, due to two consecutive below normal rainfall years in recent years, most of the horticultural crop has dried up and the farmers have incurred losses. Thus, it appears that promotion of horticulture (land use) in these conditions is not the right approach as these aquifers fill up and deplete faster in good and bad rainfall years. On the other hand, water-intensive crops even in good rainfall years may not be sustained, as rainfall fluctuations are quite normal in these regions.

Further, in this HUN, the midstream village has reported poor performance when compared to upstream as well as downstream villages. This

Table 10.1 Biophysical aspects and WSD interventions.

Hydrogeological features	Suitable interventions	Location		Present interventions	
		HUN1	HUN2	HUN1	HUN2
Very shallow basement	Suitable for on farm interventions only	Upstream and midstream	Lower portion of upstream locations, beginning of midstream locations and upper portion of downstream locations	More focus on check dams though on farm interventions are also evident	More focus on check dams through on farm interventions are also evident
Moderate-to-deep weathering and fracturing (shallow aquifer)	Artificial recharge (check dams, percolation tanks, farm ponds, etc.)	Upstream and control villages	Upper portion of upstream locations and lower portion of downstream locations	Check dams are provided everywhere with varying density; no percolation ponds	Check dams are provided everywhere with varying density; as well as percolation ponds
Deep fracture/deep weathering zone	Artificial recharge (injection wells and check dams)	Southern edge of mid and downstream; small portion in the middle of midstream locations	Middle portion of midstream locations and lower portion of downstream locations	No injection wells	One injection well provided by APFAMGS in downstream locations of HUN2
Forest fringe areas	Needs integrated treatment with croplands and alignment with existing water bodies	Mostly upstream locations	All around the HUN	Trenches in upstream locations; not integrated with crop lands	Trenches are made around most parts of forest area; no integrated crop land and existing water bodies
Conversion of wastelands to crops (land distribution)	Changes in runoff and recharge	All areas	All areas	Not considered in WSD design (unexpected); land pressure and investment support by govt. to these programs	Not considered in WSD design (unexpected); land pressure and investment support by govt. to these programs
Changes in cropping pattern	To be planned according to groundwater availability	Midstream	Midstream and downstream	Horticultural crops are promoted and sustained so far	Horticultural crops are promoted that could not be supported even with WSD
Mechanized plowing	Changes in runoff and recharge	Shifted to mechanized plowing	Shifted to mechanized plowing	WSD design did not consider these changes	WSD design did not consider these changes

has resulted in a decline in agricultural income and lower proportion of households reporting high resilience during the drought year. In fact, the midstream village has revealed lowest impact when compared to the upstream and downstream villages in this HUN. Though this goes against the given wisdom, the poor performance could be explained in terms of the hydrogeological attributes of the streams.

Similarly, the land use changes such as distribution of waste lands, mechanization, etc. have reduced the runoff in recent years. It is observed that while water bodies were filled during normal years, after the advent of these changes they are filled only during excess rainfall years. In some locations (upstream locations of HUN1), the interventions in the forest fringe areas have reduced the inflows into the water bodies, adversely affecting the cropping pattern. The farmers feel that filling of water bodies is more beneficial than groundwater recharge. According to them, once the upstream water body is filled, it not only facilitates growing of paddy, but also helps in the recharge of downstream groundwater on a wider scale. It is observed that infiltration rates are quite high in the upstream tank bed, and the recharge also includes return flows from paddy cultivation. Thus, it is necessary to integrate WSD interventions in the forest fringes with the water bodies and land use in the downstream locations. This could be another reason for the poor performance of the midstream villages.

In order to assess the possibilities for reallocation of water between streams, storing, and using more water in places where water productivity is relatively high, there is a need for a holistic approach. This we have done by estimating the production function for three different crops, viz., paddy, cotton, and groundnut. While paddy is the common crop grown under well irrigation in both the hydrological units, groundnut is grown in only HUN-1 and cotton is grown in HUN-2 only. Both groundnut and cotton are irrigated-dry crops. It may be noted that paddy continues to be the most preferred crop for the farmers in these regions, as the watershed interventions have not really helped in moving toward an alternative land use that is sustainable. This could be due to the inappropriateness of the intervention to the hydrogeology and biophysical attributes of the region. At the same time, changes in land use are short-sighted, as they did not consider the hydrogeology of the locations.

10.3.1 Discussion

In the attempts to evaluate the potential effects of differing scales and levels, it was concluded that the geographical "scale (or level) issue" in WSD is reflected by the variation in drivers associated with level outcomes, not the scale or levels themselves. The hydrological scale issue relates to where and how much of the water in the system is intercepted, as well as the interaction between groundwater and surface water, e.g., at what point does the extraction of groundwater unduly affect surface-water flows (again, level is not the driver, it is just correlated with the driver). Similarly, other factors such as land use, social and institutional commitment to WSD, and the maintenance of water-retention works also influence this.

It was also concluded that the mesolevel is reflective of the tensions between macrolevel and microlevel. There are no definitive standalone "meso" measures currently available. Hydrology can be used at a macrolevel to provide a plausible plan for the distribution of WSD, whether at the micro or meso. It can also identify the potential influence of that distribution on water availability and the likelihood of dissemination of WSD given that the whole catchment is unlikely to be covered.

Social and economic data are either collected at the macrosummary level, or typically through interviews with individuals at a microlevel and then aggregated for mesointerpretation. Thus mesolevel WSD will be evaluated with a combination of macro- and microlevel data. As such, the research challenge of using the mesolevel evaluation to recommend the "best" level of WSD implementation is unlikely to provide a precise answer.

Not only extent of area for WSD but also the quality delivery of the program will also be important factors. It must be acknowledged, however, that geographic area or level may have a significant impact on the ability to deliver, if there are insufficient resources or inappropriate institutional structures for that level. It is unlikely that mesolevel intervention will sit seamlessly on overlapping ground and surface-water catchments, especially if WSD is designed to coordinate with administrative boundaries for efficiency. In this case, the institutional scale and level become highly significant in governing WSD outcomes. Research has also shown that above microlevel economic incentives or voluntary action may have many challenges. Regardless, without adequate consideration of the hydrology, an "institutions-only" approach is unlikely to be successful.

The time scale and level is also important from the social and ecological point of view as the treated areas are expected (if managed well) to provide sustainable impacts over the longer term. How long would it take mesolevel WSD (in a normal situation) to mitigate any negative effects of WSD structures and under what conditions (e.g., quantity of rainfall)? For example, studies have shown that impacts, including externalities, are more in the regions with above 700-mm rainfall. Water tables may rise if rates of recharge exceed extraction, which over a period of time could result in higher flows downstream.

It was also found that the social-science scale without hydrological input is not enough for a definitive evaluation. Often specific hydrological conditions will tend to influence social perceptions of equity or improved livelihoods. With research efforts usually limited to the collection of survey data from relatively few villages, errors in interpretation can occur if simple WSD versus no WSD comparisons are made to demonstrate the effectiveness of WSD. The issue of selecting a control no WSD village is also problematic in that it is hard to pick similar villages at a mesolevel that are hydrologically, agriculturally, and socially compatible.

When considering the relative equity of different levels of WSD, it is also important to recognize that there is a tendency in India for people to more readily accept that water resources are unevenly distributed; that upstream and downstream areas are not naturally endowed equally in hydrological terms. Therefore, there is a need to interpret WSD outcomes along with other programs that recognize this point and promote more equitable socio-economic outcomes. This is likely to be more evident with larger land areas and populations. As has been stated: "integrated water resources management is not integrated unless water related policy is integrated with wider socio-political contexts" (Skogen, 2003).

This discussion has also demonstrated that changing the geographical area of WSD may benefit from evaluation against a strategic whole-of-catchment model to provide general guidance in regard to the investment of essentially limited funds. This could provide an assessment on the basis of broad hydrological data and an estimate of the hydrological efficiency of differing patterns and intensities of WSD at a catchment level. These patterns/intensities can then be assessed in a preliminary fashion from economic and social perspectives. The social evaluation will need to include consideration of the relative diffusion and resilience provided by alternative WSD spatial models. It must be noted though that the feasibility of doing this at this level is dependent upon basic factors such as rainfall, soil type, and the nonlinear effects of areas covered by WSD. For example, twice the geographical coverage may have more than twice the adoption impact for WSD practices. Diffusion of knowledge is likely to be compounded by greater visibility of WSD outcomes through media reporting and the perception by landholders that something "substantial" is happening, which could result in the activation of alternate and substantially wider communication networks.

From an integrated methodology perspective, perhaps the hierarchical approach adopted by the European Union for river management may ensure that each of the levels has some conceptual linkage—see Vreudenhil et al. (2010) who use five levels for a river varying from river basin to ecoelement to show how each fits into a discipline's natural level of thinking and how linkages may be created between each functional level. For the meso-WSD problem in India, one could adopt four levels (basin, sub-basin, meso, and micro). Research on each scale (e.g., hydrology, land use, institutions, and equity issues) would be at its appropriate level, which would ensure that level gaps did not occur for each scale and that there was a perspective from each scale at each level. This may go some way to answering more definitively the issue of what scales or levels are "best" for what purpose. In using such a methodology, however, it must be noted that any rescaling in WSD will require careful thought in terms of the actors involved and their changed power relationships and the requirements for fresh community-based institutions (Young, 2002). In short, the "transaction" costs of change may become a major consideration (Birner and Wittmer, 2004).

10.4 The way forward

On the whole, the assessment of the impacts validates the hypotheses that mesowatersheds could generate differential benefits at scale (upstream/downstream). However, it is clear that there is some mismatch of perceptions of the outcome of WSD in terms of the benefits and costs to other parties that relate to hydrogeology and the biophysical characteristics of the location. Therefore, these aspects, which were hitherto not considered, need to be directly addressed in order to optimize as well as sustain the watershed impacts. Important issues include:

- A hydrology-based approach of placing watersheds within the hydrological coordinates would help in understanding the upstream-downstream linkages better.
- Technical inputs need to be used for assessing surface and groundwater hydrology and their linkages in the context of biophysical attributes, while designing the watersheds. Contributions from this project have clearly demonstrated that these models and tools can be simplified with limited data demands. With training, these simple technical tools can be used at the implementation level.
- Based on the hydrogeology and biophysical aspects of the location, WSD interventions can be rationalized in terms of the type and density of interventions—it is not necessary for the entire area to be treated. In fact, some portion of the watershed could be left untreated as a buffer (donor) for the rest of the watershed. Using different scenarios pertaining to biophysical

aspects such as variations in rainfall, hydrogeology, soils, and land use, user-friendly decision support tools for watershed design could be developed.

- Achieving equity remains a difficult issue to be resolved and demands judicious planning of interventions with justice principles in mind.

References

Bardossy, A., 2007. Calibrating hydrological models for ungauged catchments. Hydrol. Earth Syst. Sci. 11 (2), 703–710.

Batchelor, C.A.K., Singh, C.H., Rama Mohan Rao, C., Butterworth, C., 2003. Watershed development: a solution to water shortages or part of the problem? Land Use Water Resour. Res. 3, 1–10.

Bergstrom, S., Graham, L.P., 1998. On the scale problem in hydrological modelling. J. Hydrol. 211 (253), 265.

Biggs, T.W., Gaur, A., Scott, C.A., et al., 2007. Closing of the Krishna Basin: irrigation, streamflow depletion and macroscale hydrology. In: IWMI Research Report 111. International Water Management Institute, Colombo. 4 pp.

Birner, R., Wittmer, H., 2004. On the 'efficient boundaries of the state': the contribution of transaction-costs economics to the analysis of decentralisation and devolution in natural resource management. Eviron. Plann. C Gov. Policy 22, 667–685.

Brown, J.C., Purcell, M., 2005. There's nothing inherent about scale: political ecology, the local trap and the politics of development in the Brazilian Amazon. 36, 607–624.

Calder, I.A., 2005. Blue Revolution, Integrated Land and Water Resource Management, second ed. Earthscan, London and Sterling, VA.

Cash, D.W., Adger, W.N., Berkes, P., Garden, L., et al., 2006. Scale and cross-scale dynamics: governance and information in a multi-level world. Ecol. Soc. 11 (4), 8.

De Groot, W.T., Tadepally, H., 2008. Community action for environmental restoration: a case study on collective social capital in India. Environ. Dev. Sustain. 10, 519–536.

Dore, J., Lebel, L., 2010. Deliberation and scale in Mekong region water governance. Environ. Manage. 46, 60–80.

Ebrahim, A., 2004. Institutional preconditions to collaboration: Indian forests and irrigation policy in historical perspective. Adm. Soc. 36, 208–242.

Enters, T., 1997. The token line: adoption and non-adoption of soil conservation practices in the highlands of northern Thailand. In: Sombatpanit, S., Zobbisch, M., Sanders, D.W., Cook, M.G. (Eds.), Soil Conservation Extension: From Concepts to Adoption. Science Publisher, Enfield, NH.

Folke, C., Pitchard, L., Berkes, F., et al., 2007. The problem of goodness of fit between ecosystems and institutions: ten years later. Ecol. Soc. 12 (1), 30.

Gibson, C., Ostrom, E., Ahn, T.K., 2000. The concept of scale and the human dimensions of global change: a survey. Ecol. Econ. 32, 217–239.

Government of India, 2008. Common Guidelines for Watershed Development Projects. http://www.eSocialSciences.com/data/articles/Document1812200890.6061823.pdf. Cited 10 March 2011.

Huchtemann, D.H., Frondel, M., 2010. Increasing the efficiency of transboundary water management: a regionalization approach. J. Water Resour. Prot. 2, 501–506.

Kerr, J., Milne, G., Chhotray, V., et al., 2007. Managing watershed externalities in India: theory and practice. Environ. Dev. Sustain. 9 (3), 263–281.

Klemes, V., 1983. Conceptualization and scale in hydrology. J. Hydrol. 65, 1–23.

Landell-Mills, N., Porras, I.T., 2002. Silver bullet or Fools' gold? A global review of markets for forest environmental services and their impact on the poor. In: A component of the International Collaborative Research Project steered by IIED. London.

Lee, K.N., 1993. Greed, scale mismatch and learning. Ecol. Appl. 3, 560–564.

Lutz, E., Pagiola, S., Reiche, C., 1994. The costs and benefits of soil conservation: the farmers' viewpoint. World Bank Res. Obs. 9, 273–295.
Merz, R., Parajka, J., Bioschl, G., 2009. Scale effects in conceptual hydrology. Water Resour. Res. 45, W09 405.
Oates, W.E., 1972. An essay on fiscal federalism. J. Econ. Lit. 37, 1120–1149.
Olsson, P., Folke, C., Galaz, V., Hahn, T., Schulz, L., 2007. Enhancing the fir through adaptive co-management: creating and maintaining bridging functions for matching scales in the Kristianstads Vatten Biosphere Reserve, Sweden. Ecol. Soc. 12 (1), 28.
Ostrom, E., 1999. Coping with tragedies of the commons. Annu. Rev. Polit. Sci. 2, 493–535.
Pagiola, S., 1999. Economic analysis of incentives for soil conservation. In: Sanders, D.W., Huszar, P.C., Sombatpanit, S., Enters, T. (Eds.), Incentives in Soil Conservation: From Theory to Practice. Oxford and IBH, New Delhi, pp. 41–56.
Pagiola, S., 2002. Paying for water services in Central America: learning from Costa Rica. In: Pagiola, S., Bishop, J., Landell-Mills, N. (Eds.), Selling Forest Environmental Services: Market-Based Mechanisms for Conservation and Development. Earthscan, London.
Pagiola, S., Platais, G., 2007. Payments for Environmental Services: From Theory to Practice. World Bank, Washington, DC.
Reddy, V.R., Syme, G.J. (Eds.), 2015. Integrated Assessment of Scale Impacts of Watershed Intervention: Assessing Hydrogeological and Biophysical Influences on Livelihoods. Elsevier, Amsterdam.
Reddy, V.R., Reddy, M.G., Galab, S., Soussan, J., Springate-Briganski, O., 2004. Participatory watershed development in India: can it sustain rural livelihoods? Dev. Chang. 35, 297–326.
Reddy, V.R., Reddy, G.M., Soussan, J., 2010. Political Economy of Watershed Management: Policies, Institutions, Implementation and Livelihoods. Rawat, Jaipur, India.
Reddy, V.R., Chiranjeevi, T., Rout, S.K., Srinivasa Reddy, M., 2015. Assessing livelihood impacts of watersheds at scale: an integrated approach. In: Ratna Reddy, V., Syme, G.J. (Eds.), Integrated Assessment of Scale Impacts of Watershed Interventions: Assessing Hydro-geological and Bio-physical Influences on Livelihoods. Elsevier Inc.
Savenije, H.H.G., 2009. The art of hydrology. Hydrol. Earth Syst. Sci. 13, 157–161.
Scoones, I., 1999. New ecology and the social sciences: what prospects for a fruitful engagement? Annu. Rev. Sociol. 28, 479–5007.
Sivapalan, M., Grayson, R., Woods, R., 2004. Scale and scaling in hydrology. Hydrol. Process. 18, 1369–1371.
Skogen, K., 2003. Adapting adaptive management to a cultural understanding of land use conflicts. Soc. Nat. Resour. 16, 435–450.
Viney, N.R., Sivapalan, M., 2004. A framework for scaling of hydrological conceptualizations based on a disaggregation- aggregation approach. Hydrol. Process. 18, 1395–1408.
Vreudenhil, H., Slinger, J., Kater, E., Thissen, W., 2010. The influence of scale preferences on the design of a water innovation: the case in Dutch river management. Environ. Manage. 46, 29–43.
Young, O., 2002. The Institutional Dimensions of Environmental Change: Fit, Interplay and Scale. MIT Press, Cambridge.

Further reading

Calder, I.A., Gosain, M.S., Rama Mohan Rao, C., et al., 2008a. Planning rainwater harvesting in India: 1, biophysical and societal impacts. Environ. Dev. Sustain. 10, 537–557.
Calder, I.A., Gosain, M.S., Rama Mohan Rao, C., et al., 2008b. Watershed development in India. 2. New approaches for managing externalities and meeting sustainability requirements. Environ. Dev. Sustain. 10, 427–440.

Farrington, J., Turton, C., James, A.J., 1999. Participatory Watershed Development: Challenges for the Twenty-First Century. Oxford University Press, New Delhi.
White, T.A., Runge, C.F., 1995. The emergence and evolution of collective action: lessons from watershed management in Haiti. World Dev. 23 (10), 1683–1698.
World Bank, 2007. Watershed Management Approaches, Policies and Operations: Lessons for Scaling-Up (Draft Report). Agriculture and Rural Development Department. World Bank, Washington, DC.

CHAPTER 11

Operationalizing IWRM concepts at the basin level: From theory to practice

M. Dinesh Kumar[a], Charles Batchelor[b], A.J. James[c]
[a]Institute for Resource Analysis and Policy, Hyderabad, Telangana, India
[b]Water Resource Management Ltd., London, United Kingdom
[c]Environmental & Natural Resource Economist, New Delhi, India

11.1 Introduction

Water of varying quality is demanded by multiple sectors and in varying quantities by multiple users within a catchment or a river basin. Allocating this water among various stakeholders involves tradeoff. Invariably, domestic and industrial sectors have the first call on the available water. These sectors largely contaminate more water than the amount they consume, unlike agricultural sectors where water is consumed in evapotranspiration or lost to saline aquifers. Therefore, the appropriate treatment of water used by domestic and industrial sectors will enable the reuse of water by another sector. Environmental flows, on the other hand, often improve the quality of water, and this water too may be used by another sector when needed (Kumar et al., 2017).

Water management has time and quality dimensions, in addition to quantity dimension. By incorporating all three dimensions while allocating water, conflict among sectors can be minimized. This would not necessarily eliminate difficult, but essential, tradeoffs between agricultural and environmental sectors. These tradeoffs will affect a catchment's hydrological processes and outcomes accordingly. As water scarcity looms large, the paradigm of water management should shift from fresh water for all uses and users to "timely supply of adequate amounts of water of acceptable quality" to each user. This will encourage water recycling within catchments, and the use of alternative water supplies for agriculture within catchments.

It is widely argued that integrated water resources management at the level of river basins, which gives due consideration to water demand management as well as supply augmentation, hydrological system integrity

Current Directions in Water Scarcity Research, Volume 1
https://doi.org/10.1016/B978-0-12-814851-8.00011-2
© 2019 Elsevier Inc.
All rights reserved.

(surface water-groundwater interactions, interconnection between catchment land use, runoff, and water quality), and meaningful community role in management decision making, is the most cost-effective, socially viable, and ecologically sound strategy for managing water, when freshwater resources are scarce, and there are multiple uses of water competing with each other (Jønch-Clausen, 2004; van der Zaag and Savenije, 2014).

The challenge is in operationalizing the concept of IWRM through appropriate coordinating institutions at the river basin level. As discussed by Biswas (2008, p. 12), "...operationally it has not been possible to identify a water management process at a macro- or meso-scale which can be planned and implemented in such a way that it becomes inherently integrated, however this may be defined, right from its initial planning stage and then to implementation and operational phases." For river basins that are falling entirely within the provincial territories, this mammoth task is in the realm of the state water agencies, which have to create such coordination mechanisms. For interstate river basins such as Krishna, Cauvery, Godavari, and Mahanadi, it is in the realm of the Union government.

The basic concepts and principles of IWRM (such as economic efficiency, social equity, and ecological sustainability) are widely known and accepted (Jønch-Clausen, 2004; van der Zaag and Savenije, 2014). So are the key coordination mechanisms (UNESCO, 2008). But the fundamental question is in identifying and sequencing the various technical and institutional processes for getting them followed. In this regard, the real gray areas, which academics and researchers have not systematically worked on, are: (i) the nature of scientific/technical studies and inputs that would be required for planning IWRM in a basin context; (ii) the institutional model for implementing IWRM at the level of river basins, including the institutional design principles and institutional structure; and, (iii) the various institutional processes that need to be initiated for affecting implementation. Among all these, institutional design principles are of paramount importance, as poor institutional design would lead to institutional ineffectiveness.

In this chapter, an attempt is made to: (1) lay out the institutional design principles for setting up effective water management institutions; (2) discuss the overall institutional regime for implementing integrated water resources management actions at the state level and the institutional structure; and, (3) describe the key scientific and institutional processes that are necessary to operationalize them at the level of river basins. While doing this, we would also highlight some of the water management outcomes that are "integrated" in nature.

11.2 General plan of action for implementing IWRM

11.2.1 Institutional capability building for water management

11.2.1.1 Design principles for setting up of new water management organizations

In India, there are no separate agencies for water resources management covering planning, design, and implementation functions either at the union level or at the provincial level. At the union level, technical agencies such as Central Water Commission and Central Ground Water Board perform resource investigation and evaluation functions to quantity the amount of water available in river basins and underground formations, respectively.

Further, the same agency performs multiple functions. For instance, the water resources departments (erstwhile irrigation department) at the provincial level perform water resources evaluation and planning; undertake water resource development activities for surface water; and also do allocation of water to various sectors, including industries and drinking water supply. They are also responsible for flood control. Similarly, the state pollution control boards undertake water quality monitoring of aquatic resources (rivers, lakes, and groundwater), and are also responsible for enforcement of pollution control norms they set up. These create situations of conflict of interest and reduce their effectiveness (Kumar, 2006).

Since water is a state subject in India, setting up of organizations for carrying out the line functions in water resource management sector and nurturing them will be paramount to building institutional capabilities for water management at the state level. The line functions include water resource monitoring, water resource evaluation and planning, execution of water development and management schemes, water allocation, and water resources management. This apart, there is a need for an organization that can deal with the more complex issues. Some of them are deciding water allocation priorities across sectors, evolving criteria for allocation of water for different sectors in different regions, determining private/cooperative use rights, and fixing water tax and pollution tax.

The institutional design principles for IWRM, as formulated by Kumar (2006) and later on elaborated by Kumar (2016), are as follows: (1) the operational boundaries of the governance units created for effective management should match with the hydrological system boundaries; (2) clear distinction between water development and water resources management functions; (3) institutions responsible for water allocation or regulating water use have to be different from those providing water-related services—viz., irrigation water supply and environmental management; (4) institutions responsible for water quality monitoring will be different from the ones managing water quality; (5) the institution responsible for investment in water quantity management and WRM should also be enforcing norm and regulations on water use; and, (6) the governance/management structure should be such that it integrates local water management interventions with larger basin management actions and promote involvement of user groups/communities in the governance and management.

11.2.1.2 State water development and management board

The State Water Development and Management Board will have the mandate of provide the overall governance of water resources in the State. To be set up through an act passed by the State Legislator. The Board is to be headed by a full-time Chairperson, who will be a highly respected "water engineer" or economist with a long career record with the State or Central Government. He/She will be directly appointed by a "Committee of State Legislatures" chaired by the Chief Minister. The candidate will be an official of cabinet rank. The Chairperson will have a 5-year tenure. The Board is not to be in the administrative control of any government department.

The Secretaries of all ministries dealing with water directly will automatically become its members, apart from senior professionals—environmental lawyers, academicians, water engineers, economists, and ecologists—social workers—who are to be nominated by the Governor. The nominated members will have a 3-year term.

The Chairperson can constitute a search committee to find and appoint the member-secretary, who will be the Director of the SWDMB Board. The Director and a group of senior professionals from hydrology, water technology, agricultural sciences, economics, management, and law will consist of the Board will constitute the executive wing of the Board, who will be guided by the Chairperson.

The Board will have the following objectives:

- Help the Government formulate State Water Policy and Water Law
- Review State Water Policy periodically
- Create appropriate organizational/institutional structures and mechanisms for governance of water resources for river basins and regions.
- Evolve criteria for allocation of water from river basins among various sectors of use along the guidelines of State Water Policy

- Evolve criteria for establishing private/cooperative property rights in water, including groundwater, canal water, and other surface water resources for different users in a particular sector.
- Evolve criteria for fixing water taxes for different sectors, viz., irrigation, drinking and domestic, industry and recreation, and pollution tax for different regions
- Evolve pricing structures for efficient use of water in irrigation, industry, and domestic use
- Represent the respective state governments in situations of interstate water disputes in the interstate water tribunals.

The operating funds for the SWDMB are to come from the State's annual budgetary allocations plus the annual service fee paid by the basin authorities/regional water councils.

11.2.1.3 Water resources monitoring and planning unit

Water resource monitoring and planning are important line functions in water management sector. However, this is a highly neglected area in India. While reliable scientific data are the crux of good water resource planning, they are often lacking. While data required for planning are generated through systematic monitoring, monitoring is also required to assess the impact of water management interventions on the socioeconomic system as well as the hydrologic system. There are several issues involved in water resource monitoring.

These issues point to the need for a separate institution to perform the line function of monitoring and planning. Since the same agency is carrying out planning and monitoring, the data requirements for realistic planning could be worked out, and accordingly monitoring system could be designed and implemented. It can be constituted as a scientific/technical wing of the State Water Development and Management Board. The unit should also co-opt specialists/experts from social science fields such as economics and sociology.

The unit can also evolve the criteria, decide upon, and observe the parameters for assessing the performance of other line agencies, which are engaged in implementing water development and management schemes. Some of them are basin yields, silt load of water in the reservoir and rate of siltation, quality of water in the rivers, streams, and aquifers, and the pollution load. Some others are per-capita water supplies for domestic uses, their quality and reliability, duration of supply, irrigation intensities in terms of area irrigated, number of irrigation, rate of recovery of water charges, water utilization efficiencies, conveyance efficiencies, and economic outputs.

This means that the unit should collect the range of socioeconomic data apart from the hydrology-related data. The collection, compilation, and periodic publishing of such data will put pressure on the line agencies to perform better. For instance, publishing of data on irrigation performance and rate of recovery will put pressure on the farmer organizations and the irrigation bureaucracy to perform better.

An important task to be performed by the monitoring and planning unit will be to carry out resource evaluation, which can form the basis for detailed planning. It will undertake the resource evaluation for the major river basins of the state concerned and carry out prospective planning. The prospective planning will include the future demands of water in different basins and the potential future supplies and the ways to meet or manage the demands including interbasin transfer of water. The River Basin Organizations will have to obtain the data on basin's water availability and balance from the planning and monitoring unit.

The unit can collaborate with NGOs and other civil society organizations for collection and compilation of water resource-related data. This will help strengthen the capabilities of these civil society organizations in analyzing water management problems and issues and orient them about the state of affairs with regard to water management vis-à-vis the resource condition, access to and control over the resource by different sections of the communities. The government of Rajasthan has a state water resources planning department, which has similar mandate.

11.2.2 Allocation of tradable water use rights

In order to create incentive for efficient use, the water users should be confronted with opportunity cost of using water. To induce opportunity cost of using water from the natural system (rivers, lakes, surface reservoirs, and aquifers), establishment of tradable water use rights among the user groups and users is crucial (Howe et al., 1986; Rosegrant and Binswanger, 1994). Equity and sustainability are important considerations for establishing institutional regimes for water use. Hence, realistic assessment of total water resource availability, renewable freshwater, and economically accessible water resource in different regions/basins will be the cornerstone of establishing tradable property rights regimes. This is an extremely difficult task.

This task is to be performed by the planning and monitoring unit of SWDMB. In this section, we will discuss how the water use rights shall be operationalized for both surface water and groundwater resources.

As regards groundwater, in the deep alluvial areas, the total groundwater storage is many times more than the groundwater, which is accessible economically. The economically accessible groundwater is much more than the replenishable groundwater. Since sustainability considerations override economics, only the renewable portion of the water should be brought under the property rights regimes. But, there are certain practical considerations in evolving the criteria, which will differ for different regions (Kumar, 2006).

In alluvial areas, in view of the problem of excessive withdrawal (like in north Gujarat, central Punjab and western Rajasthan), the total volumetric use rights to be allocated every year should be much less than the renewable groundwater. This will help reverse the process of mining over a period of time. On the other hand, in certain canal command areas, due to the problem of underdevelopment of groundwater, the actual annual groundwater use rights could be much higher than the renewable groundwater till a good balance is achieved in the groundwater ecology of the region (Kumar, 2006).

In hard rock areas of India, the economically accessible groundwater will be less than the renewable groundwater resources as these regions have very little static groundwater. As groundwater is trapped in the fractures, fissures, and weathered zones, not all the recharge can be extracted using the available technologies. Another factor, which needs to be considered while establishing private property rights systems, is correlative right. In hard rock areas, well interference is a common phenomenon. When farmer pumps out water from his well for a long time, large draw down is created, which will bring down the water levels in the neighboring wells also. For an individual, the ability to access groundwater from the common pool, which he/she owns under the rights system, is related to the amount of water his/her neighbors are pumping out from the same pool. Therefore, the rights are correlative. Due to this phenomenon, the actual "abstraction quota" that can be permitted to any farmer and which does not deprive the neighbors of their rights will be much less (Kumar, 2006).

In order to internalize this phenomenon of well interference and make the rights "nonrelative" or exclusive, the communities can devise mechanisms.

The allocation of volumetric use rights in surface water should be determined primarily by two important considerations: the annual runoff, which is the renewable surface water (x), and the amount of water needed for ecosystem services (y). The amount of water for ecosystem services will depend on the range of ecological parameters including the water requirements of riverine ecosystem, watersheds, the assimilating the pollution load in the river. The amount of water that can be allocated for various competitive uses annually therefore will be $(x) - (y)$.

But for all practical purposes, the surface water (use) rights allocated to irrigation agency, drinking water supply agency, and industrial corporation in any region will be decided by the total volume of water that can be harnessed by the existing storage and diversion head works. While these considerations automatically factor in the ecological sustainability concerns of the water scarce regions, it will not be so for water-abundant regions. The import of surface water to such regions will have to be controlled to maintain ecological balance. Therefore, the rights to use canal water will have to be limited to encourage higher rate of withdrawal of groundwater.

Water rights systems exist in many developed and some developing countries, and in certain cases these rights are tradable (Rosegrant and Binswanger, 1994). For instance, in Australia, the water rights allocated to different users are tradable (Connell, 2011). For maintaining environmental flow in the river, which is a public good, surface water rights are purchased by the government from the market through trading. In Spain, the river basin organization allocates water rights for both surface water and groundwater among various users, including irrigation agencies, industries, and farmers (private irrigators). In the Middle Western United States, groundwater management districts exist in states such as Kansas and California, where water rights of individual appropriators of groundwater are allocated volumetrically and enforced.

11.2.2.1 Intersectoral allocation of rights

Once the quantum of resources to be brought under private property regimes is assessed for different basins and regions, the SWDMB is to evolve criteria for allocating water rights among different sectors of use and among different users and user groups within each sector. Hence, there are two major steps involved: allocation across sectors; and allocating rights within sectors among the users.

One important criterion for allocation of water use rights across different sectors could be the per-capita renewable freshwater availability in each of the four regions. In those regions where per-capita renewable fresh water is quite low, lion's share of the rights to use the available water will have to be allocated for basic survival needs (as defined by P. Glieck (1997)) and domestic food security.

Nevertheless, some allocation will have to be made to sustain the existing economic activities such as industries and recreational services and cash crop production. Nevertheless, the percentage allocation for these will be much less. The objective will have to be discouraging such economic activities in the regions that the water ecosystem cannot support in the long run. Therefore, the criteria for future allocation will also have to be based on the existing use patterns.

In regions, where per-capita renewable freshwater is high, larger share of the rights will be allocated to economic production purposes such as irrigation of cash crops, industries, and recreational purposes.

An important sector of water use, which will not come under the purview of markets, will be water for ecosystem services. The government will have to invest in procuring that much amount of water from the river basin organizations in larger public interest. Investing in procurement of water for ecosystem services will be of much higher priority in the water-scarce regions.

11.2.2.2 Intrasectoral allocation of rights among users/user groups

When rights are allocated across sectors, a major task for the SWDMB will be to allocate water use rights among individuals (private property rights) and user groups (cooperative property rights). According to the variation in criterion for allocation of available water across sectors, criterion for allocation within sectors will vary remarkably across regions. For instance, food security being the overriding concern in a water-scarce region, the allocation of water for agriculture in water-scarce region will be primarily based on considerations such as family size. The criterion such as "total cultivable land holding" will be only secondary. On the other hand, in water-abundant region, the criteria for allocation of water within agriculture sector will involve total cultivable holding as a primary criterion.

Similarly, in absolutely water-scarce regions, the criterion for allocation of water for domestic uses may involve only consideration such as "basic survival need" and "family size." While in water-abundant regions, allocation may also involve considerations such as water for recreational services, water for gardening, etc. apart from "per capita water for domestic purpose" and "family size." Here it is important to mention that water for basic survival (50 lpcd) is much less than the water requirements for drinking and domestic purposes.

Once the Board evolves appropriate criteria for allocation of rights, it becomes the responsibility of the respective River Basin Organization to define and establish the rights of individuals, groups, and agencies. They are discussed in the respective sections on river basin organizations under the specific action plan.

11.2.3 Strengthening legal regimes

The State should have a comprehensive water law. The law should provide for:

- Recognition and enforcement of private/cooperative, tradable property rights in groundwater and canal water through appropriate institutional structures such as RBOs;
- Setting up of river basin organizations, State Water Development and Management Board, and Water Resources Monitoring and Planning Organization;
- Setting up of organizations for management of irrigation management up to the system level with appropriate legal status, including organization of water users;
- River Basin Organizations to have statutory powers to decide on the allocation of water across sectors; prices of water and sanction on certain uses of water; and,
- Involvement of multiuser stakeholder forums in governance and management of water

The State Water Development and Management Board shall be responsible for drafting the Water Law. The law should enable communities to protect and manage the water resources and for that purpose. It should allow them to frame and enforce rules and regulations locally regarding its use that are socially viable and hydrologically meaningful, within the existing institutional regimes, rather than trying to police. The agency should initiate institutional processes to activate public debate on the "draft document" and get maximum public and political opinion on the pros and cons of implementing the same. After receiving this, the draft bill could be considered for introduction in the assembly for making legislation. The Water Law should be widely publicized for community awareness and education. Every user of water should be aware of the law governing the use of water in the State, especially the provisions in the law for protection of private property rights and the penalties for actions that lead to violation of such rights.

The State Water Development and Management Board can launch programs aimed at educating the public about the provisions of the Water Law.

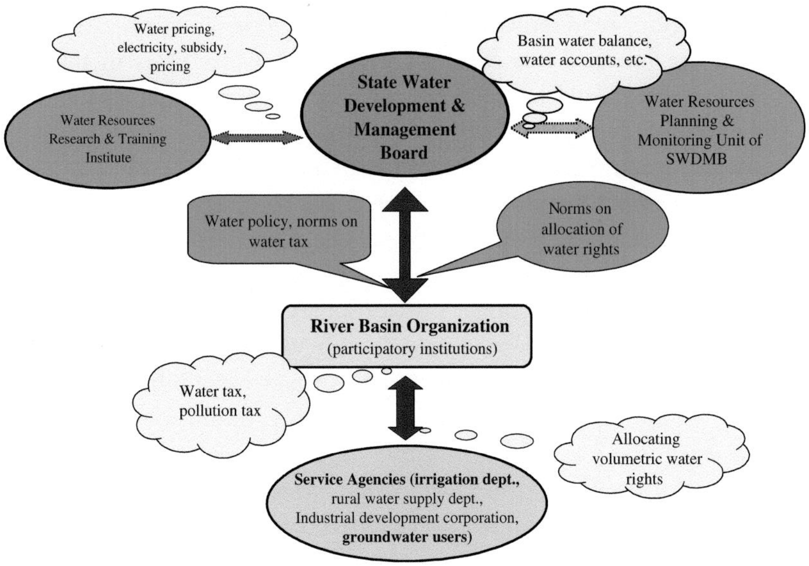

Fig. 11.1 Institutional regime at the state level for integrated water resources management.

The overall institutional regime for governance and management of water resources under the proposed paradigm is given in the flow diagram (Fig. 11.1).

11.3 Specific action plan for different regions

The action plans for addressing the water management challenges involve several steps. They are:

1. *Designation and Constitution of River basin Organizations*: the State Water Development and Management Board (SWDMB) shall designate river basin organizations. The first step will be to demarcate the exact boundaries of the basin vis-à-vis the administrative boundaries. The SWDMB shall then initiate constitution of Governing Councils for these basin organizations. It should have adequate representation from civil society organizations, and widely respected individuals and organizations concerned with/about water problems in the region. It should include the district dairy unions, credible NGOs working in the region, social reformers, naturists, and champions of community development and environment.

The Board can also initiate institutional processes to form *Stakeholder Forums*. The forum can work toward realizing the objective of effective, user-centered management of water in these basins. This shall comprise progressive farmers or representatives of farmer organizations, leading industrialists from the region, representatives of grass root women's organizations who represent the interests and concerns of women water users in the basin. Besides these, it should have officials of various government agencies such as the irrigation department, forest department, GWSSB, and water supply wing of major municipal corporations as members.

The *Governing Councils* can then select the chief executive of the *Management Units* of the respective RBOs. The Chief Executive will subsequently set up a group of professionals to carry out the day-to-day management of the RBO under his/her supervision. The Chief Executive shall be a water resource professional with vast experience working on water management issues in the State. The management unit should consist of water technologists, hydrologists, geologist, economist, ecologists, and political scientists.

2. *Setting up of database of water users, socioeconomic profiles, and water accounts*: Database on the physical and social aspects of water use is as much a prerequisite for management of river basins as the database on the physical aspects of water availability like the basin water balance. Developing this database will be the responsibility of the RBO. The specific database requirements for basin management would include the different types of users of water, their socioeconomic profiles, and the water accounts. The water accounts will show how much water is being diverted and consumptively used in different sectors. The basic socioeconomic data required for setting up the database are available with a number of organizations such as the revenue department, Industrial Commissionerate, Agricultural Directorates, and Animal Husbandry departments. Therefore, the task before the RBO will be in compiling these data at the level of river basin. Based on this database, water accounting can be carried out.
3. *Water resource mapping*: the RBO shall take up the exercise of complete mapping of the water resources in the basin. This will form the basis for basin planning. This exercise should include mapping of drainage patterns, determining surface flows, demarcation of aquifers, their recharge and discharge areas, mapping groundwater levels and quality, evaluation of recharge, and abstraction. The RBO can obtain the necessary data from the water resource monitoring and planning unit of SWDMB.
4. *Mapping of the existing water harvesting/storage/abstraction structures*: data on the existing water harvesting, storage, and abstraction structures can be obtained from water resources department, groundwater/minor irrigation department, rural development department/watershed

development agency and state remote-sensing application centers (SAC). The water resources department can provide data on all minor, medium, and major irrigation schemes. The groundwater/minor irrigation department can provide data on their irrigation tube wells. The Agricultural Directorate can provide data on private wells. The SAC can provide data on land use pattern.

5. *Mapping of areas and villages affected by serious water quality problems*: excessively high levels of fluoride in groundwater are a menace in many regions of India—Gujarat, Rajasthan, Haryana, Telangana, Karnataka, and Chhattisgarh are some of them (GOI, 2010). Therefore, one of the basin management objectives will be to reduce the fluoride content in groundwater to permissible levels. The RBO has to take up complete mapping of the fluoride-affected areas. The State Water Supply and Sewerage Boards or Public Health Engineering Departments can act as the nodal agency to set up this database. It can carry out detailed investigations on the extent of impact of fluorides on human and animal life. The purpose is to get a comprehensive understanding of not only the degree of fluoride contamination of groundwater, but also the extent of physiological and biological and toxicological impact on the fluorotic individuals.
6. *Mapping of the existing rural economic institutions*: a large number of grass root organizations are required to implement the local water management interventions planned for the basin. Local institution development is a painstaking process, and needs investments in terms of human resources, time, and money. Where local institutions are active, they can be motivated to play an important role in local water management. This is subject to the condition that they have large stake in water management.
7. *Evolving basin management objectives and development of basin plans*: analysis of water balance, water accounts, and the needs and interests of the stakeholders can form a sound basis for evolving basin management objectives. Based on the management objectives, a basin plan can be developed. Water balance studies have to be taken up for this as well. The plan should indicate different physical interventions that are possible in the basin, and the extent to which they could contribute to augmenting water supplies. For instance, it should indicate the potential for watershed management, wastewater treatment systems, local water harvesting, large-scale artificial recharging and conjunctive

management options, and their potential impacts. It should also indicate the technological interventions for managing the demand for water in different water use sectors. The basin management plans are not blue prints for direct action but will throw indications on the directions for developing more realistic local management plans that can be taken up for implementation, and their potential impacts on the overall water supply and demand balance.

There are several good examples of development of basin management plans from Australia, Spain, and France. In the case of Murray Darling basin in Australia, the basin management plan is a statutory document, which the basin states will have to comply by while undertaking land and water management interventions in the respective jurisdictions (Connell, 2011). In Spain river basin, authorities exist for all the major basins of the country, known as *Confederacion Hydrographica Del* (the Basin name) and they carry out hydrological planning for the respective basins (*source*: "The Jucar River Basin District," prepared by Ministry of Agriculture and Fisheries, Govt. of Spain).

8. *Setting up of subbasin organizations*: many of the river basins in India that are falling entirely within the country's territory are too large—to name a few, Cauvery, Godavari, Krishna, and Narmada. They cut across different agroecological subzones. Apart from the variations in soils and climate, there are remarkable variations in geology, hydrology, vegetation, and overall water availability situation. Hence, it is difficult to work out detailed local management plans from the basin plan at least in the case of these two basins. Keeping in mind the broader basin management objective, specific subbasin planning has to be carried out. Sub-Basin Organizations (SBOs) have to be created for this purpose. Constitution of the SBO will be the mandate of the RBO.
9. *Preparation of specific management plans for subbasins*: the objective will be to match the demand and supplies, and arrest groundwater depletion. This will be the responsibility of the SBO. The following analysis would be carried out to arrive at subbasin plans.
   - The scale and potential impact of watershed management interventions and local water harvesting structures in the upper catchment areas and plains, respectively, on water balance.
   - Identification of villages and watersheds.
   - The scale and potential impacts of conjunctive management activities such as recharging of deep aquifers using the excess runoff diverted from the river, and their potential impacts on groundwater hydrology.

- The scale and potential of demand management interventions: change in water use technology, change in cropping patterns, efficient conveyance system, wastewater treatment and reuse systems, etc., and target the regions/localities for their adoption.

Scientific water balance studies are required for evaluation and planning of water management interventions. The studies should assess the catchments for estimating the availability of surface runoff, base flows, and groundwater using hydrological (simulation) models for various rainfall conditions and compare the water availability against the estimated future water demands from various sectors. The studies should then also consider various water management scenarios for the subbasin; evaluate the impacts of the key interventions planned under those scenarios to improve the water balance of the subbasin; and analyze their overall impact on the water resource condition—stream flow regimes, quality of the runoff water, groundwater recharge, and changes in groundwater levels. The models should be such that they capture the groundwater-surface interactions adequately. The specific management plans for the subbasins will aggregate to constitute the basin management plan.

10. *Determining sustainable levels of groundwater abstraction in the basin*: simulation models of groundwater for the entire basin need to be developed. Such models are available in the market. They can simulate the natural recharge and artificial recharge conditions under different rainfall regimes and different groundwater abstraction and surface water diversion scenarios under the basin management plan. The groundwater-level trends could be analyzed for different degrees of abstraction. The abstraction levels suggested by the scenario, which creates permissible draw down, can be adopted as the sustainable level of abstraction for a given rainfall regime. However, the permissible annual draw down for the basin will be decided by the overall basin management objective. If the basin management objective is to raise the groundwater levels over a period of time, then the permissible draw down will be negative and accordingly, the sustainable abstraction figures will come down significantly.

11. *Technical assistance from scientific institutions to RBOs*: since the RBOs will have to focus on governance and management issues in water, they cannot develop into full-fledged scientific institutions. But scientific skills are required for carrying out detailed resource mapping, periodic water balance studies, water accounting, and basin planning. These skills will not be available with the RBO. The RBO, therefore, will have to enter into technical collaboration from scientific agencies such as the Central Ground Water Board, Space Application Centre, and the newly promoted Water Resources Research and Training Institute. The areas for technical collaboration are:

- Delineating different aquifers spatially—horizontally and vertically
- Identifying areas suitable for artificial recharge in shallow and deep aquifers.
- Designing conjunctive management systems and artificial recharge structures

For perspective planning of river basins, the RBO can seek inputs from the water resource monitoring and planning unit of SWDMB.

12. *Identification of agencies for local institutional development*: appropriate institutional arrangement for implementing the management actions is as important as developing a good management plan. The RBO and the SBOs will have to identify agencies for putting in place appropriate institutional arrangements at the local level for microplanning and their implementation. These agencies will have two primary roles. (i) Local institutional development, and (ii) integrating them with the RBOs. This apart, the agencies will also have to help the local institutions so formed in preparing their management plans. However, in areas where grass root institutions already exist, the effort will have to be only in motivating them and developing and channelizing their potential.

13. *Promotion effective local watershed management institutions*: in areas where local institutional capacities do not exist, the agencies can focus on the following: (i) Creating awareness on the need for water conservation and management, and (ii) Social Organization, (iii) Developing the technical and organizational capabilities for water management. There are three distinct typologies for local water management institutions, depending on the overall focus.

First: in the upper catchment areas of the subbasins or large watersheds, as the interventions will be watershed based, watershed management committees can be set up. Watershed boundaries can form the basis for constituting such organizations. If a particular watershed falls within a village boundary, the watershed management committees shall be formed at the village level. The Committees shall involve all water users within the village administrative boundary. If the watershed boundary extends beyond the village administrative boundaries, then users from more than one village have to be included in the watershed management committee.

In the plains, where local harvesting activities such as construction of percolation tanks, and other decentralized recharge systems are to be carried out, water management committees shall be promoted. They can be formed at the level of a village. In situations where boundaries of the water systems like tanks and ponds extend beyond village boundary, water management institution should include all users from the villages concerned.

Finally, in the lower plains, where no supply management interventions are possible, water users' associations shall be promoted. Next step will be developing the technical and managerial skills of the newly formed organizations. The technical skill development shall include water management activities, water harvesting, artificial recharge, construction and postconstruction management such as desilting and catchment protection.

In areas where community-based, local institutions are present, they can be motivated. Once these local institutions are fully developed, they need to be integrated with the higher institutions in the hierarchy.

The PIAs shall be engaged in educating the local water management institutions about the new property rights/management regimes that the users are to be brought under. They will also educate the users about the enabling laws that are newly introduced for protection of private/cooperative property rights.

The local management institutions should be legal entities as transfer of funds is required. They can be registered as cooperative societies.

14. *Promotion of electricity users' cooperatives*: the state electricity utilities can facilitate formation of Electricity Users' Cooperatives (EUCs). In areas, where electricity is not consumed in the farm sector, or the consumption is negligible, such cooperatives are not required. The SEBs can go for bulk supply of electricity to the EUCs in these areas but have to supply meters to them. The EUCs will have to install meters in each well owner's pump. The SEBs will sell electricity to the EUCs. They will meter it at the transformer kept for each cluster. The EUCs will have to pay for the consumption as per the meter reading at the transformer level. It will be the responsibility of the EUCs to monitor the consumption of every individual user and recover the charges. The information on electricity consumption will be a critical input for the local water management organization for monitoring groundwater use and understand as to what extent they are able to enforce property rights locally.

15. *Promotion of WUAs for irrigation management*: Though water users associations are promoted under the participatory irrigation management program, they have so far made only negligible impacts on the ways in which irrigation systems are managed. There are three reasons for this.

First: the number of farmer organizations so far created is very low due to the low level of priority irrigation management enjoys within the department. Second: the involvement of WUAs in irrigation management is limited to the outlet and below due to which the farmers have little control over the water. Lack of assurance from the irrigation department to the farmers on either meeting the minimum requirements, or otherwise adequately compensating for the losses arising out of nonprovision of irrigation water is the third, but the most important reason.

There is a need for setting up a separate wing within the irrigation department to implement the irrigation management transfer. This can be called a Farmer Management Support Cell along the lines of the Water and Sanitation Management Organization (WASMO) of Gujarat and Water Supply and Sanitation Organization (WSSO) of Maharashtra. This FMSC has to have good social, managerial orientation as well as the ability to appreciate technology-related issues. The FMSC has to run as a project of the irrigation department. Such a flexible institutional framework can provide the much-needed flexibility in administration.

The FMSC can comprise mainly professional with background in rural management, social work, and social and political sciences and can be headed by a professional in the rank of Chief Engineer. The mandate of the FMSC is to enable formation of water users' associations, and build their technical, organizational, and managerial capabilities for managing irrigation. For this, it can take up the following activities:

- Publicity campaigns on irrigation management and turnover policy through its field offices;
- Regular meetings with progressive farmers and NGO representatives on irrigation management; and,
- Identification of NGOs/voluntary organizations for social mobilization

NGOs and other voluntary organizations so identified will then be responsible for setting up appropriate forms of organization of farmers in the commands of the schemes. The FMSC can organize "training of trainers" for the staff of the NGOs and other civil society organizations, which are engaged in promoting the WUAs. The topics for training should include management of irrigation systems, irrigation water management, concepts of farmer management of irrigation, and roles and responsibilities of farmer organizations in irrigation management at various levels in the irrigation system.

An alternative will be to use the social mobilizing skills of the dairy unions. In such a case, promoting of water users' associations at the primary, secondary, and tertiary levels will be the responsibility of the union.

The NGOs can promote water users' associations comprising farmers who own land under each outlet as the Tertiary Committees. They can then set up a federation of such Tertiary Committees at the minor level, by drawing two representatives from each Tertiary Committee. This can be called a VSA Committee. In order for promoting WUAs, the NGOs can take up the following activities. 1. Social mobilizing, and 2. Training on technical and organizational aspects of irrigation management for capacity building. The training requirements for various levels in the organizational hierarchy of farmers will be different.

In states like Gujarat, the district dairy unions might be appropriate for managing the system at the level of distributary and could be equipped to manage 4–5 distributaries commanding a total irrigated area of 40–50,000 ha. The district unions can motivate the village dairy cooperatives to take over the management of minors. The village dairy cooperatives can organize the farmers under each outlet and set up the Tertiary Committees (TC). Since the district dairy unions already have attained social mobilizing skills, it will be easy to prepare them for local institutional development.

This is one component of the irrigation modernization. The next step is technological innovation and is discussed in the following paragraph.

16. *Technological Innovations as part of modernization*: The modernized scheme should have robust hydraulic designs to facilitate automatic control and measurement of flow rates and/or water levels at various control points along the hydraulic system. The control points at various levels in the hydraulic systems also correspond to the interface between different levels in the organizational/institutional hierarchy created. Since the irrigation system from distributary to the lower levels is to be turned over to the FOs, infrastructure modernization should cover distributary and below.

The options are:

- *Improved control of main and branch canals*: replacement of old control structures with overflow gates or composite gates with overflow characteristics to provide good water level control in the branch canals. Full automatization of these canals using dynamic control is also needed.
- *Automatic flow measurement and control in distributary, minors, and watercourses*: in this case, the flow rate into the distributary, minors, and watercourses is

kept constant irrespective of the fluctuations in water levels in the supply canals. This uses automatic electric gates to maintain a constant head on a flume.

- *Modernization of distributary, minors, and watercourses:* in order to maintain constant flow rate at the turn out to field channels, at the minor outlets, and the head regulators, water levels in the supply channels like water courses, minors, and distributary need to be kept constant. Therefore, cross regulators such as long crested weirs, hydraulic gates, etc. are to be installed in the distributary, minors, and watercourses. Watercourses, which are unlined, can be lined.
- *Modernization* of field channels: field channels are open ditches and excessive losses take place during conveyance of water due to seepage, as soils are sandy. Therefore, low-pressure pipes can be installed.

This is one of the most controversial interventions in irrigation management. Worldwide, irrigation agencies have been investing in interventions such as lining of canals under irrigation system modernization on the pretext that this would result in water saving. However, recent evidence suggests that this may not be the case always, and on the contrary in many situations, more water would get consumed. The reason for this is that the water that seeps through unlined canals or canals with broken lining actually recharges groundwater. Preventing this seepage actually makes more water available for supplying to the field, thereby increasing the crop consumptive use (Perry, 2007).

17. *Development of microplans for watersheds*: development of microplans will be the responsibility of the local water management institutions. The promoters of the local institutions could help facilitate the process through exercises such as Participatory Rural Appraisal and orienting them about the subbasin and plans. The idea is to make sure that the local management plans fits very well within the larger basin management plan. Once the microplans are developed, the subbasin plans and the basin plan will have to be reviewed and revised to incorporate the location specific changes and deviations that are likely to occur. This can give a complete picture of the proposed management interventions for the basin.

Unfortunately, the past approach has been to develop watershed management plans at the level of microwatersheds based on hydrological assessment of these micro catchments and then aggregate them at the level of large catchments, without considering the "scale effects" (Batchelor et al., 2002; Gupta, 2011; Kumar et al., 2008) and watershed development (Batchelor et al., 2002; Syme et al., 2012).

18. *Enable support from scientific agencies for preparation of technical plans*: if the local management systems are complex, the local institution will require inputs for preparation of technical plans. For instance, the design and preparation of estimates wastewater treatment systems can sometimes be very complex, depending on the type of pollutants present in the effluents from urban areas and industries. The RBO should enable support to these local management institutions from technical agencies such as the National Environmental Engineering Research Institute, Nagpur.
19. *Allocation of groundwater and surface water use rights*: once the agency has complete idea about the quantum of surface water and groundwater, which can be sustainably extracted including the new water generated through management interventions, it can start allocating the use rights among the users, and user group organizations. At the basin level, the rights to use groundwater will be allocated to the local water management institutions, which in turn will allocate the rights among their members. In allocating water rights, the agency will follow the criteria decided by the SWDMB for inter- and intrasectoral water allocation. Further, water right allocation for groundwater will be correlative to surface water use rights. In other words, the total entitlement of an individual user or user group organization from surface sources and aquifers will be equal to the total volumetric right of that user of user group organization from the entire basin.

At the practical level, the groundwater rights will be allocated mainly to the watershed committees, water management committees, and (ground) water users' associations, industrial clusters, Panchayats, Municipalities. The volumetric rights will depend on the sectors these institutions cater to, and the number of users within the sectors, based on inter- and intrasectoral water allocation criteria. Similarly, the rights to use water from surface schemes will be allocated to irrigation department, and other line agencies, which have already invested in appropriating the resource or those who wish to appropriate, depending on the sectors they cater to. The norm used for water allocation in this case will be same as that for groundwater. The irrigation department can subsequently transfer the rights to the farmers' organizations and municipal water authorities. However, if the municipal authorities have already established their infrastructure for water transport, then the RBO could directly allocate the rights to them. The farmer organization at the highest level in the hierarchy can transfer the rights to the lower level.

The water use rights are likely to vary from year to year based on the amount of water available in the basin in form of surface water storage and recharge. Though the criteria for intersectoral allocation can be kept the same for the region, the criteria for a particular user within the sector have to vary. The best example to illustrate this point is the Cartagena in Spain where rights of irrigators in Segura basin to use water (concessions) imported are decided on annual basis.[a] The volumetric allocation was found to be varying from a mere 6 MCM to a maximum of 132 MCM over a 30-year period from 1979–80 to 2015–16, depending on the amount of water available in the donor basin for transfer (*source*: crcc.es).

20. *Fund generation through sale of water and collection of pollution tax*: the RBOs need huge amount of financial resources to invest in water resource management projects, including those for improving water quality. In addition, it needs funds to take care of its own management cost, and transaction costs of local water management institutions and NGOs. A major source of resource generation for the RBOs will be the fund generated through allocation of water rights in the form of water taxes and pollution tax. The norms for fixing water tax and pollution tax will be as decided by the State Water Development and Management Board. However, the water used to meet basic survival needs could be made free of charge.

The taxes to be paid for the purchase of surface water use rights by the farmers are however likely to be much higher than those for groundwater. This is to recover the capital investment that irrigation department has made for building the infrastructure such as the dams and the canal systems.

China started introducing resource fee for water delivered to farmers in irrigation commands on volumetric basis sometime in 2002, while only domestic and industrial water supply services were charged on volumetric basis till then. The irrigation fee collected from the farmers goes directly to the agency, instead of the state finances. It also charges a water supply tariff

[a] In *Comunidad de Regantes del Campo* de *Cartagena (Community of Irrigators in De Campo Cartagena)*, the total irrigation service area was 40,000 ha. The farmers pay an average of around 0.32 euros per m^3 of water. These concessions are linked to the size of plot under cultivation, and the quantum varies from year to year depending on the amount of water available from interbasin transfers. Water from the donor basin (Tagus) is dispensed on a monthly basis, and farmers are informed a month in advance of the maximum quantum of water they are likely to receive per unit area of the cultivable land. The rationing had a huge impact on the efficiency of water use (*source*: www.crcc.es).

for hydraulic infrastructure, an urban water supply tariff, a wastewater collection and treatment tariff, and a pollutant discharge fee (Shen and Wu, 2017).

21. *Negotiation of canal water use rights between WUAs and Urban Water Utilities*: for an individual user or a Water Users' Association, the volumetric water use right is fixed for a particular year. Hence, there could be tremendous scope for trading (across sectors, and also within sectors) due to the potential gap between demand and supplies. In drought years, the amount of water that will be allocated for municipalities is likely to fall short of the actual requirements as is evident from the past. If the municipality or corporation or the rural water supply board needs extra water, it could negotiate with the WUAs. The terms of the agreements reached through such negotiations will mainly include the amount of rights WUAs could transfer and the price they would charge. The negotiations can take place through the irrigation department, which will act as mediator and arbitrator if situation calls for. Based on the terms of the negotiations, the irrigation department can reallocate water from the system.

22. *Fund allocation for implementing water management activities*: a major chunk of the resources generated through the sale of water rights and collection of pollution tax (from municipalities and industries) needs to be diverted for implementing the management activities. Thus, it will be routed back to the local water management institutions as implementation funds. The annual fund allocation should be based on the financial outlay for the microplans developed by the local water management institutions and the time frame required. The local management institutions will also include urban water councils, which engage certain agencies to treat wastewater, and the financial resources for this will be made available by the RBO from the (pollution) tax collected from urban utilities and industries, which dispose of untreated effluents.

23. *Agency Intervention for large water management projects*: water resource management interventions in the semiarid river basins are of two types. *First*: those interventions, which involve large geographical areas and technical complexities. *Second*: local water harvesting structures, which are confined to small geographical areas.

One important technological option for regions like north Gujarat and western Rajasthan is impounding of excess runoff from their river basins during excessively wet years and diverting it into decentralized water storage systems. Another option is to use water from large water systems for recharging the aquifers. In Gujarat, water from the Sardar Sarovar project is being used to recharge the alluvial aquifers in north and central Gujarat by diverting water from the main canal into dry river beds and decentralized storage systems (Kumar and Perry, 2018).

All these are large projects and cannot be implemented by the local communities alone. Therefore, there is a need to involve agencies such as the state water resources departments and Central Ground Water Board in planning, designing, and implementing these projects. Nevertheless, the local water management institutions could implement the smaller components of these projects.

24. *Implementation of watershed management/local water harvesting activities*: the small water harvesting interventions include watershed management, check dams, percolation ponds with recharge tube wells, and surface water storage tanks. They can be remodeled, deepened, and could be equipped with recharge tube wells for directly putting the runoff water into the deeper aquifers.

Watershed management activities could be possible in the upper catchment areas that receive good amount of rainfall and which have steep topography. The local institutions themselves can construct these structures, with little technical supervision from the personnel appointed by the NGOs. For special structures, the RBOs could enable the assistance of the technical staff of water resources department or rural water supply department or state groundwater department, or experts hired from the private sector.

25. *Monitoring of cropping patterns*: the local management institutions shall periodically monitor the cropping patterns in their villages and watersheds in the case of groundwater-irrigated areas. The WUAs will monitor the cropping patterns in the case of canal commands in different seasons. The data can then be fed back to the higher-level institutions. These will help to understand the existing water use priorities and to predict future water uses in the irrigation commands and the basin.

26. *Enforcement of groundwater use rights among tube well companies and well owners*: the enforcement of volumetric use rights among the individual users within the villages will be the responsibility of the local water management institutions. The enforcement of rights among the tube well companies, and owners of tube wells and wells fitted with electric motors will be easy. This is because the data on electricity consumption can be obtained from the EUCs. From the electricity consumption figures, hours of electricity use can be obtained. The local water institution can employ staff to measure the discharges of individual wells for every season. The figures of hours of pumping, coupled with discharge figures can give more or less accurate figures of groundwater withdrawal. In the case of diesel wells, as the hours of running the pump are not known, one way to tackle the problem is to ask the well owner to record their own use in every season. The local institution could crosscheck using cropping pattern data. However, every user group or user will have to carry out water budgeting in the beginning of the season. This is to make sure that use does not exceed the allocated water right.

The RBO or the SBOs shall conduct or arrange special training programs for local institutions to build their capability to do water budgeting and accounting. The local institutions can decide on the penalties for excessive use and impose them so as to minimize the incidence of free riding.

In northwestern China, two pilot projects are currently underway to control groundwater overexploitation, one with a groundwater quota system enforced on farmers and the other with a tiered groundwater pricing. Studies show that the quota system had a higher impact in reducing groundwater abstraction (Aarnoudse et al., 2016).

27. *Monitoring of groundwater levels and water quality*: monitoring of groundwater levels and quality of surface and groundwater resources is paramount to assessing the resource condition at any given point of time. Thus, periodic monitoring can help analyze the impacts of water resource management interventions at the basin as well as local level. This will also help to understand the overall effectiveness of these interventions. The primary responsibility of monitoring will be vested with the local water management institutions.

In the rural areas, the local institutions need to be equipped with simple water-level recorders and water quality testing kits for this purpose. In the case of groundwater, water-level monitoring and water quality monitoring (WQM) shall be carried out four times a year, including one premonsoon and one postmonsoon. In the case of surface water, the monitoring will have to be more frequent to ensure that the effluents that are discharged into the streams and lakes are of acceptable quality and the wastewater treatment systems are functioning as per the requirements. In urban and industrial areas, periodic monitoring of water quality performed by the State Pollution Control Boards of rivers and lakes will give indications on the effectiveness of water quality management interventions upstream.

The data generated from monitoring can be directly used by the local institutions for (1) understanding their resource conditions; and, (2) analyzing the effectiveness of the local institutions. One of the interventions for which monitoring would be very crucial to analyze effectiveness is use of microirrigation systems. The key question will be the nature of their actual impact, i.e., whether it is on irrigation return flows or on reduction in non-beneficial evaporation, and therefore the actual change in evapotranspirative demand of water. The main point of enquiry is whether the use of the system actually results in real water saving at the catchment/aquifer level or not. Accordingly, the management strategy will have to be modified. They will be critical to institutional capability building of the local institutions. But the data will have to be eventually shared with the higher-level institutions for analyzing the basin-level impacts of a large number of local management interventions. This can feed into making necessary amendments in the basin management plans.

28. *Demonstration of on-farm water management practices in farmers' fields*: the farmer management support unit (FMSU) of the irrigation department can collaborate with Water and Land Management Institute to carry out demonstration of on-farm water management practices in farmers' fields in canal commands. The demonstration will include small border irrigation, furrow irrigation, and farm leveling. This will provide opportunities for farmers to compare these practices against the conventional irrigation practices in terms of their impact on irrigation water use and motivate them to take up such practices in their fields. The demonstration can be arranged before the starting of the crop season in several locations.

29. *Demonstration farms for land and water management practices; water use technologies*: the efforts of institutions such as WALMI have so far concentrated their efforts on only canal-irrigated areas. No attention is paid to groundwater-irrigated areas. But, there is a tremendous scope for improving the physical and economic efficiencies of water use through land and water management practices even in well commands.

The basin organization or the subbasin organization can collaborate with state agricultural universities to establish demonstration plots of organic farming in the fields of progressive farmers. The team from the agricultural universities can constantly provide scientific inputs to the farmers on preparation of biofertilizers through efficient use of the available biomass; and, their efficient use in the farms for proper management of soil organic matter and moisture.

The basin organization or the subbasin organizations can help microirrigation companies to set up demonstration plots for their cost-effective technologies in sample villages across the basin. This will provide farmers with first-hand experience of these technologies vis-à-vis their advantages against conventional technologies such as easy adaptability, and low maintenance requirements apart from water saving, and increase in crop yields.

30. *Monitoring of local water management activities and their impacts*: the basin organization and the subbasin organizations shall monitor the water management interventions being taken up by the local institutions in order to analyze their performance and effectiveness. Performance evaluation is critical to reviewing and redefining the objectives. This will include close monitoring of the physical activities and the "enforcement of volumetric water use rights," performance of wastewater treatment systems, and the hydrological monitoring work carried out by the local institutions. The basin or the subbasin organization can set up independent committees to assess the performance of the local institutions. They can also use the in-built mechanisms to check free riders in certain instances. For example, to what extent, the volumetric use rights are enforced can be understood by examining the electricity consumption figures that can be obtained from the electricity users' co-operatives or the State Electricity Board. The basin organization can also crosscheck the data (groundwater levels, groundwater quality, etc.) gathered by the local institutions with those compiled by the State Water Resources Monitoring and Planning Board.

In northern China, prepaid energy meters are used to enforce a quota system for groundwater, volumetrically defined on the basis of land holding and livestock holding size of farmers, introduced for controlling abstraction. The swipe card transactions are transmitted through the internet to a Digital Water Resource Information Centre. This center meticulously records the number of units consumed by each farmer based on the swipe card transactions (Wang et al., 2017).

As regards performance of the agencies treating wastewater, the RBO can analyze the data generated by the state pollution control board on water quality of rivers, lakes, and groundwater in order to assess the performance of the wastewater treatment systems set up by the service providers to improve the quality of the effluents.

31. *Periodic water accounting and water balance exercises*: as the situation with regard to water availability, water diversion, and water use changes annually, the RBO should carry out periodic water balance and water accounting studies of the basin. They are essential for determining sustainable levels of groundwater abstraction and surface water utilization, and the volumetric water use rights. They change from time to time with variations in rainfall, runoff and recharge, and with changes in water allocation policies. The water accounting will give clear ideas as to what extent the actual water use by different sectors, deviates from the intersectoral water done by the basin authority. This will be a direct indicator of the effectiveness of market instruments such as tradable private property rights on demand management.

The basin authority can also use the data to evaluate the changes in overall basin efficiency with changes in water allocation patterns, the "third-party effects" of water transfers and thereby taking necessary precautionary measures. The institutional processes resulting from the change in the institutional paradigms and the dynamics of interaction between various institutions are summarized in the flow diagram (Fig. 11.2).

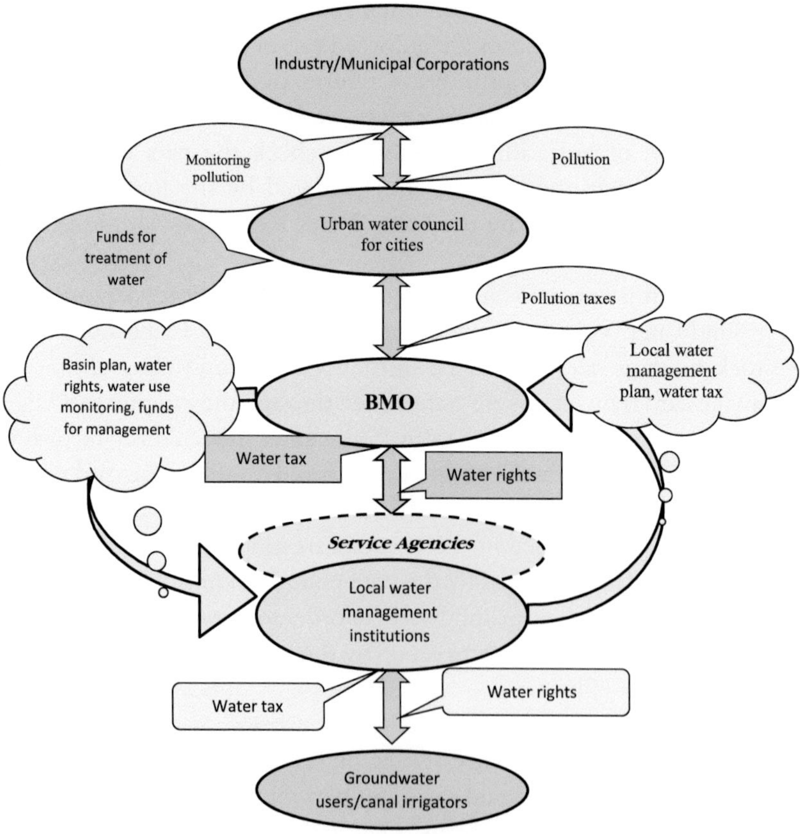

Fig. 11.2 Institutional arrangements for integrated water resources management in river basins. *(Adapted from Kumar, M.D., 2010. Managing Water in River Basins: Hydrology, Economics, and Institutions. Oxford University Press, New Delhi.)*

11.4 Conclusions

Several countries have adopted IWRM principles in their water policies and programs (Jønch-Clausen, 2004). However, too little has been achieved on the ground in terms of operationalizing the concept and implementing the practices (Biswas, 2008). One of the biggest reasons for this limited success is the inability to develop a model of IWRM to fit the socioeconomic context, and design appropriate institutions to implement and operate it, following sound institutional design principles for effective water resources management. While IWRM envisages taking cognizance of the interaction between various components of the hydrological system that affect the effective water availability, viz., catchments, groundwater and surface water, and water quantity and quality, the institutions designed to implement water resources management actions should have the capability to plan, design, and execute WRM activities at the basin level in an integrated manner considering these factors, and monitor the basin performance in terms of progress toward achieving water resources management goals set. Often, the various processes and actors (institutions) and the desirable level of coordination between various institutions at various levels to achieve integrated resource management are difficult to foresee. Problems are also faced with simulating various physical processes (land use changes, increase in groundwater withdrawal, import of surface water for irrigation through canals, introduction of microirrigation) that truly capture the hydrologic system characteristics.

Similarly, while IWRM envisages taking cognizance of the interaction between various socioeconomic systems and ecosystem that demand water from the hydrological system, the institution that is designed to cater to various water demands should have the capability to handle intersectoral water allocation, especially when water scarcity becomes acute, with minimum conflicts. But the agencies concerned with water allocation often fail to predict the likely changes in water demand of different sectors affected by behavioral change induced by the changes in water allocation and pricing policies, and the resultant changes in the hydrological system characteristics, and their overall impacts on the various users of water. These are some of the key challenges in operationalizing IWRM concept. In this chapter, we have laid out sound design principles for effective water management institutions; and discussed the overall institutional framework required for affecting integrated water resources management in the Indian context, and the scientific and institutional processes that need to be initiated to overcome some of these challenges in implementing IWRM concepts at the basin level. The discussion of the action plan for IWRM suggests that implementation of IWRM has to be an evolving process and there can't be a standard blue print that can be implemented at one time.

References

Aarnoudse, E., Wei, Q., Bluemling, B., Herzfeld, T., 2016. Groundwater quota versus tiered groundwater pricing: two cases of groundwater management in north-west China. Int. J. Water Resour. Dev. https://doi.org/10.1080/07900627.2016.1240069.

Batchelor, C., Singh, A., Rao, M.S.R.M., Butterworth, J., 2002. Mitigating the potential unintended impacts of water harvesting. In: Paper Presented at the IWRA International Regional Symposium 'Water for Human Survival', 26–29 November, 2002, Hotel Taj Palace, New Delhi.

Biswas, A.K., 2008. Integrated water resources management: is it working? Int. J. Water Resour. Dev. 24, 5–22. https://doi.org/10.1080/07900620701871718.

Connell, D., 2011. Water reform and the federal system in the Murray-Darling basin. Water Resour. Manage. 5 (15), 3993–4003.

Glieck, P.H., 1997. Human population and water: meeting basic needs in the 21st century. In: Pachauri, R.K., Qureshi, L.F. (Eds.), Population, Environment and Development. Tata Energy Research Institute, New Delhi, pp. 105–121.

Government of India, 2010. Ground Water Quality in Shallow Aquifers of India, Central Ground Water Board, Ministry of Water Resources. Government of India, Faridabad, India.

Gupta, S., 2011. Demystifying 'tradition': the politics of rainwater harvesting in rural Rajasthan. Water Alternatives 4 (3), 347–364.

Howe, C.W., Schurmeier, D.R., Douglas Shaw Jr., W., 1986. Innovative approaches to water allocation: the potential for water markets. Water Resour. Res. 22 (4), 439–445.

Jønch-Clausen, T., 2004. Integrated Water Resources Management (IWRM) and Water Efficiency Plans by 2005 Why, What and How? Global Water Partnership. (January 2004).

Kumar, M.D., 2006. Water Management in River Basins: A Study of Sabarmati River Basin in Gujarat. Ph.D. Thesis Submitted to Sardar Patel University, VV Nagar.

Kumar, M.D., 2016. Status and institutional arrangements of river basin management in India: some analytical insights. In: Presented at the EU-India Water Partnership Capacity-Building Workshop on River Basin Management Planning and Governance, Park Hotel, New Delhi, 14–15 June, 2016.

Kumar, M.D., Perry, C.J., 2018. What can explain groundwater rejuvenation in Gujarat in recent years? Int. J. Water Resour. Dev. https://doi.org/10.1080/07900627.2018.1501350.

Kumar, M.D., Ankit Patel, A., Ravindranath, R., Singh, O.P., 2008. Chasing a mirage: water harvesting and artificial recharge in naturally water-scarce regions. Econ. Polit. Wkly 43 (35), 61–71.

Kumar, M.D., Prathapar, S.A., Saleth, R.M., Reddy, V.R., Narayanamoorthy, A., Bassi, N., 2017. New 'water management paradigm': outdated concepts? Econ. Polit. Wkly. 52 (49), 89–94.

Perry, C., 2007. Efficient irrigation, inefficient communication, flawed recommendations. Irrig. Drain. 59, 367–378.

Rosegrant, M., Binswanger, H.P., 1994. Markets in tradable water rights: potential for efficiency gains in developing country water resource allocation. World Dev. 22 (11), 1613–1625.
Shen, D., Wu, J., 2017. State of the art review: water pricing reform in China. Int. J. Water Resour. Dev. 33 (2), 198–232.
Syme, G., Reddy, V.R., Pavelik, P., Croke, B., Ranjan, R., 2012. Confronting scale in watershed development in India. Hydrogeol. J. 20 (5), 985–993.
United Nations Educational, Scientific and Cultural Organization (UNESCO), 2008. IWRM Guidelines at River Basin Level, Part 2. Guidelines for IWRM Coordination, Paris, p. 1.
van der Zaag, P., Savenije, H.H.G., 2014. Principles of Integrated Water Resources Management. UNESCO-IHE Institute for Water Education, Delft. (October 2014).
Wang, X., Shao, J., van Steenbergen, F., Zhang, Q., 2017. Implementing the prepaid smart meter system for irrigated groundwater production in northern China: status and problems. Water 9, 379. https://doi.org/10.3390/w9060379.

Further reading

Kumar, M.D., 2010. Managing Water in River Basins: Hydrology, Economics, and Institutions. Oxford University Press, New Delhi.
Murray Darling Basin Authority (MDBA), 2009. Issues Paper: Development of Sustainable Diversion Limits for the Murray-Darling Basin. MDBA, Canberra. http://www.mdba.gov.au/files/publications/sustainable-diversion-limits-issues-paper-12-11-09.pdf.

CHAPTER 12

Conclusions and areas for future research

M. Dinesh Kumar[a], V. Ratna Reddy[b], A.J. James[c]
[a]Institute for Resource Analysis and Policy, Hyderabad, Telangana, India
[b]Livelihoods and Natural Resource Management Institute, Hyderabad, Telangana, India
[c]Environmental & Natural Resource Economist, New Delhi, India

12.1 Summary

Catchment management and basin management are complex and process oriented. The process of planning, designing, and implementation requires a multidisciplinary approach that integrates biophysical, socioeconomic, and governance aspects. The experiences regarding catchment management and basin management vary across countries. This book is an attempt to compile experiences with regard to catchment assessment, planning and management, and river basin planning and management across various countries, including those from developed countries and some countries in South Asia. It first presents the findings of the review of international experiences and then presents several case studies from India.

The review of international experiences covers the following: (1) the scientific tasks of carrying out hydrological assessments of the catchment (to quantify water availability in terms of quantity and quality), and analyzing the water allocation and water uses in different sectors, viz., agricultural use, urban use, industrial use, livestock use, and environment; (2) technical task of planning catchment management interventions and preparing river basin planning; and, (3) institutional processes of organizing catchment management and river basin management actions.

As regards the scientific task of catchment assessment, understanding the land-use and hydrology interactions, "interaction between upstream water withdrawals and alterations in downstream flows" and "surface water-groundwater interactions" are of prime importance, in addition to quantifying the amount of surface runoff from precipitation, base flows, and groundwater recharge. The former set of activities can tell us what kind

Current Directions in Water Scarcity Research, Volume 1
https://doi.org/10.1016/B978-0-12-814851-8.00012-4
© 2019 Elsevier Inc.
All rights reserved.

of land and water-based interventions can bring about alterations in hydrological processes of the catchment and to what extent, which is essential for arriving at sound catchment management decisions.

As regards water accounting, it is important for river basins that are already developed and from which water is allocated to estimate consumptive use of water in various sectors against the water supplied to examine whether current water uses are efficient from social, economic, and environmental points of view, and to identify the leverages for improving water management either through water saving or through supply augmentation. For instance, the actual water consumption by paddy crop against the total volume of water applied through irrigation in paddy fields is important, as a large part of the "nonconsumptive fraction" of the water applied in the field will recharge groundwater and will be available for reuse. Similarly, estimating consumptive water use in thermal power plants against the actual water required for running the plant including that for cooling is important for getting actual water accounts as the water used for nonconsumptive uses will be available for reuse after treatment.

The collection of case study papers in this volume demonstrates some of these scientific tasks through case studies from selected river basins/sub-basins/watersheds in India, viz., Krishna, Luni, and Narmada. In the case of Krishna, a sub-basin named Warna was chosen for the analysis. In the case of Luni, the entire basin was chosen for water accounting work. In the case of Narmada, a watershed of size nearly 450 sq. km of one of the sub-basins, viz., Orsang, was chosen for catchment assessment and planning. Two other sub-basins of Narmada, viz., Kundi and Hathni, were chosen for understanding the "scale effects" in hydrology, and this involved analysis of the downstream impacts of intensive watershed development upstream in the form of runoff change.

In the case of the large watershed in south Gujarat, the analysis (presented in Chapter 5), apart from assessing various components of the watershed hydrology (total rainfall, evapotranspiration, surface runoff, and base flow), helped us ascertain that catchment management interventions there should be capable of regulating the flows as there was sharp difference between dry-year flows and wet-year flows. Hydrological simulation using SWAT showed that the creation of permanent storage structures for watering crops in the catchment would reduce the flows into a reservoir downstream, whose storage is committed for irrigation, in drought years. In the absence of a stream-gauging point, the availability of reservoir mass balance data for a large reservoir at the end of the watershed, which acted as a hydraulic response unit, enabled calibration and validation of the model for its robustness in computing rainfall data with high degree of accuracy. The chapter also drew lessons for the approach and methodology to be chosen for hydrology-based watershed planning, including that of project area selection, hydrological assessment, and intervention planning.

Analysis of time series data on groundwater level trends (between pre- and postmonsoon) of two sub-basins, viz., Kundi and Hathni (Chapter 5), showed that watershed development led to increase in recharge, as indicated by higher average value of "recharge fraction" in both the sub-basins post-watershed development. Further analysis for two watersheds (one treated and the other control) in Hathni sub-basin, done to nullify the effect of rainfall difference, showed an increased difference in average "recharge fraction" between the two watersheds postinterventions (1996), with higher difference in recharge fraction during high rainfall years in the case of the treated watershed indicating the effectiveness of recharge structures during high rainfall years. Sub-basin level analysis for Kundi also showed reduction in streamflows postwatershed development. Such an outcome was the result of an increase in withdrawal from wells (followed by augmented recharge), as suggested by an analysis of groundwater balance of one treated and one untreated microwatershed. In the treated, not only the annual groundwater recharge, but annual draft was also higher. The study highlighted the need for considering groundwater-surface water interactions in watershed development and basin planning.

The scale of analysis of watershed interventions assumes importance in capturing the biophysical linkages. Chapter 10 demonstrated that changing the geographical unit of analysis of WSD against a strategic "whole-of-catchment" model would help evaluation process to provide general guidance in regard to the investment of essentially limited funds. This could provide an assessment on the basis of broad hydrological data and an estimate of the hydrological efficiency of differing patterns and intensities of WSD at a catchment level. The assessment of the impacts validated the hypotheses that mesowatersheds could generate differential benefits at scale (upstream/downstream). However, it was clear that there is some mismatch of perceptions of the outcome of WSD in terms of the benefits and costs to other parties that relate to hydrogeology and the biophysical characteristics of the location. Therefore, these aspects, which were hitherto not considered, need to be directly addressed in order to optimize as well as sustain the watershed impacts.

In the case of Luni basin, the water accounting exercise (Chapter 7) showed that with the total consumptive use of water exceeding the renewable water resources, the basin's annual flows were overappropriated in most years, while groundwater resources were mined. The study found that the key intervention that is effective for managing the basin water resources is water demand management in agriculture, involving technologies and practices to reduce consumptive water use in crop production, while groundwater banking would be a good option in the abnormally wet years though those are very rare in the basin area.

A water accounting exercise for one of the sub-basins of Krishna river basin (Warna) was undertaken for the purpose of understanding water tenure in the sub-basin. The exercise used annual streamflow data for the gauging station at the drainage outlet of the sub-basin, historical data on inflows, and annual storage changes in a large reservoir located inside the basin and data on renewable groundwater resources and estimated annual groundwater storage change, and estimates of consumptive water use in various sectors. The analysis (Chapter 6) showed that the amount of water consumed in the sub-basin for various consumptive uses (irrigation, domestic use, industry, and livestock use) is far less than the total volume of water generated from the rainfall-rich upper catchments of the sub-basin. Further, water from a reservoir build on Warna river is also exported from the sub-basin to irrigate the command area in the districts of Kolhapur and Sangli, outside the basin. Yet the sub-basin has large amount of excess flows, which is quite critical to maintaining the overall water balance of Krishna river basin, which is on the verge of closure. The uniqueness of the sub-basin was that the climate (rainfall and humidity), land use, and water use patterns changed dramatically from the upper catchment to the middle and lower catchments.

The findings of the water accounting study undertaken for Kanpur Metropolitan Area (Chapter 9) suggests that there is not much physical risk in meeting existing or future water demand of in terms of water quantity. However, water quality is an issue as the infrastructure required to treat raw water and wastewater is absent. This scenario poses a major threat to the river ecology and poses a risk to livelihoods of the communities dependent on it. The economic risk for the industries alone is estimated to be around US$ 540 million per annum. To mitigate these risks, it is suggested that various parastatal agencies involved in water supply and sewerage services work in a coordinated fashion to strictly enforce pollution control norms for the industries. The water accounting approach can be replicated in other major cities along river Ganga to assess their water risk and water security challenges.

In Chapter 8, we have also demonstrated how water resources planning using the concept of IWRM (Integrated Water Resources Management) can be undertaken at the district level, through case study of one district in Rajasthan falling in Luni river basin for which a water accounting exercise was carried out. A lot of the data on renewable water resources, both surface and underground, were deduced from the data generated for Luni basin water accounting. The IWRM planning exercise determined the extent to which various agricultural water demand management interventions can be undertaken in the basin and their overall impact in terms of real water saving and on reducing the water demand-supply imbalance. It also quantified the total amount of runoff water that can be captured from the basin for groundwater banking.

The international literature review of institutional processes in catchment management and river basin management showed that there are several institutional models for watershed management, catchment management, and river basin management, with varying objective functions and mechanisms to achieve them. In the case of the National River Authority of England and Wales, protection of water environment through Catchment Management Plans was the prime objective.

The experience of watershed management (WSM) in South Asian countries has shown that WSM is an effective technology for addressing the broader issues of farm productivity and food security as well as soil and water conservation in arid and semiarid regions. The experiences of these countries in terms of technical as well as institutional aspects provide important lessons for countries like Afghanistan, which are just started implementing WSM. It is argued that there is an immediate need for creating policy environment to initiate WSM program, with a national policy and objectives (Chapter 3). A common framework for implementing the WSM programs needs to be provided on priority, given the potential of WSM in strengthening the natural resource base and ensuring food security.

In the case of Murray Darling Basin Authority of Australia, the main objectives were limiting surface and groundwater diversions from the basin for environmental sustainability; ensuring that sub-plans for each state are in line with the basin plan; and promoting water trading to maximize the social and economic benefits. The MDBA develops and implements the basin plan, which has legislative backing and is an enforceable document to be followed by all basin states while preparing their own plans. In the case of French river basin organizations that were active during 1970–92, the main objective was water resource management in major river basins, by promoting efficient water use and pollution reduction through financial incentives to local agencies and private players.

The Catchment Management Agencies in South Africa implements the National Water Act and management of water at local level. These are participatory institutions in a decentralized format, but at the same time are also statutory bodies. The governing board of CMA is to have representation from present and potential future water users, local and provincial government, and environmental interest groups. Their objectives are planning of WRM at catchment scale; registration of water users; collection of water charges collection; water authorization and licensing; and, ensuring poor communities' access to water for domestic and productive needs.

The institutional mechanisms used for water management also differ. The French RBOs were financial investment institutions with no statutory powers. Under this set up, there were separate sets of agencies for performing various functions, the RBO for water quality monitoring and regulation, and separate (private) agencies for water supplies, and for setting up and running wastewater treatment plants. The RBO regulator, which is responsible for managing river water quality through financial incentives, ensured that private agency, which is responsible for investing in and running WWT plants, treats the effluent to conforms to water quality standards through water quality monitoring.

In the case of Murray Darling basin authority, the market instrument of "tradable water rights" was successfully used during years of drought to achieve higher economic efficiency of water use in agriculture to maintain the economic outputs at the level of normal years, with a 50% cut in water allocation. The establishment of separate "water entitlement for environment" at the basin level and creation of environmental water holder enabled securing water for environmental flows, though it created resentment among irrigators as it reduced their overall allocation.

The National River Authority of England and Wales was a centralized agency with range of regulatory powers on private and public bodies. The CMP formed the basis for NRA action within each catchment. The CMPs were consultative with respect to the planning process, with no ability to enforce on land users, developers, or local planning authorities. The NRA could do little to incorporate water management decisions in land-use planning as it had limited formal authority to influence land use, while the CMP had no immediate impact on statutory authorities.

In the case of CMAs of South Africa, the shift from a centralized management approach to a decentralized participatory approach enabled water users from poor communities to discuss directly with the concerned agency on water resources management. The mismatch between catchment boundaries and local administrative boundaries led to poor coordination between CMAs and the water supply agencies.

The experience of countries like India in implementing integrated water resources management is almost nil. While the IWRM concepts and operational principles are largely understood, most often, the agencies are confronted with the question of where to start and where to end, i.e., whether the process should start with scientific studies or crafting of institutions. Another question is how to initiate the institutional processes. In one of the chapters (Chapter 11), we have therefore discussed the overall institutional regime for implementing integrated water resources management actions; described the key scientific and institutional processes that are necessary to operationalize them at the level of river basins, and in the same order and sequence in which they should be initiated. We have also highlighted some of the water management outcomes.

12.2 Areas for future research

In many regions of the developing world, reliable hydrological data essential for taking water resources development and management decisions are often sparse due to lack of sufficient number of precipitation and stream-gauging stations, weather stations, and groundwater monitoring stations spread well across river basins and catchments.

There are two major exercises that water managers have to undertake to arrive at sound management decisions. They are (1) hydrological assessment of catchments to know the catchment yield (the amount of runoff generated from the catchment plus the groundwater recharge from precipitation) and (2) water accounting to know how much of the available annual yield is consumed for various purposes (ET from both natural and cultivated vegetation in the catchment, evaporation from soils and water bodies) and how much is held in storages (surface and underground) and how much is discharged from the drainage outlet.

The hydrological assessment of catchments would help carry out water budgeting, and planning of water resources development and management interventions and water allocation. Catchment assessments if done at various scales will also help minimize the tradeoffs in watershed developments occurring due to scale effects of hydrology (Chapter 2 of this volume). The water accounting exercise would help us estimate the consumptive use of water in various sectors in volumetric terms in a year or in a season, and assess the total amount of water that remains unutilized in the basin, and how much of the water generated in the basin is consumed for beneficial uses (evapotranspiration) and nonbeneficial uses (evaporation from barren land, uncultivated areas within crop land, swamps, and reservoirs) (Chapter 1 of this volume and Molden et al. (2003)). This forms the basis for deciding on the nature of intervention that can be taken up in the basin for improving water management.

The most crucial data that are required for carrying out hydrological assessment of catchments and water accounting, but are often very hard to obtain, are of streamflow or river discharge, which is the net of the total runoff generated in the catchment minus the water effectively utilized upstream through the help of reservoirs and diversion systems, and water held in small and large reservoirs. If streamflow data for some gauging stations are available, then the virgin runoff can be estimated by doing backward calculation, on the basis of data on total effective diversions upstream of the gauging point. The streamflow data for the last gauging point in a catchment itself can give strong hints on whether there is scope for taking up more water resource development interventions in the catchment. Availability of sufficient flows at the last gauging point in all years is a strong indication of the fact that the catchment is "open," and further development is possible unless these flows are committed for downstream uses.

The data, which are less scarce are rainfall as many regions in the developing countries, barring some countries of Africa, now have reasonably high density of rain-gauging stations, which record not only daily rainfall data but also have real-time monitoring data using automatic gauges. If we have good network of rain-gauging stations in a catchment that can measure the annual precipitation accurately, then the catchment water yield (the sum of runoff, groundwater recharge, and soil moisture storage) can be indirectly estimated if we know how much water is lost in evapotranspiration during the year. The catchment yield will be total precipitation-evapotranspiration (which also includes evaporation from barren soils and swaps).

Evapotranspiration from cultivated land can be estimated using the widely accepted Penman–Monteith equation, if we know the crop growing season, the date of sowing and harvesting and have the weather data for the nearest weather station. But estimation of evapotranspiration from the natural vegetation cover poses a major scientific challenge. Since the natural vegetation (forest) cover consists of a mix of large and small trees, grasses and shrubs, neither estimation using empirical methods nor actual physical measurement (using lysimeters) in the field would be practical. Since the extent of natural vegetation cover is more for water-rich catchments in tropical regions, arriving at realistic estimates of annual and seasonal flows to assist water management decisions would depend heavily on the accuracy of estimation of evapotranspiration.

In some situations, even when streamflow data are available (like in the case of tributaries of many large rivers originating from the western Ghats of India such as Krishna and Cauvery), it might be important to get estimates of the volume of water consumed annually by the natural vegetation that covers large areas. The reason is that in many situations, we may have to consider altering the vegetation cover as one of the water management solutions for the catchment, to alter the total runoff volume or to reduce the sediment load in the runoff water. Since the direct method of using lysimeters for field measurement cannot be employed for estimating the actual ET, in such situations, indirect methods can be employed wherein the instantaneous evapotranspiration from the forest cover is estimated using satellite remote-sensing data for the exposure time with the help of SEBAL (Soil Energy Balance) model, and a coefficient called "evapotranspiration fraction"—which is the ratio of instantaneous ET and reference ET (ET_0)—is estimated. In this case, ET_0 for the exposure period is estimated using Penman–Monteith equation based on weather data. Using the ET fraction and the total reference evapotranspiration for the vegetation for a given period, the actual ET can be estimated as $ET_0 \times$ Evapotranspiration Fraction (Roby and Jose, 2018).

However, for using such hybrid models, it is necessary to have weather data for the nearest location so that reliable methods like Penman–Monteith equation can be employed to estimate reference evapotranspiration (ET_0).

Another major area of research in hydrology, which can immensely benefit water resources development planning and evolution of water management approaches, is the groundwater-surface water interactions. In developing countries, detailed hydrological investigations groundwater-surface water interactions are rare and often thumb rules are used. For instance, groundwater recharge estimates of official agencies of India for hard rock hilly regions are highly unreliable due to the nonscientific methods used for estimation of "groundwater discharge into streams," an important variable crucial for estimating the utilizable groundwater resources in shallow aquifers. In the absence of gauging stations that are capable of measuring the lean season flows at appropriate scales, some simplistic norms are often employed to estimate the nonmonsoon (lean season) base flow, and the estimated value is deducted from monsoon recharge into the aquifer to obtain utilizable recharge.

In many river basins, the small upper hilly catchments contribute majorly to the total annual runoff. Here again, the baseflows, which is the groundwater discharge into streams, is a major portion of the total runoff, and this include outflow during monsoon and as well as nonmonsoon period. In any catchment, as the rainfall increases, the runoff and base flow increase. But the baseflow as a fraction of the runoff reduces. This is because at lower magnitudes of rainfall, the baseflow component of the total runoff is generally very high for hilly catchments, and at higher magnitudes of rainfall, the surface runoff component increases. But such data are nonexistent for most river basins in developing countries. Estimating the "base flow coefficients" (ratio of total base flow and total runoff) of a catchment for different rainfall regimes is of great relevance in research on catchment hydrology, as it can help arrive at realistic estimates of utilizable groundwater resources from precipitation. For this, partitioning of base flow and surface runoff components of the total runoff from precipitation in these catchments using hydrographs is required. Since the effect of base flow is not visible in small microcatchments, setting up gauging stations at appropriate locations is an important starting point.

Data of stream discharge (along with quantum of precipitation) obtained from monitoring of major (runoff-generating) rainfall events, followed by analytical process of partitioning base flow, relationships (statistical regression models) can be established between: (1) rainfall and runoff and (2) rainfall and base flow coefficient.

If the "lean season base flow" is known (through gauging) for different magnitudes of rainfall in the catchment (for the same years for which data on total runoff are available), then the base flow during the monsoon can be estimated for different values of rainfall, using the regression model for "rainfall-baseflow coefficient." This, when added to monsoon recharge (estimates), gives value of total infiltration during monsoon. Relationship can then be established between rainfall and total infiltration for any given catchment for different magnitudes of rainfall. Once this is done, the three models, viz., rainfall-runoff model, rainfall-base flow coefficient model, and "rainfall-total infiltration model" can be employed to compute the total utilizable recharge as "Estimated total infiltration-Estimated total base flow" (GSDA & IRAP, 2015). Such investigations and analysis need to be carried out for hilly and mountainous catchments in different hydrological settings.

References

GSDA (Groundwater Survey and Development Agency) & IRAP (Institute for Resource Analysis and Policy), 2015. Using Technology to Ensure Ground Water Safety and Security in a Tribal Block of Chandrapur, Maharashtra. Final Report submitted to UNICEF-Mumbai, January.

Molden, D., Murray-rust, H., Sakthivadivel, R., Makin, I., 2003. A water productivity framework for understanding and action. In: Kijne, J., Barker, R., Molden, D. (Eds.), Water productivity in agriculture limits and opportunities for improvement, comprehensive assessment of water management in agriculture. CABI Publishing in Association with International Water Management Institute, UK.

Roby, T.J., Jose, K., 2018. Estimation of evapotranspiration in a tropical wildlife sanctuary using GIS and remote sensing techniques. For. Res. Eng.: Int. J. 1(3). 00013.

Index

Note: Page numbers followed by "*f*" indicate figures, "*t*" indicate tables, "*b*" indicate boxes, and "*p*" indicate plates.

A

Adaptive Environmental Assessment and Management (AEAM), 37–38
Afghanistan
  lessons for, 79–81
  policy imperatives, 81
  watershed impacts
    equity in benefits, 73–74
    hydrological impacts, 71–73
    range land rehabilitation, 71
    reforestation, 71
    vegetative cover, 71
  WSM, 56–57, 63–64
Agricultural water demand, 230*t*, 231
Agroecology, 43–44
Anthropogenic factors, 55
Aquifer geometry, watershed development, 288–289

B

Bagmati Integrated WSM Project (BIWMP), 69
Bangladesh, WSM, 63
Basin-level water accounting, 184
Basin management, 13–14, 331
Basin organization, 324–325
Basin water accounting, 15
Bench terraces, 64
Bhutan, WSM, 61–62
Biological oxygen demand (BOD), 252
Bottom-up approach, 278
Bunds, artificial embankments, 64

C

Canal-based water distribution, 239–240
Catchment, 276
  closure of, 280, 284
  governance of, 282
  groundwater, 277–278
  hydrological assessment of, 337–338
  modeling part of, 285
  socioeconomic context, 285–286
  surface-water, 292
Catchment assessment
  catchment management decisions, knowledge use, 25–26
  integrated catchment management
    institutional, legal/regulatory measures, 27–29, 30–36*t*
    physical/technological aspects, 26–27
  land use/land cover and catchment yield and soil erosion changes, 23–24
  scientific task of, 331–332
Catchment assessment and management planning (CAMP) study, Gujarat
  catchment choice, 107
  hydrological assessment, 110–111
  hydrological modeling, 103–104
  IWMP watersheds, choice of (*see* Sukhi catchment, IWMP watersheds)
  objectives, 106–107
  SWAT (*see* Soil and Water Assessment Tool (SWAT))
  village-level information
    cropped area, 120
    elevation ranges and steep topography, 117
    forest and degraded forest, 118
    ground water levels, interannual variability in, 119
    irrigated area, 120
    kharif shift to double cropping, 118
    postmonsoon depth to water level, 120
    water harvesting structures, 119
    water-level fluctuation, 120
    wells, 118
    years of drought, groundwater recharge, 120
  villagers' perception of water problems, 120–121
  watershed characteristics, 117
  watershed management, challenge for, 121–122
Catchment assessment and planning, 2–3, 13
  in South Gujarat, western India, 14–15
  theoretical and conceptual aspects, 23

Catchment characteristics, modeling
basic steps in, 123*b*
data collection, 122–123
final model run, 126–128, 127*b*, 127*f*
global datasets, 123, 124*b*, 124–125*f*
improved datasets, 123–126, 125–126*f*
Catchment closure, 72
Catchment functioning, 47
Catchment hydrology, 13
land use and land cover, changes in, 46
modeling tools, 47
state of knowledge, 40–41
Catchment management, 331–333, 335
and basin management, 13–14
decisions, knowledge use, 25–26
economic outcomes, 47
inadequate scientific knowledge, 12–13
integrated
institutional, legal/regulatory measures, 27–29, 30–36*t*
physical/technological aspects, 26–27
land and water resources, 47
Catchment Management Agencies (CMAs), 28–29, 40, 336–337
Catchment management planning (CMP), 47–48
in Britain, 48
in India
agroecology, 43–44
central funds, 42
forest hydrology, 43
KAWAD project, 42
microwatershed level, 43
plan allocation, 41–42
soil moisture regime, 42
water demand, 42–43
treatment technologies, 26
Catchment planning
hydrological considerations, 48
implementation outcomes, 38–40
in India
catchment hydrology, state of knowledge, 40–41
in IWMP (*see* Integrated watershed management program (IWMP))
physical and socioeconomic impacts, 44–45
watershed management institution sustainability, 45–46
integrated crop, rainfall runoff, soil erosion, and economic models, 29–37
process, 37–38
Central Ground Water Board (CGWB), 228
China, resource fee, 320–321
Clean Ganges Programme and Sustainable Cities Mission, 273
Climate extremes, 1
CMAs. *See* Catchment Management Agencies (CMAs)
CMP. *See* Catchment management planning (CMP)
Community development committee (CDC), 77
Compensation approach, 283
Conveyance losses, irrigation, 239–240
Crop production, 190
Curve number method, 224

D

Deep percolation losses, 10–11
Demonstration effect, 283
Department of land resources (DoLR), 48, 75, 107
Department of Soil Conservation and Watershed Management (DSCWM), 76–78
Distributary
flow measurement and control, 317
modernization of, 318
District Development Committee (DDC), 77
District soil conservation officers (DSCOs), 77
District water management agency (DWMA), 75–76
DoLR. *See* Department of land resources (DoLR)
Domestic water
demands, 219
supply, 245
unplanned disposal of, 246
use in Luni river basin, 202–207
Downscale basin-level IWRM plans, 213–216

balancing demand and supply, 229–230
scale effects and economic viability, 232–233
water availability-demand gap, 230–231
water management interventions, 231–232
challenges of, 216
district water management options, 234–235
irrigated area reduction, 236
microirrigation and mulching, 235
irrigation
conveyance losses and piped water systems, 239–240
water use hydrology of, 237
livestock population in Luni river basin, 238
microirrigation, 240–241
mulching, 241–243
rainfall and climate variability, 215
water demands, 214, 216, 222
domestic, 219
industrial, 221–222
irrigation, 216–219
livestock, 219–221
water supplies, 214–215, 223, 229
groundwater resources, 226–228
surface water resources, 224–225
Drinking water security, 245
Drip irrigation, 210–211, 233
DSCWM. *See* Department of Soil Conservation and Watershed Management (DSCWM)

E

Effective rainfall, 160–161
Effluent Treatment Plant (ETP), 268
Electricity Users' Cooperatives (EUCs), 315
Employment assurance scheme (EAS), 89
European Union Special Partnership Programme (EU-SPP), 213
European Water Framework Directive (EWFD), 38
Evapotranspiration (ET), 136, 338–339
Evapotranspiration fraction, 339

F

Farmer management support cell (FMSC), 316
Farmer management support unit (FMSU), 324
Flagship national watershed development scheme, 103–104
FMSC. *See* Farmer management support cell (FMSC)
Forest cover, 339
Forest hydrology, 13, 43
Future water requirement, 264–265

G

Ganga River Basin, 16
Ganges barrage construction, 265
Geographic information system (GIS), 90
Geohydrology
of Luni river basin, 187–190
Warna river basin, 154–155
Government of India (GOI), 103–105
Gram Panchayats (GP), 233
Groundwater
abstraction
in Luni river basin, 188–189
sustainable levels of, 313
in Kanpur metropolitan area, 252–254
levels and water quality monitoring, 323
level trends, time series data on, 333
resources, 170–171*t*, 226–228
tradable water use rights allocation, 305
use rights allocation, 319
Groundwater banking, 231, 233
Groundwater-surface water interactions, 13, 41, 339–340

H

Harvesting cost, 233
Hathni sub-basin, 89–96, 97*f*, 333
Hilkot participatory WSM program, 70
Hydroecology, 88–89
Hydrogeology, 286–289
Hydrological processes, 278–279
Hydrological simulation models, 13–14
Hydrology
of Luni river basin, 187–190
Warna river basin, 153–154, 154*f*
watershed development, 284–285, 292

Hydrology-based watershed management
application
hydrological assessment, 141–144
intervention planning, 121–122, 141
project area selection, 141–142
approach and methodology, 139–140
Hyperarid river basin, 15

I

ICM. *See* Integrated catchment management (ICM)
IGNP. *See* Indira Gandhi Nehar project (IGNP)
India
catchment management planning
agroecology, 43–44
central funds, 42
forest hydrology, 43
KAWAD project, 42
microwatershed level, 43
plan allocation, 41–42
soil moisture regime, 42
water demand, 42–43
drinking water security, 245
hydrological research, 280
integrated water resources management in, 337
meso-WSD problem in, 293–294
water resources management in, 301
watershed impacts and equity, 67–68
watershed institutions and implementation, 74–76
WSD policy, 275
WSM, 48, 56, 58–60
institutions, sustainability of, 45–46
physical and socioeconomic impacts, 44–45
Indira Gandhi Nehar project (IGNP), 183, 198
Industrial growth, Kanpur metropolitan area, 250–252
Industrial production, 190–192
Industrial water demands, 221–222
In situ moisture conservation, 65–66
In situ water harvesting, 2
Institutional approach, 279
Integrated catchment management (ICM)
goals of, 39
institutional, legal/regulatory measures, 27–29, 30–36*t*
interventions, 29, 37, 44
physical/technological aspects, 26–27
Integrated water management interventions, 233
Integrated water resource management (IWRM), 13, 16, 213, 299–300, 327–328, 335
action plan for implementation
institutional design principles for, 301–302
State Water Development and Management Board, 302–303
concepts and principles of, 300
in India, 337
interventions, river basins, 15–16
monitoring and planning unit, 303–304
operationalizing concept of, 300
Integrated watershed management
agricultural intensification and groundwater abstraction, 10
catchment assessment, 8
hydraulic interconnectedness, 9–10
lack of integration, 10
microwatershed, 8–9
thrust, 7–8
watershed planning, 8
water use efficiency, 9
Integrated watershed management program (IWMP), 75
catchment management planning, in India
agroecology, 43–44
central funds, 42
forest hydrology, 43
KAWAD project, 42
microwatershed level, 43
plan allocation, 41–42
soil moisture regime, 42
water demand, 42–43
PMKSY, 103–104
water harvesting structures, 119
Intensive water-harvesting activities, 49
Intrabasin water transfers, 216
Irrigation
components of, 194

conveyance losses and piped water systems, 239–240
groundwater suitability for, 190
management, 315
water consumption through, 201–202
water demands, 216–219
water requirement, 161, 169–171
water use hydrology of, 237
IWMP. *See* Integrated watershed management program (IWMP)
IWRM. *See* Integrated water resource management (IWRM)

J

Jawaharlal Nehru National Urban Renewal Mission (JNNURM), 256

K

Kanpur metropolitan area (KMA), 245–249, 273
climate in, 249
demographic changes, 249
groundwater, 252–254
industrial growth, 250–252
major findings from desktop study, 270–273
methodology, 247–248
surface water, river Ganges, 252
urban water management, institutional framework in, 260–261
wastewater treatment, 257–259
water accounts of, 263–264
estimated water demand, 261–262
estimated water supply and use, 262–263
water quantity and quality management, 16
water supply management challenges, 264
economic risks, 267–270
physical risks, 264–266
water supply system, 254–256
Krishna river basin, 15, 334
Kundi sub-basin, 89–96, 98–99, 98*f*, 333

L

Land degradation, 55
Land use-hydrology interactions, 2–3
Large-scale water resources development, 187–188
Livelihoods, watershed development, 284–285
Livestock
farming, 190
population in Luni river basin, 238
water demands, 219–221
water use in Luni river basin, 207–209
Local institutional development
agencies for, 314
Local water management activities
monitoring, 325
Longitudinal and cross-sectional analysis, 90
Luni river basin, 15–16, 159, 184–186, 210–211, 333–334
domestic water use in, 202–207
drainage map of, 184, 185*p*
evaporation from reservoirs, 198–201, 201*t*
hydrology and geohydrology, 187–190
imported water, 198, 199*t*
livestock population in, 238
livestock water use in, 207–209
methodology and analytical procedure, 193–195
rainfall in, 186–187
rainfall-runoff relationship estimation, 196–198
replenishable groundwater estimation, 198, 199*t*
socioeconomic drivers of water use, 190–192
virgin flows estimation, 195–196
water accounts, 209–210
water consumption through irrigation, 201–202, 203–205*t*, 206*f*

M

Mapping, 90
Mesoapproach, 282
Microcatchment, 8–9, 172–173
Microcatchment management programs, 24
Microirrigation (MI) system, 232, 235, 240–241
Microplanning, 50
Microwatershed, 8–9, 14, 276, 279–280

Minors
flow measurement and control, 317
modernization of, 318
Model scenarios
catchment-level scenarios, 129–131
characteristics, 128, 128*b*
description of, 130, 130*t*
impact on yields, 137
important and uncertain impacts, 129
village-level scenarios
description, 134–136, 135*t*
IWMP watersheds, 132
Kevdi village, RWH structures, 133–134, 135*t*
microcatchments, 133, 134*f*
SWAT subcatchments, 132, 133*f*
water balance components, 136–138
Modernization, 317
of distributary, minors, and watercourses, 318
of field channels, 318
Mulching, 210–211, 232, 235–236, 241–243
cost of, 233
Multiannual storage of water, 216
Multitier ridge-to-valley sequenced approach, 105–106
Murray Darling basin, 312–313
Murray Darling Basin Authority (MDBA) of Australia, 335–336

N

Namami Ganga (Clean Ganges) program, 261
Narmada river basin, 14
district falling in, 87, 88*f*
groundwater development, 87–88
Hathni sub-basin, 89–96, 97*f*
hydrological changes
annual rainfall, moving average of, 93–95, 95*f*
pre- and postmonsoon well water levels, 95
recharge fraction, 95–96, 96*f*
rejected recharge, 95
on stream flow at sub-basin level, 98–99, 98*f*, 99*t*
water balance at microwatershed level, 97–98, 98*t*, 98*f*
watershed intervention effect, 96, 97*f*
intensive rainwater harvesting activities, 87–88
Kundi sub-basin, 89–96, 98–99, 98*f*
land use, changes in, 92–93, 94*t*
lines of analysis, 89–90
Narmada Master Plan, 88–89
objectives, 89
policy implications, 100
socioeconomic changes, 90–92
sustainable basin management, 88
tools and techniques, 90
tributaries, 89
National Green Tribunal (NGT), 267
National Institute of Hydrology, 172–173
National River Authority (NRA), 28
of England and Wales, 336
National Water Act, 28–29, 39, 48, 336
Natural resource management
Bangladesh, 63
Nepal, 76–77
policies, Sri Lanka, 62
Nepal
watershed impacts and equity, 69
watershed institutions and implementation, 76–78
WSM, 58, 60–61
Netravati river basin, 172–173
NGOs, 316–317
NGT. *See* National Green Tribunal (NGT)

O

On-farm water management practices, 324
Operating funds, SWDMB, 303

P

Pakistan
watershed impacts and equity, 69–70
WSM, 61
Participatory decision-making processes, 69
Participatory WSM, 58, 63, 69–71, 73, 76–80
Payment for environmental services (PES), 283
Pebble bunding, 64–65

PES. *See* Payment for environmental services (PES)
PIAs. *See* Project implementing agencies (PIAs)
Piped water supply, Kanpur city, 254–255
Piped water systems, 239–240
Plastic mulching, 235, 236*t*
Polluters pay principle, 268
Ponds, rejuvenation of, 231
Potential evaporation (PE) values, 200
Pressurized irrigation, 236
Project implementing agencies (PIAs), 42, 48, 76

R

Rainfall
  downscale basin-level IWRM plans, 215
  in Luni river basin, 186–187
Rainfall-baseflow coefficient, 340
Rainfall-runoff relationship, 3–4, 173, 224–225
  estimation of, 196–198
Rain water harvesting (RWH), 66–67
Rajasthan, 183–184
  IWRM plan for Pali district (*see* Downscale basin-level IWRM plans)
  Luni river basin (*see* Luni river basin)
RBMPs. *See* River Basin Management Plans (RBMPs)
RBOs. *See* River basin organizations (RBOs)
Recharge fraction, 95–96, 96*f*
Reference evapotranspiration, 339
Regulatory risk, water supply management, 267–268, 272
Renewable surface water resources, 197–198
Renewable water resources, 265–266
Replenishable groundwater estimation, 198, 199*t*
Reputational risk, water supply management challenges, 269–270, 272
Reservoir evaporation, 159, 161–162, 198–201, 201*t*
Risk assessment framework, 248, 264
River basin
  IWRM interventions, 15–16
  management, 13–14, 331, 335
  water accounting, 13
River Basin Management Plans (RBMPs), 28–29
River basin organizations (RBOs)
  designation and constitution of, 309
  fund generation, 320
  local institutional development agencies, 314
  technical assistance from scientific institutions to, 313
  training programs for local institutions, 323
  water resource mapping, 310
River Ganges. *See also* Kanpur metropolitan area (KMA)
  clean and protect, 261
  pollution of, 268
  surface water, 252
  water quality monitoring of, 253*t*
River valley projects
  decision making in, 5–7
  micro watersheds, 7
Rock and spill drains, 64
Rural domestic water, 206

S

Self-help groups (SHGs), 75
Sewage treatment, 246
Sewage Treatment Plants (STP), 257–258, 259*t*
Sewerage system, 257–259, 271
Simulation models, 111
Smart Cities Mission, 261
Social impact assessment, 50
Socioeconomic drivers, Luni river basin, 190–192
Socioeconomic system
  Warna river basin, 157–158, 157*f*
  watershed management, 55–56
Soil
  conservation, 64–65
  in Luni river basin, 185–186
  technology for, 275–276
Soil and Water Assessment Tool (SWAT), 111–113
  features, 112*b*
  primary data, 113–116, 116*t*
  secondary data, 113–116, 114–115*t*

Sri Lanka
watershed impacts and equity, 70–71
WSM, 62–63
Stakeholder forums, 309–310
State-level nodal agency (SLNA), 75–76
State Water Development and Management Board (SWDMB), 302–303, 308
State Water Policy, 213
Stone walls, 64
STP. *See* Sewage Treatment Plants (STP)
Stream interventions, 289
Sub-basin
Hathni, 89–96, 97*f*, 333
Kundi, 89–96, 98–99, 98*f*, 333
organizations, 312, 324–325
specific management plans for, 312
water tenure
cropping systems, 178–179
land use, 178
lower catchment, 177
total drainage area, 176–177
upper catchment, 177–178
urban population, 180
water quality deterioration, 179
water supply schemes, 180
water table and groundwater outflows, 178–179
wide spatial heterogeneity, 176–177
Sukhi catchment, IWMP watersheds, 107
DEM and SWAT sub-basins, 107
features of, 110*b*
field visit, 107
location, 108–109*f*, 130
upper Orsang catchment, 107, 110*f*
Surface water
resources, 224–225
river Ganges, 252
tradable water use rights allocation in, 305–306
use rights allocation, 319
Surface-water catchment, 292
Surface-water hydrology, 285
Sustainable livelihoods (SL) framework, 67
SWAT. *See* Soil and Water Assessment Tool (SWAT)
SWDMB. *See* State Water Development and Management Board (SWDMB)

T

Tanks, rejuvenation of, 231
Theisson Polygon method, 225
Topography, Warna river basin, 150–151, 151*f*
Tradable water use rights allocation
groundwater, 305
intersectoral allocation of rights, 306–307
intrasectoral allocation of rights, users/user groups, 307–308
in surface water, 305–306
water rights systems, 306
Traditional water bodies, construction of, 231

U

UMA-OYA upper WSM project, 70–71
Urban piped water leakage, prevention of, 232
Urban water management, 260–261
Urban water supply, 246, 321
User groups (UGs), 75–77
Uttar Pradesh Pollution Control Board (UPPCB), 267–268

V

Village Development Committee (VDC), 77
Village-level cluster approach, 75–76
Village-level heterogeneities, 286
Village Watershed Committees (VWCs), 42, 48
Virgin flows
estimation, Luni river basin, 195–196
water accounting, 159–160

W

Warna river basin
agriculture, 175–176
annual renewable water resource, 174
basin outflows, reservoir storage and water utilization pattern
annual streamflow at Samdoli gauging station, 162–163, 163*f*
effective water diversion, 164
effective water utilization, 164, 165*t*
gravity irrigation, 166

ground water availability and draft, 166, 167*t*
inflows, 163–164
irrigated cropping pattern, 168, 168*t*
lift irrigation, 164–166
live storage, 163–164, 163*f*
storage change, 164, 165*t*
climate and agroecology, 153
evaporation loss, 174
geology and geohydrology, 154–155
groundwater balance, 173–174
hydrology, 153–154, 154*f*
outflow, 176
rainfall in, 151–153, 152*f*, 152*t*
socioeconomic system, 157–158, 157*f*
surface water accounts, 175
topography and geography, 150–151, 151*f*
total annual groundwater replenishment, 173
total surface inflow, 172
water accounts of, 174, 174*f*, 175*t*
water audit reports, 176
water consumption, 175
in domestic sector, 168–169, 169*t*
in irrigation, 169–172, 171*t*
by livestock, 169, 170*t*
water resource system
groundwater, 155–157
surface reservoirs and irrigation projects, 155, 156*t*
Wastewater
generation, from KMA, 266, 271
and treatment, 257–259
Wastewater return flows, 247–248
Wastewater treatment systems, 319, 323, 325–326
Water-abundant regions, 307
Water accounts/accounting, 10–11, 332, 334, 337–338
analysis and sources, data for, 161–162
basins in Western Ghats, 172
consumptive use, 149–150, 159
hydrological stress, impacts of, 149–150
hydrological system, 149–150
for Kanpur Metropolitan Area, 334
of KMA, 247, 263–264
estimated water demand, 261–262
estimated water supply and use, 262–263
of Luni river basin, 209–210
methodology, 15
methodology and analytical framework
for blue water, 159
consumptive use of water in urban area, 160
EWD, 160
irrigation, 160
irrigation application, 161
return flows, 159
storage change, 159–160
sum of irrigation dosage, 160–161
total inflow, 159
virgin flows, 159–160
microcatchments, 172–173
storage change, 149–150
water governance and management, 150
water security challenges, KMA, 16
in western Maharashtra (*see* Warna river basin)
Water availability-demand gap, 230–231
Water balance estimation, 90
Water conservation, 275–276
Water consumption, Warna river basin, 175
in domestic sector, 168–169, 169*t*
in irrigation, 169–172, 171*t*
by livestock, 169, 170*t*
Watercourses
flow measurement and control, 317
modernization of, 318
Water demands, 214, 216, 222
domestic, 219
downscale basin-level IWRM plans, 214, 216, 222
domestic, 219
industrial, 221–222
irrigation, 216–219
livestock, 219–221
industrial, 221–222
irrigation, 216–219
livestock, 219–221
Water harvesting structures (WHSs), 93
Water imports, 216
Water management, 299
for Afghanistan, 14
ecological characteristics, 1

Water management (*Continued*)
hydrological and topographical characteristics, 1
institutional capability building for, 301–304
institutional mechanisms for, 336
integrated watershed management
agricultural intensification and groundwater abstraction, 10
catchment assessment, 8
hydraulic interconnectedness, 9–10
lack of integration, 10
microwatershed, 8–9
thrust, 7–8
watershed planning, 8
water use efficiency, 9
interventions of, 231–232
river valley projects
decision making in, 5–7
micro watersheds, 7
scale effects in, 3–5, 14
socioeconomic characteristics, 1
in south Asian countries, 14
water accounting, 10–11
Water pollution, 266, 269
control norms, 267–270
Water quality, 266, 272, 299
problems in areas and villages, 311
Water quality monitoring (WQM), 323
Water-related risks, 248
Water resource(s)
augmentation, 216
of city of Kanpur, 247
development and management, 337–338
mapping, 310
planning, 335
Water resources monitoring and planning unit, 303–304
basin management objectives, 311
basin plan development, 311
cropping pattern monitoring, 322
Electricity Users' Cooperative promotion, 315
existing water harvesting/storage/abstraction structures, 310
groundwater abstraction, sustainable levels of, 313
groundwater and surface water use rights allocation, 319
groundwater levels and water quality monitoring, 323
groundwater use rights enforcement, 322
land and water management practices, 324
large water management projects, 321
local institutional development agencies, 314
local water management activities monitoring, 325
main and branch canals control, 317
microplan development for watersheds, 318
modernization, technological innovations, 317
on-farm water management practices, 324
periodic water accounting and water balance exercises, 325
river basin organizations, designation and constitution of, 309
rural economic institutions, 311
specific management plans for subbasins, 312
strengthening legal regimes, 308–309
subbasin organizations, 312
technical assistance from scientific institutions to RBOs, 313
technical plan preparation, 319
tradable water use rights allocation
establishment of, 304–305
groundwater, 305
intersectoral allocation of rights, 306–307
intrasectoral allocation of rights, users/user groups, 307–308
in surface water, 305–306
water rights systems, 306
water management activities, fund allocation, 321
water quality problems, areas and villages, 311
water resource mapping, 310
water sale and pollution tax collection, fund generation, 320
watershed management institutions, 314

watershed management/local water harvesting activities implementation, 322
water users, socioeconomic profiles, and water accounts database, 310
WUAs
promotion for irrigation management, 315
and urban water utilities, canal water use rights negotiation, 321
Water resource system, Warna river basin
groundwater, 155–157
surface reservoirs and irrigation projects, 155, 156*t*
Water rights systems, 306
Water savings, 216
Water-scarce regions, 307
Water security challenges, 245
Watershed
definition, 55–56
microplans development for, 318
in south Gujarat, 332–333
treatment, 99
Watershed development (WSD), 21, 55–56, 294–295
aquifer geometry, 288–289
at catchment/landscape scale, 105
closure of catchments, 280
ecological/environmental and social aspects, 281
ecological/environmental aspect, 281
history of, 275
hydrology and livelihoods integration, 284–285
impact assessment and monitoring, 105
impact of, 275
institutional approach, 279
institutional issues, 283–284
internalizing externalities, 282–283
interventions management, 281
local water loss, 104
mesoapproach, 282
micro- to mesoscale shift, 282
microwatershed approach, 279–280
Narmada river basin (*see* Narmada river basin)
National watershed guidelines, 104–105
policy, 276
political preference, 22
rainwater harvesting structures, 104
scale aspects, 276–286
social aspect, 281
socioeconomic benefits, 21–22
socioeconomic context, 285–286
socioeconomic implications, 286–294
technology for soil and water conservation, 275–276
water-related consequences, 105–106
Watershed Development Projects, common guidelines, 105–106
Watershed hydrology simulation
alternative rain water harvesting options, 138
nonwater-based livelihoods, 139
water demand exceeds supply, 138
water productivity, 139
watershed inflows exceed outflows, 138
Watershed impacts and equity
socioeconomic and natural resource impacts, 67
in South Asia, 67, 68*t*
Afghanistan, 71–74
India, 67–68
Nepal, 69
Pakistan, 69–70
Sri Lanka, 70–71
sustainable livelihood framework, 67
Watershed institutions and implementation
India, 74–76
Nepal, 76–78
Watershed interventions, 287–288, 291
management, difficulty of, 281
scale of analysis of, 333
Watershed management (WSM), 335
catchment assessment and planning (*see* Catchment assessment; Catchment planning)
characteristics of, 282
hydrological, biophysical, and socioeconomic systems, 55–56
in India, 48
institutions, sustainability of, 45–46
physical and socioeconomic impacts, 44–45
institutions, 314
positive hydrological impacts, 49

Watershed management (WSM) (*Continued*)
programs, 49–50, 279, 283
projects, 276–277, 279
scale effects in, 16
in South Asia
in Afghanistan, 56–57, 63–64
Bangladesh, 63
Bhutan, 61–62
evolution, 58, 59*t*
in India, 56, 58–60
natural resource, 57–58, 62
Nepal, 58, 60–61
Pakistan, 61
Sri Lanka, 62–63
technical interventions of, 64 (*see also* Watershed structures)
Watershed management interventions, 145–146
Watershed management projects
hydrological assessment, 141–144
intervention planning, 121–122, 141
project area selection, 141–142
Watershed structures
in situ moisture conservation, 65–66
rain water harvesting, 66–67
soil conservation, 64–65
Water supply, 214–215, 223, 229
degradation of, 245
downscale basin-level IWRM plans, 214–215, 223, 229
groundwater resources, 226–228
surface water resources, 224–225
system, 254–256
Water supply management challenges, 264
economic risks, 267
regulatory risk, 267–268
reputational risk, 269–270
physical risks, 264
future water requirement, 264–265
renewable water resources availability, 265–266
water quality, 266
Water-surplus river basin, 15
Water tenure
definition, 149
in sub-basin
cropping systems, 178–179
land use, 178
lower catchment, 177
total drainage area, 176–177
upper catchment, 177–178
urban population, 180
water quality deterioration, 179
water supply schemes, 180
water table and groundwater outflows, 178–179
wide spatial heterogeneity, 176–177
Water use efficiency (WUE) improvement measures, 25–26
Water use hydrology, irrigation, 237
Water Users' Association (WUAs)
promotion for irrigation management, 315
and urban water utilities, canal water use rights negotiation, 321
Whole-of-catchment model, 278, 293, 333
World Bank-assisted Neeranchal Watershed Management Project, 106–107
World Wildlife Fund (WWF), 248
WSD. *See* Watershed development (WSD)
WSM. *See* Watershed management (WSM)
WSM Division (WMD), 62
WUAs. *See* Water Users' Association (WUAs)

Printed in the United States
By Bookmasters